国家出版基金项目
NATIONAL PUBLICATION FOUNDATION

有色金属理论与技术前沿丛书

中空结构无机功能材料

INORGANIC FUNCTIONAL MATERIALS WITH HOLLOW INTERIORS

刘小鹤　马仁志　李星国　著
Liu Xiaohe　　Ma Renzhi　　Li Xingguo

中南大学出版社
www.csupress.com.cn

中国有色集团

Introduction

This book aims to present some recent advances in the exciting research field of inorganic functional materials with hollow interiors as well as the insight into the morphological and structural evolution among them. We have selected our related publications dedicated to the subject covering layered hydroxide and oxide nanocones, inorganic nanotubes, cobalt and nickel oxide nanorings, rare-earth oxide and oxysulfate hollow spheres, transition-metal oxide and chalcogenides hollow spheres, ferrite and metallic and polypyrrole hollow spheres. It surely will be a great relief and pleasure for us if this book provides a candid glimpse to the active and exciting research area of inorganic functional materials with hollow interiors. We sincerely hope that this book may be available for reference to established scientists and scholars.

About the Authors

Liu Xiaohe received his Ph. D from the Central South University in 2005, Changsha. He became an Associate Professor and then Full Professor at the School of Resources Processing and Bioengineering of Central South University in 2007 and 2012, and distinguished professor of the Shenghua Scholar Program of Central South University in 2011. He joined the School of Materials Science and Engineering of Central South University in 2014. His research interests focus on, layer-structured materials, novel unilamellar nanosheets, and hollow structure of nanomaterials, as well as darifying the close relationship between the microstructure of as-prepared products and their properties. He has already published more than 70 papers in SCI journals, such as *Advance Materials*, *Angewandte Chemie International Editon*, *ACS Nano*, *Advanced Functional Materials*, *Journal of the American Chemical Society*, etc. which have been cited more than 1600 times, and H-index has reached 26. He has received Hunan Province Natural Science Award (2010) and Outstanding Youth Fund in Hunan Province (2013).

Ma Renzhi received his B. S. and Ph. D both from Tsinghua University in 1995 and 2000, Beijing. He has been a principal researcher at the National Institute for Material Science (Japan). He joined the School of Materials Science and Engineering of Central South University in 2014. He has accomplished a series of international cutting-edge research work, especially the systematic research on the intercalation/exfoliation chemistry of inorganic layered compounds, single-layer nanosheets and hollow structure of nanomaterials. He has already published more than 130 papers and have been cited for more than 7000 times. The H-index of his citations is 45. He co-holds 22 authorized Japan patents. He was selected as an Innovative Talent offered by Recruitment Program of Global Experts (One-Thousand-Talents Plan) in China. He received 2009 Award for Distinguished Lectureship at Asian International Forum for Young Scientists of the Chemical Society of Japan, and "Chemical Physics Letter" 2003—2007 Most Cited Paper Award.

Li Xingguo received his Ph. D from Tohoku university in 1990, Japan. He is the Professor and Director of New Energy and Nano Materials Laboratory, Chairman of the Inorganic Chemistry Institute at Peking University. He is also a Guest Professor of Hiroshima University (Japan). He is an editor for the *Chinese Journal of Inorganic Materials*, *the Chinese Journal of Inorganic Chemistry*, *the Chinese Journal of Process Engineering*, *Journal of Chinese Rare Earth Society*, *Journal of Rare Earth*, *Journal of Applied Chemistry and the Chinese Journal of Functional Materials*. His research interests focus on hydrogen storage materials, hydrogen generation and purification, battery electrode materials and nano materials. He has already published more than 250 papers and contributed chapters to three books published by Nova Science Publishers, Inc. and American Scientific Publishers. He has received the National Distinguished Young Investigator Fund (2000), Lectureship Award of Japan Research Institute of Material Technology (2002) and GM Foundation Science & Technology Achievement Award (2005).

Academic Committee

National Publication Foundation / Series of Theoretical and Technological Frontiers of Nonferrous Metals

Director: Wang Dianzuo   Academician of Chinese Academy of Sciences
                        Academician of Chinese Academy of Engineering

Members: (in alphabetical order)

Yu Runcang	Academician of Chinese Academy of Engineering
Gu Desheng	Academician of Chinese Academy of Engineering
Zuo Tieyong	Academician of Chinese Academy of Engineering
Liu Yexiang	Academician of Chinese Academy of Engineering
Liu Baochen	Academician of Chinese Academy of Engineering
Sun Chuanyao	Academician of Chinese Academy of Engineering
Li Dongying	Academician of Chinese Academy of Engineering
Qiu Dingfan	Academician of Chinese Academy of Engineering
He Jilin	Academician of Chinese Academy of Engineering
He Jishan	Academician of Chinese Academy of Engineering
Yu Yongfu	Academician of Chinese Academy of Engineering
Wang Xuguang	Academician of Chinese Academy of Engineering
Zhang Wenhai	Academician of Chinese Academy of Engineering
Zhang Guocheng	Academician of Chinese Academy of Engineering
Zhang Yi	Academician of Chinese Academy of Engineering
Chen Jing	Academician of Chinese Academy of Engineering
Jin Zhanpeng	Academician of Chinese Academy of Sciences
Zhou Kesong	Academician of Chinese Academy of Engineering
Zhou Lian	Academician of Chinese Academy of Engineering
Zhong Jue	Academician of Chinese Academy of Engineering
Huang Boyun	Academician of Chinese Academy of Engineering
Fuang Peiyun	Academician of Chinese Academy of Engineering
Tu Hailing	Academician of Chinese Academy of Engineering
Zeng Sumin	Academician of Chinese Academy of Engineering
Dai Yongnian	Academician of Chinese Academy of Engineering

Editorial and Publication Committee

National Publication Foundation / Series of Theoretical and Technological Frontiers of Nonferrous Metals

Director

Luo Tao   Professor of Engineering
            General Manager of China Nonferrous Metal Mining (Group) Co. , Ltd

Deputy Directors

Qiu Guanzhou   Academician of Chinese Academy of Engineering
Tian Hongqi   Professor
              Deputy president of Central South University
Yin Feizhou   Professor of Editorship
              Deputy Director of Hunan Provincial Bureau of Press & Publication
Zhang Lin   Professor of Engineering
              Chairman of Daye Nonferrous Metals Group Holdings Co. , Ltd

Deputy Executive Director

Wang Haidong Professor   Wang Feiyue Professor
              President of Central South University Press

Commissioners

Su Renjin	Wen Yuanchao	Li Changjia	Peng Chaoqun
Chen Canhua	Hu Yemin	Liu Hui	Tan Ping
Zhang Xi	Zhou Ying	Wang Yiye	Yi Jianguo
Li Hailiang			

Preface

Nowadays, nonferrous metals have become the important material basis for the development of national economic, science and technology, national defense and so on, and are the key strategic resources for enhancing the comprehensive national strength and national security. As a nonferrous metals production superpower, the nonferrous metals research field in China has made great progress, particularly in the development and utilization of complex and low-grade nonferrous metals resources.

The nonferrous metals industry in China has developed rapidly in the recent 30 years. The production has been the maximal in the world year after year. The science and technology of nonferrous metals is playing an increasingly important role in the national economic construction and the modernization of national defense. At the same time, the contradiction between the shortage of nonferrous metals resources and the economic development is increasingly serious. The dependence on foreign resources has been increasing year by year, which seriously affects the healthy development of our national economy.

With the economic development, high-quality mineral resources have proven to be nearly depleted. This makes China face the crisis of critical shortage of the total supply of nonferrous metals. Furthermore, when the complex and low-grade mineral resources or secondery raw materials, which have the characteristics of difficult exploration, difficult mining, difficult processing and difficult metallurgy, gradually become the main resources, a huge challenge will be exerted on the traditional science and technology, such as geology, mining, mineral processing, metallurgy, materials, processing, and enviroment. The nonferrous metals industry and related industries are facing the competition for the survival of the crisis due to low-quality resources. The development of nonferrous metals industry urgently needs new theories and new technologies to adapt to the characteristics of the resources. The publication of nonferrous metals books that own complete system, leading level and mutual integration will improve the independent innovation ability of the nonferrous metals industry and promote the application of new theories and new technologies of high-efficiency, low consumption, non-pollution and comprehensive utilization of the nonferrous metals resources. It will play a vital role in ensuring the sustainable development of the nonferrous metals industry in China.

As a major national publication project, the *Series of Theoretical and Technological Frontiers of Nonferrous Metals* plan to publish 100 kinds of books, which cover materials, metallurgy, mining, earth science, mechanical and electrical engineering and so on. The authors of these books include academicians who work in the field of nonferrous metals, chief scientists of major national research projects, Yangtze River Scholars, winners of China National Funds for Distinguished Young Scientists, winners of the National Excellent Doctoral Dissertation, talented persons selected by major national talent programs, leading experts in large-scale research institutes of nonferrous metals and key enterprises.

The National Publication Foundation established by the state is to encourage and support the publication of outstanding public service works that represent the highest level of the national academic publication. The *Series of Theoretical and Technological Frontiers of Nonferrous Metals* will aim at the frontiers of nonferrous metals and grasp the latest development of nonferrous metals at home and abroad. They will comprehensively, promptly and accurately reflect the new theories, new technologies and new applications of nonferrous metals, and explore and capture the highly valuable research achievements, thus owning high academic values.

The Central South University Press has devoted long-term effort to publishing books of nonferrous metals. A lot of very effective work has been done in the course of publishing the *Series of Theoretical and Technological Frontiers of Nonferrous Metals*. This will vigorously promote the publication of outstanding scientific and technological works in the nonferrous metals industry, and play a direct and significant role in cultivating personnel of the nonferrous metals disciplines in the universities, research institutes and enterprises of China.

Wang Dianzuo
Dec 2010

Foreword

This book aims to present some recent advances in the exciting research field of nanotubes, especially the designed synthesis of various functional materials with hollow interiors as well as the insight into the morphological and structural evolution among them. It is generally believed that, for any inorganic functional material with a layered structure, nanotubes could be formed under appropriate conditions (chemical vapor deposition, hydrothermal synthesis, etc.). In fact, many kinds of inorganic nanotubes made of carbon, boron nitride (BN), transition-metal halides, oxides, and chalcogenides, all possessing natural or artificial layered structures, have successfully been produced based on a "rolling-up" mechanism, analogous to a paper scroll.

Compared with the layered counterparts, the formation of nanotubes from nonlayered materials requires much more effort to assemble atoms or small particles into a tubular structure during crystallization. As a result, the preparation of nanotubes from nonlayered materials is often subjected to the use of sacrificial templates such as nanotubes, nanowires, nanorods, and porous membranes. Although the sacrificial templating method has been proved to be a facile and efficient approach for the growth of tubular structures, sometimes the final product might be disrupted during the templates removing processes.

In addition to cylindrical nanotubes, layered materials might be able to form conical structures with hollow interiors by the same rolling-up mechanism. Derived from the peculiar conical feature, nanocones exhibit special electronic, mechanical, and field-emission properties. However, apart from carbon and boron nitride systems, there are few reports focusing on conical structures originating from the rolling-up of layered materials. On the other hand, a large variety of layered materials could be exfoliated/delaminated into single-layer nanosheets by controlling layer-to-layer interaction through soft chemical procedures. Even nanotubes can be unwrapped/unravelled into individual sheets under a similar scenario. For example, carbon nanotubes could be unzipped/exfoliated to fabricate graphene sheets or ribbons. In general, layered oxide and hydroxide crystals adopt a lamellar or plate-like morphology. The question arises as to, besides carbon and boron nitride (BN), whether conical structures can be formed by the rolling-up of layered oxides and hydroxides, and if the nanocones can be further unwrapped/exfoliated into single-layer nanosheets. The pursuit of the answer to this intriguing question will be very important in revealing the formation mechanism of nanocones as well as the energy balance between nanocones and nanosheets.

Another interesting morphology for inorganic materials, hollow spheres with nanometer or micrometer dimensions, are attractive in various fields, such as controlled release capsules, artificial cells, selective catalysis, chemical storage and sensors, photonic crystals, biomedical diagnosis and therapy. Numerous chemical and physicochemical strategies such as the kirkendall diffusion effect, spray pyrolysis, ostwald ripening, template-assisted synthesis and chemically induced self-transformation have been employed for the design and controlled fabrication of various micro/nanospheres with hollow interiors. Traditionally, template-assisted synthetic strategies including hard and soft templates have been demonstrated to be the most effective and versatile approach. In this regard, various templates, such as hard ones (e.g., silica, polymer, and carbon spheres) and soft ones (e.g., vesicles, emulsions, and surfactant), have been extensively utilized to fabricate hollow spheres. In a typical procedure, the surfaces of the hard templates are coated with a designed material (or its precursor) to form core-shell structures. Subsequent removal of the core templates via calcination at elevated temperatures in air or selective etching in an appropriate solvent generates hollow spheres. The procedures are somewhat technically complicated, which to some extent limits their wide application prospects. More recently, biomolecules, as life's basic building blocks inherent with special structures and fascinating functions for assembling purpose, have also been generally utilized as templates for the design and synthesis of various micro/nano hollow spheres.

Under such a research background, we have selected our related publications dedicated to the subject covering layered hydroxide and oxide nanocones, inorganic nanotubes, cobalt and nickel oxide nanorings, rare-earth oxide and oxysulfate hollow spheres, transition-metal oxide and chalcogenides hollow spheres, ferrite and metallic and polypyrrole hollow spheres. The published information of all the original papers is listed in the end of the references and marked with asterisk. It surely will be a great relief and pleasure for us if this book provides a candid glimpse to the active and exciting research area of inorganic functional materials with hollow interiors.

The authors would like to express our sincere thanks to the many wonderful students who have contributed to the research and made the publication of this book possible. They are Ph.D students Yan Aiguo and Shi Rongrong; Master course students Zhang Ning, Wu Hongyi, Yi Ran, Zhou Huifen, Li Yongbo, Jing Jinghui, Wang Jingying, Gao Guanhua, Chen Gen, Ma Wei, Zhang Dan, Yuan Peng, Ji Deli, Zhu Yanjun, Liu Tao, Wan Hao, Chen Fashen, Wang Duo, Chen Chen, Zhong Yishun. We would also like to thank Profs. Li Junhui and Chen Limiao, and Dr. Jingling Pan (Central South University),

Prof. Jia Baoping (Changzhou University), Prof. Zhang Haitao (Institute of Process Engineering, Chinese Academy of Sciences), Prof. Shi Youguo (Institute of Physics, Chinese Academy of Sciences), Prof. Zhang Shoubao (Qufu Normal University), Prof. Wang Zhong (General Research Institute for Nonferrous Metals) for their kind help and passionate cooperation. This work was the outcome of several research projects supported by the National Natural Science Foundation of China, Hunan Provincial Key Science and Technology Project of China, Hunan Provincial Natural Science Foundation of China and the Shenghua Scholar Program of Central South University.

We are also very grateful to Prof. Qiu Guanzhou of Central South University (Academician of Chinese Academy of Engineering), Prof. Li Yadong of Tsinghua University (Academician of Chinese Academy of Science), as well as Prof. Sasaki Takayoshi (University of Tsukuba, Fellow of National Institute for Materials Science, Japan), and Prof. Bando Yoshio (Chief Operating Officer of International Center for Materials Nanoarchitectonics, Fellow of National Institute for Materials Science, Japan).

Li Xingguo
June 2015

Contents

Chapter I Layered Hydroxide and Oxide Nanocones 1

Layered Cobalt Hydroxide Nanocones: Microwave-Assisted Synthesis, Exfoliation and Structural Modification 3

A General Strategy to Layered Transition-Metal Hydroxide Nanocones: Tuning the Composition for High Electrochemical Performance 17

High-Yield Preparation, Versatile Structural Modification, and Properties of Layered Cobalt Hydroxide Nanocones 34

Layered Zinc Hydroxide Nanocones: Synthesis, Facile Morphological and Structural Modification, and Properties 59

Chapter II Inorganic Nanotubes 75

Selective and Controlled Synthesis of Single-Crystalline Yttrium Hydroxide/Oxide Nanosheets and Nanotubes 77

Rational Synthetic Strategy: From ZnO Nanorods to ZnS Nanotubes 96

Selective Synthesis and Magnetic Properties of Uniform CoTe and $CoTe_2$ Nanotubes 110

Shape-Controlled Synthesis and Properties of Manganese Sulfide Microcrystals via a Biomolecule-Assisted Hydrothermal Process 120

Conversion of Metal Oxide Nanosheets into Nanotubes 132

Chapter III　Cobalt and Nickel Oxide Nanorings　　147

Cobalt Hydroxide Nanosheets and Their Thermal Decomposition to Cobalt Oxide Nanorings　　149
Rationally Synthetic Strategy: From Nickel Hydroxide Nanosheets to Nickel Oxide Nanorolls　　166

Chapter IV　Rare-Earth Oxide and Oxysulfate Hollow Spheres　　179

Controllable Fabrication and Optical Properties of Uniform Gadolinium Oxysulfate Hollow Spheres　　181
Hollow Spherical Rare-Earth-Doped Yttrium Oxysulfate: A Novel Structure for Upconversion　　197
Controlled Fabrication and Optical Properties of Uniform CeO_2 Hollow Spheres　　213
General Synthetic Strategy for High-Yield and Uniform Rare-Earth Oxysulfates ($RE_2O_2SO_4$, RE = La, Pr, Nd, Sm, Eu, Gd, Tb, Dy, Y, Ho, and Yb) Hollow Spheres　　225

Chapter V　Transition-Metal Oxide and Chalcogenides Hollow Spheres　　241

Controllable Fabrication and Magnetic Properties of Double-Shell Cobalt Oxides Hollow Particles　　243
Shape-Controlled Synthesis and Characterization of Cobalt Oxides Hollow Spheres and Octahedra　　257
Shape-Controlled Synthesis and Properties of Dandelion-Like Manganese Sulfide Hollow Spheres　　274
Nickel Dichalcogenides Hollow Spheres: Controllable Fabrication, Structural Modification and Magnetic Properties　　286
Biomolecule-Assisted Hydrothermal Synthesis and Properties of Manganese Sulfide Hollow Microspheres　　300

Biomolecule-Assisted Hydrothermal Synthesis and Electrochemical Properties of Copper Sulfide Hollow Spheres   311

Chapter VI   Other Hollow Spheres   323

Hollow Metallic Microspheres: Fabrication and Characterization   325

Selective Synthesis and Properties of Monodisperse Zn Ferrite Hollow Nanospheres and Nanosheets   339

Controllable Fabrication of SiO_2/Polypyrrole Core-Shell Particles and Polypyrrole Hollow Spheres   351

Chapter I

Layered Hydroxide and Oxide Nanocones

Layered Cobalt Hydroxide Nanocones: Microwave-Assisted Synthesis, Exfoliation and Structural Modification

Abstract: Layered cobalt hydroxide nanocones intercalated with dodecyl sulfate (DS) ions could be successfully synthesized through the rolling process of lamellar structures under microwave-assisted conditions, which underwent exfoliation in formamide into unilamellar cobalt hydroxide nanosheets. Layered cobalt hydroxide nanocones could also be converted into CoOOH and Co_3O_4 nanocones retaining original morphological features, respectively.

1 Introution

Layered materials have drawn immense attention because of their distinctive properties and their wide range of practical and potential applications, such as anion/cation exchangers, selective separation membranes, catalysis, adsorbents, chemical or biosensors, solid-state nanoreactors, and molecular sieves.[1-5] It is generally believed that layered materials might be able to form tubular structures by a rolling mechanism. Various layered materials, such as carbon, boron nitride (BN), transition-metal halides, oxides, and chalcogenides, could roll up or fold up into tubular forms, for example nanotubes.[6-8] In particular, conical structures with hollow interiors, namely nanocones/nanohorns formed from carbon and boron nitride, have also been discovered.[9-12] Due to the conical feature, nanocones/nanohorns might have special electronic, mechanical, and field-emission properties. However, apart from carbon and boron nitride systems, there are few reports of conical structures originating from the rolling-up of layered materials. On the other hand, layered materials could be exfoliated/delaminated into unilamellar nanosheets by controlling layer-to-layer interaction through soft chemical procedures.[13-15] In particular, unilamellar nanosheets, typically about one nanometer in thickness and several tens of nanometers to several micrometers in lateral size, can curl or fold up into nanotubes/nanoscrolls.[16-18] Very recently, carbon nanotubes could be unzipped/exfoliated to fabricate graphene sheets and ribbons.[19-21] The question arises as to whether conical structures be formed by the rolling-up of layered materials other than carbon and boron nitride, and if they can be further unwrapped/exfoliated into unilamellar nanosheets. The answer will be very important in revealing the formation mechanism of nanocones/nanohorns as well as the energy balance between nanocones/nanohorns and nanosheets.

Layered cobalt hydroxide has received enormous attention in recent years on the basis of its unique catalytic, magnetic and electrochemical properties.[22] It is well-known that layered cobalt hydroxides have two polymorphs: $\alpha - Co(OH)_2$ and $\beta - Co(OH)_2$. The $\alpha - Co(OH)_2$ consists of stacked layers intercalated with various anions (such as CO_3^{2-}, NO_3^-, Cl^-) and water molecules in the interlayer gallery, which thus has a larger interlayer spacing (> 0.70 nm) than that of $\beta - Co(OH)_2$ (0.46 nm) without guest species.[23,24] We have recently demonstrated hexagonal microplatelets of layered α - cobalt and β - cobalt hydroxides could be selectively synthesized by homogeneous precipitation using hexamethylenetetramine (HMT) as an alkaline reagent.[23] Herein, we present layered cobalt hydroxide nanocones intercalated with dodecyl sulfate (DS) ions that can be formed by a facile microwave-assisted method in which the surfactant sodium dodecyl sulfate (SDS) is used as a structure-directing agent. Furthermore, unilamellar cobalt hydroxide nanosheets can then be obtained by direct exfoliation of these nanocones in formamide. By using layered cobalt hydroxide nanocones as the precursor, cobalt oxyhydroxide (CoOOH) and cobalt oxide (Co_3O_4) nanocones can also be obtained by oxidation in alkaline solution and thermal decomposition, respectively. This feature offers a vast opportunity to rationally design related nanostructures based on layered hydroxides.

2 Experimental Section

All chemicals used were of analytical grade and were purchased from Wako Chemical Reagents Company (Japan). They were used without further purification. Milli - Q water was used throughout the experiments.

2.1 Synthesis. In a typical procedure, $CoCl_2 \cdot 6H_2O$ (1 mmol), HMT (3 mmol), and SDS (5 mmol) were charged into a Teflon-lined autoclave of 100 mL capacity. The autoclave was filled with Milli-Q water up to 80% of the total volume, then sealed and microwave-heated at 100 ℃ for 1 h under magnetic stirring. After the reaction finished, the green precipitate was filtered, washed with water and ethanol, and finally dried at 60 ℃ for 5 h. The resulting product (30 mg) was mixed with formamide (50 mL) in a conical beaker, which was tightly capped after purging with nitrogen gas. Then, the mixture was agitated in a mechanical shaker at 120 r/min for 24 h, yielding a translucent green colloidal suspension. To remove possible unexfoliated nanocones, the suspension was further treated by centrifugation at 4000 r/min for 20 min. To obtain the CoOOH nanocones, 50 mg corresponding green product was dispersed in NaOH solution (0.5 M, 100 mL) under magnetic stirring for 6 h in an ambient environment. Layered cobalt hydroxide nanocones were also annealed in air at 700 ℃ for 2 h to prepare Co_3O_4 nanocones.

2.2 Characterizations. XRD data were collected by a Rigaku RINT – 2000 diffractometer with monochromatic Cu_{K_α} radiation ($\lambda = 0.15405$ nm). The morphology of the synthesized products was examined using a JEOL JSM – 6700F field-emission scanning SEM. TEM was performed on a JEOL JEM – 3100F energy-filtering (Omega type) transmission microscope. UV/Vis absorption spectra were recorded using a Hitachi U – 4100 spectrophotometer. Thermogravimetric-differential thermal analysis measurements (TG-DTA) were carried out using a Rigaku TGA – 8120 instrument in a temperature range of 25 – 900 ℃ at a heating rate of 1 ℃/min.

3 Results and Discussion

Figure 1(a) shows a typical scanning electron microscopy (SEM) image of as-prepared product obtained by using 3 mmol HMT and 5 mmol SDS at 100 ℃ for 1 h. A large quantity of conical nanostructures was found in the product. A typical transmission electron microscopy (TEM) image [Figure 1(b)] also clearly shows that as-prepared nanocones possess conical structures with hollow interiors. These nanocones have an average bottom diameter of about 400 nm, tip diameter of about 20 nm, and length of up to 2 μm. The wall thickness ranges from a few to several tens of nanometers. The inset in Figure 1(b) shows the SAED pattern taken on a few nanocones, which can be indexed to in-plane or [001] zone-axis diffraction of hexagonal cobalt hydroxide with a lattice constant of $a = 0.31$ nm. Figure 1(c) depicts a typical TEM image of an individual nanocone. A high-resolution transmission electron microscopy (HRTEM) image (inset) shows the layered structure; the interlayer spacing is measured at about 2.4 nm. The layered structure was also confirmed by X-ray powder diffraction (XRD). The basal reflection series [Figure 1(d)] corresponds to an interlayer distance of 2.4 nm. Taking into account the green color of the product, the nanocones may be identified as DS^- – intercalated α – type cobalt

Figure 1

(a) SEM and (b) TEM images of layered cobalt hydroxide nanocones. The inset in (b) shows an SAED pattern taken on nanocones: 1) 100, 2) 110. (c) TEM and (inset) HRTEM images of an individual nanocone. (d) XRD pattern of layered cobalt hydroxide nanocones intercalated with DS ions.

hydroxide.[23]

The chemical composition of as-prepared products is estimated to be $Co(OH)_{1.75}DS_{0.25} \cdot 0.4H_2O$ based on thermogravimetric and differential thermal analysis, as shown in Figure 2. Figure 2(a) shows the thermal behavior of layered cobalt hydroxide nanocones intercalated with dodecyl sulfate (DS) ions in a temperature range of 25–900 ℃. The gradual mass loss below 100 ℃ can be attributed to the evaporation of the adsorbed water on the nanocone surface. The consecutive stage of weight loss in the range 100–170 ℃ was 4.5%, which can be attributed to the removal of interlayer water molecules. The DTA curve shows two corresponding endothermic peaks. The weight-loss profile exhibits a well-defined decrease between 170 ℃ and 800 ℃, which can be mainly attributed to combustion of DS and dehydroxylation. First, the combustion of organic components in DS to form inorganic sulfate, might be accounted for a huge exothermic peak between 250℃ and 350℃. Accompanying with that, the dehydroxylation and oxidation of some Co cations may also occur. After that, the removal of inorganic sulfate may be responsible for the loss around 700℃. The final product of thermal decomposition at 800℃ is Co_3O_4 with a weight loss of 41.7%. As

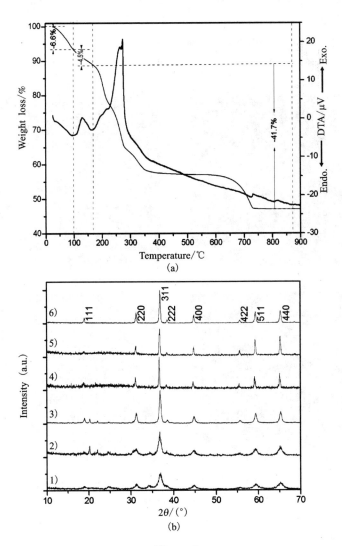

Figure 2
(a) TG and DTA curves of layered cobalt hydroxide nanocones recorded at a heating rate of 1℃/min in air. (b) The evolution of XRD patterns of products obtained by calcination of as-prepared nanocones at 1) 400℃, 2) 500℃, 3) 600℃, 4) 700℃, 5) 800℃, and 6) 900℃ for 2 h in air atmosphere, respectively.

shown in Figure 2(b), layered cobalt hydroxide nanocones intercalated with DS ions could be completely converted into Co_3O_4 (JCPDS 42 – 1467) by calcination at 700 ℃ for 2 h in air atmosphere. On the basis of the results, the chemical composition of layered cobalt hydroxide nanocones was estimated to be $Co(OH)_{1.75}DS_{0.25} \cdot 0.4H_2O$. Excluding adsorbed water on the nanocone surface (6.6%), the measured total weight loss is 49.5%, namely (4.5% + 41.7%)/(100% − 6.6%) = 49.5%. The calculated weight loss from suggested formula $Co(OH)_{1.75}DS_{0.25} \cdot 0.4H_2O$ into Co_3O_4 is approximate 49.7%, which is consistent with the measured value.

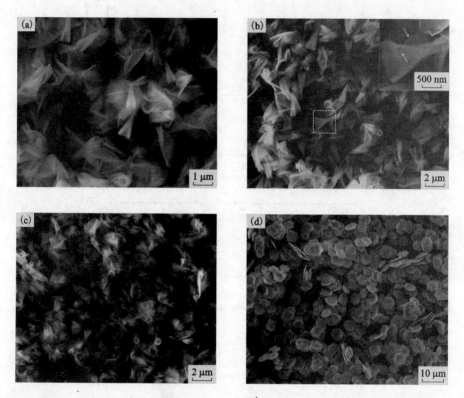

Figure 3   SEM images of as – prepared products synthesized at different reaction time and temperature

(a) 100 ℃, 10 min; (b) 100 ℃, 30 min; (c) 80 ℃, 1 h. The inset in (b) is a high-magnification SEM image obtained from the area marked by a quadrangle. (d) SEM image of as-prepared product obtained in the absence of surfactants at 100 ℃ for 1 h.

The morphology and size of the products strongly depend on the synthetic parameters, such as reaction time, temperature, and surfactant. Figure 3(a) shows a typical SEM image of as-prepared product obtained at 100 ℃ for 10 min. Apart from a few nanocones, a large quantity of lamellar structures with rolled-up or curled edges can be clearly observed. When the reaction time was prolonged to 30 min, the product was mainly composed of nanocones [Figure 3(b)], indicating that the lamellar structures gradually curled into nanocones with longer reaction time. The inset in Figure 3(b) is a higher magnification SEM image of an individual nanocone, which evidently exhibits the curling and folding up of an individual lamella onto itself to form a nanocone. The reaction temperature also influences the morphology of final products. When the reaction temperature was reduced to 60 ℃, only lamellar structures are obtained, whilst the product mainly consisted of nanocones at temperatures higher than 80 ℃ [Figure 3(c)]. In particular, surfactant SDS seems to play a crucial role in the formation of layered cobalt hydroxide nanocones. A SEM image [Figure 3(d)] shows the as-prepared product synthesized at 100 ℃ for 1 h without SDS surfactant. Hexagonal platelets with average lateral size of about 5 μm and thickness of several tens of nanometers were produced, and they are identified as β - Co(OH)$_2$ without intercalating any anionic species on the basis of XRD result. Based on the experimental results, we assume that DS$^-$ – intercalated lamellar structures with few layers are firstly formed, and then tend to curl up at the edge, producing a conical angle (θ, ca. 10° to 60°) rather than a tubular structure owing to their morphological features and relatively lower energy barrier.[25] The lamellar structures gradually roll up along the conical angle, which may further grow and finally form nanocones under suitable conditions, as shown in Figure 4.

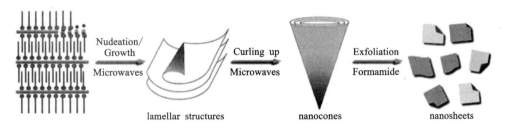

Figure 4　Schematic illustration of the formation and exfoliation of layered cobalt hydroxide nanocones intercalated with DS ions

By dispersing the layered cobalt hydroxide nanocones in formamide, a translucent green colloidal suspension was formed. Figure 5(a) shows a typical photograph of the colloidal suspensions. Clear Tyndall light scattering was discerned for the suspension,

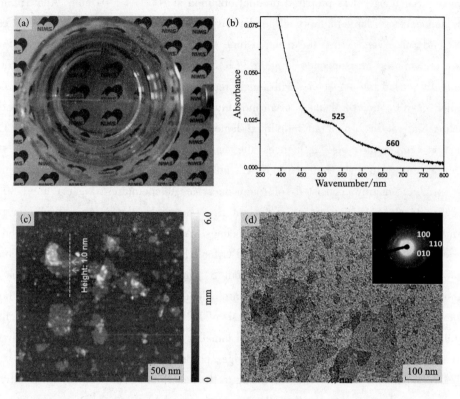

Figure 5

(a) Photograph of a colloidal suspension of the exfoliated cobalt hydroxide nanosheets. The suspension is side illuminated to demonstrate the Tyndall scattering effect. (b) UV/Vis absorption spectra of colloidal suspensions of the exfoliated nanosheets. (c) Tapping-mode AFM image of the exfoliated nanosheets deposited on a silicon substrate. Arrows indicate a height of 1.0 nm. (d) TEM image of the exfoliated cobalt hydroxide nanosheets. Inset: SAED pattern taken from an individual nanosheet.

indicating the presence of abundant exfoliated nanosheets dispersed in formamide. The UV – Vis spectrum of the colloidal suspension is shown in Figure 5(b). In the visible region, a broad absorption band centered at about 525 nm is observed, which is the typical $^4T_{1g}(F) \rightarrow ^4T_{1g}(P)$ transition of Co^{2+} octahedrally coordinated by weak-field ligands.[26] Furthermore, weak peak at around 660 nm corresponding to the absorption features of Co^{2+} with a tetrahedral coordination geometry can also be detected.[24] The UV/Vis results can be regarded as the evidence for the maintenance of octahedral and tetrahedral coordination of cobalt ions in the exfoliated nanosheets, indicating that the α – type host layer architecture of cobalt hydroxide was retained during the exfoliation process. A tapping-mode atomic force microscope (AFM) image [Figure 5(c)] shows sheet-like objects with lateral dimensions of several hundred nanometers. These nanosheets are irregular in shape, suggesting breakage or fracture of the nanocones during the delamination process. The thickness of the nanosheets was measured to be about 1.0 nm, which is very similar to that previous observed for layered double hydroxide (LDH) nanosheets (ca. 0.8 nm).[13a] The slightly larger value might be due to either the DS adsorption or the tetrahedral coordination on each side of the hydroxide plane.[24] A typical TEM image [Figure 5(d)] indicates very faint but homogeneous contrast of the nanosheets, reflecting their ultrathin and uniform thickness. The inset shows a typical SAED pattern taken from an individual nanosheet, and is compatible with the in-plane hexagonal unit cell ($a = 0.31$ nm).

More interestingly, layered cobalt hydroxide nanocones could be transformed into other related structures (such as CoOOH and Co_3O_4) that retain their original morphological features, which would endow their potenional application in various fields. The layered cobalt hydroxide nanocones can be gradually oxidized into CoOOH in alkaline medium. After treatment in alkaline solution, the color of the product turns from green to dark brown, which also indicates the transformation of hydroxide into oxyhydroxide.[27] Figure 6(a) depicts a typical SEM image of as-prepared CoOOH, which reveals that the initial conical structure was maintained. A TEM image reveals the CoOOH nanocones with hollow interior [Figure 6(b)]. The average bottom diameter of the CoOOH nanocones is smaller than that of the layered cobalt hydroxide nanocones, which is caused by the removal of DS ions and the resultant decrease in interlayer spacing. The inset shows an SAED pattern taken from a mass of the nanocones, which can be indexed as the CoOOH structure. Figure 6(c) shows a typical HRTEM image of a selected area of an individual CoOOH nanocone. The interlayer spacing is measured to be 0.44 nm, which agrees well with the separation between (003) lattice planes of CoOOH.

Co_3O_4 could also be obtained by calcination of layered cobalt hydroxide nanocones. The product was composed of many pores, but still maintained the original conical framework [Figure 6(d)] through pyrolysis and dehydration. The upper inset shows the SAED pattern, revealing the satisfactory crystallinity of Co_3O_4 nanocones. The lattice spacings in the HRTEM observation for Co_3O_4 nanocones shown in lower inset of Figure 5(d) are measured to be about 0.29 nm and 0.24 nm, which are consistent with the values of {220} and {311} lattice planes of spinel Co_3O_4, respectively.

Figure 6

(a) SEM, (b) TEM and (c) HRTEM images of CoOOH nanocones obtained by reacting layered cobalt hydroxide nanocones in a 0.5 M NaOH solution for 6 h. The inset in (b) shows an SAED pattern taken on nanocones. (d) TEM image of Co_3O_4 nanocones obtained by calcination of layered cobalt hydroxide nanocones in air at 700 ℃ for 2 h. The upper right and lower left insets in (d) are an SAED pattern and a HRTEM image of Co_3O_4 nanocones, respectively.

4 Conclusion

In summary, we have developed a simple and reliable synthetic strategy for production of layered cobalt hydroxide nanocones by using HMT as an alkaline reagent and SDS as both a surfactant and structure-directing agent. The results demonstrate that conical structures with hollow interiors might form through a rolling process of lamellar hydroxide. Unilamellar cobalt hydroxide nanosheets can be obtained by direct exfoliation of layered cobalt hydroxide nanocones in formamide. Furthermore, layered cobalt hydroxide nanocones can be transformed into CoOOH and Co_3O_4 nanocones by oxidation in alkaline solution and thermal-decomposition, respectively. The synthetic strategy presented herein may provide an effective route to synthesize and rationally design other inorganic nanocones and nanosheets, which would be helpful in understanding the energy equilibrium between folding-up and unwrapping of layered structures. These conical structures with hollow interior can also be expected to bring new opportunities for further fundamental research, and also for technological applications in catalysts, solid-state sensors, and as anode materials in lithium-ion rechargeable batteries.

References

[1] (a) R. Schöllhorn. Inclusion compounds (Eds.: J. L. Atwood, J. E. D. Davies, D. D. MacNicol)[M]. London: Academic Press, 1984. (b) A. Clearfield. Role of ion exchange in solid-state chemistry [J]. Chemical Reviews, 1988, 88(1): 125 – 148. (c) M. Ogawa, K. Kuroda, Photofunctions of intercalation compounds [J]. Chemical Reviews, 1995, 95(2): 399 – 438.

[2] (a) Y. Du, D. O'Hare, Observation of staging during intercalation in layered α – cobalt hydroxides: a synthetic and kinetic study [J]. Inorganic Chemistry, 2008, 47(24): 11839 – 11846. (b) F. Geng, H. Xin, Y. Matsushita, R. Ma, M. Tanaka, F. Izumi, N. Iyi, T. Sasaki. New layered rare-earth hydroxides with anion-exchange properties [J]. Chemistry-A European Journal, 2008, 14(30): 9255 – 9260.

[3] G. Centi, S. Perathoner. Catalysis by layered materials: a review [J]. Microporous and Mesoporous Materials, 2008, 107(1): 3 – 15.

[4] M. J. Manos, V. G. Petkov, M. G. Kanatzidis. $H_{2x}Mn_xSn_{3-x}S_6$ ($x = 0.11 - 0.25$): a novel reusable sorbent for highly specific mercury capture under extreme pH conditions [J]. Advanced Functional Materials, 2009, 19(7): 1087 – 1092.

[5] C. Gérardin, D. Kostadinova, N. Sanson, B. Coq, D. Tichit. Supported metal particles from LDH nanocomposite precursors: control of the metal particle size at increasing metal content [J]. Chemistry of Materials, 2005, 17(25): 6473 – 6478.

[6] R. Tenne, A. K. Zettl, Nanotubes from inorganic materials [M]. Berlin: Springer Berlin Heidelberg, 2001: 81 – 112.

[7] G. R. Patzke, F. Krumeich, R. Nesper, Oxidic nanotubes and nanorods-anisotropic modules for a future nanotechnology [J]. Angewandte Chemie International Edition, 2002, 41(14): 2446 – 2461.

[8] C. N. R. Rao, M. Nath. Inorganic nanotubes [J]. Dalton Transactions, 2003, 1: 1 – 24.

[9] (a) S. Iijima, M. Yudasaka, R. Yamada, S. Bandow, K. Suenaga, F. Kokai, K. Takahashi.

Nano-aggregates of single-walled graphitic carbon nano-horns [J]. Chemical Physics Letters, 1999, 309(3): 165 – 170. (b) C. M. Yang, H. Noguchi, K. Murata, M. Yudasaka, A. Hashimoto, S. Iijima, K. Kaneko. Highly ultramicroporous single-walled carbon nanohorn assemblies [J]. Advanced Materials, 2005, 17(7): 866 – 870. (c) S. P. Economopoulos, G. Pagona, M. Yudasaka, S. Iijima, N. Tagmatarchis. Solvent-free microwave-assisted bingel reaction in carbon nanohorns [J]. Journal of Materials Chemistry, 2009, 19(39): 7326 – 7331.

[10] C. Cioffi, S. Campidelli, C. Sooambar, M. Marcaccio, G. Marcolongo, M. Meneghetti, D. Paolucci, F. Paolucci, C. Ehli, G. M. A. Rahman, V. Sgobba, D. M. Guldi, M. Prato. Synthesis, characterization, and photoinduced electron transfer in functionalized single wall carbon nanohorns [J]. Journal of the American Chemical Society, 2007, 129(13): 3938 – 3945.

[11] (a) L. Bourgeois, Y. Bando, W. Q. Han, T. Sato. Structure of boron nitride nanoscale cones: ordered stacking of 240 and 300 disclinations [J]. Physical Review B, 2000, 61(11): 7686 – 7691. (b) C. Zhi, Y. Bando, C. Tang, D. Golberg, R. Xie, T. Sekiguchi. Large-scale fabrication of boron nitride nanohorn [J]. Applied Physics Letters, 2005, 87(6): 063107.

[12] H. Terrones. M. Terrones. Curved nanostructured materials [J]. New Journal of Physics, 2003, 126.1 – 126.37.

[13] (a) R. Ma, Z. Liu, L. Li, N. Iyi, T. Sasaki. Exfoliating layered double hydroxides in formamide: a method to obtain positively charged nanosheets [J]. Journal of Materials Chemistry, 2006, 16(39): 3809 – 3813. (b) T. Sasaki. Fabrication of nanostructured functional materials using exfoliated nanosheets as a building block [J]. Nippon Seramikkusu Kyokai Gakujutsu Ronbunshi, 2007, 115(1): 9 – 16. (c) R. Ma, K. Takada, K. Fukuda, N. Iyi, Y. Bando, T. Sasaki. Topochemical synthesis of monometallic (Co^{2+} - Co^{3+}) layered double hydroxide and its exfoliation into positively charged Co(OH)$_2$ nanosheets [J]. Angewandte Chemie International Edition, 2008, 47(1): 86 – 89.

[14] S. Ida, D. Shiga, M. Koinuma, Y. Matsumoto. Synthesis of hexagonal nickel hydroxide nanosheets by exfoliation of layered nickel hydroxide intercalated with dodecyl sulfate ions [J]. Journal of the American Chemical Society, 2008, 130(43): 14038 – 14039.

[15] T. W. Kim, E. J. Oh, A. Y. Jee, S. T. Lim, D. H. Park, M. Lee, S. H. Hyun, J. H. Choy, S. J. Hwang. Synthesis of hexagonal nickel hydroxide nanosheets by exfoliation of layered nickel hydroxide intercalated with dodecyl sulfate ions [J]. Journal of the American Chemical Society, 2008, 130(43): 14038 – 14039.

[16] (a) G. B. Saupe, C. C. Waraksa, H. N. Kim, Y. J. Han, D. M. Kaschak, D. M. Skinner, T. E. Mallouk. Nanoscale tubules formed by exfoliation of potassium hexaniobate [J]. Chemistry of Materials, 2000, 12(6): 1556 – 1562. (b) R. E. Schaak, T. E. Mallouk. Prying apart ruddlesden-popper phases: exfoliation into sheets and nanotubes for assembly of perovskite thin films [J]. Chemistry of Materials, 2000, 12(11): 3427 – 3434.

(c) Y. Kobayashi, H. Hata, M. Salama, T. E. Mallouk. Scrolled sheet precursor route to niobium and tantalum oxide nanotubes [J]. Nano Letters, 2007, 7(7): 2142-2145.

[17] L. M. Viculis, J. J. Mack, R. B. Kaner. A chemical route to carbon nanoscrolls [J]. Science, 2003, 299(5611): 1361-1361.

[18] R. Ma, Y. Bando, T. Sasaki. Directly rolling nanosheets into nanotubes [J]. The Journal of Physical Chemistry B, 2004, 108(7): 2115-2119.

[19] (a) L. Jiao, L. Zhang, X. Wang, G. Diankov, H. Dai. Narrow graphene nanoribbons from carbon nanotubes [J]. Nature, 2009, 458(7240): 877-880. (b) L. Jiao, X. Wang, G. Diankov, H. Wang, H. Dai. Facile synthesis of high-quality graphene nanoribbons [J]. Nature Nanotechnology, 2010, 5(5): 321-325.

[20] A. G. Cano-Márquez, F. J. Rodríguez-Macías, J. Campos-Delgado, C. G. Espinosa-González, F. Tristán-López, D. Ramírez-González, D. A. Cullen, D. J. Smith, M. Terrones, Y. I. Vega-Cantú. Ex-MWNTs: graphene sheets and ribbons produced by lithium intercalation and exfoliation of carbon nanotubes [J]. Nano Letters, 2009, 9(4): 1527-1533.

[21] D. V. Kosynkin, A. L. Higginbotham, A. Sinitskii, J. R. Lomeda, A. Dimiev, B. K. Price, J. M. Tour. Longitudinal unzipping of carbon nanotubes to form graphene nanoribbons [J]. Nature, 2009, 458(7240): 872-876.

[22] (a) D. L. Bish, A. Livingstore. The crystal chemistry and paragenesis of honessite and hydrohonessite: the sulphate analogues of reevesite [J]. Mineralogical Magazine, 1981, 44 (335): 339-343. (b) M. Kurmoo. Hard magnets based on layered cobalt hydroxide: the importance of dipolar interaction for long-range magnetic ordering [J]. Chemistry of Materials, 1999, 11(11): 3370-3378. (c) L. Cao, F. Xu, Y. Y. Liang, H. L. Li. Preparation of the novel nanocomposite $Co(OH)_2$/ultra-stable Y zeolite and its application as a supercapacitor with high energy density [J]. Advanced Materials, 2004, 16(20): 1853-1857. (d) M. Oshitani, H. Yufu, K. Takashima, S. Tsuji, Y. Matsumaru. Development of a pasted nickel electrode with high active material utilization [J]. Journal of The Electrochemical Society, 1989, 136(6): 1590-1593.

[23] Z. Liu, R. Ma, M. Osada, K. Takada, T. Sasaki. Selective and controlled synthesis of α-cobalt and β-cobalt hydroxides in highly developed hexagonal platelets [J]. Journal of the American Chemical Society, 2005, 127(40): 13869-13874.

[24] R. Ma, Z. Liu, K. Takada, K. Fukuda, Y. Ebina, Y. Bando, T. Sasaki. Tetrahedral Co(II) coordination in α-type cobalt hydroxide: Rietveld refinement and X-ray absorption spectroscopy [J]. Inorganic Chemistry, 2006, 45(10): 3964-3969.

[25] P. C. Tsai, T. H. Fang. A molecular dynamics study of the nucleation, thermal stability and nanomechanics of carbon nanocones [J]. Nanotechnology, 2007, 18(10): 105702.

[26] (a) M. A. Ulibarri, J. M. Fernández, F. M. Labajos, V. Rives. Anionic clays with variable valence cations: synthesis and characterization of cobalt aluminum hydroxide carbonate hydrate

$[Co_{1-x}Al_x(OH)_2](CO_3)_{x/2} \cdot nH_2O$ [J]. Chemistry of Materials, 1991, 3(4): 626-630. (b) S. Velu, K. Suzuki, S. Hashimoto, N. Satoh, F. Ohashi, S. Tomura. The effect of cobalt on the structural properties and reducibility of CuCoZnAl layered double hydroxides and their thermally derived mixed oxides [J]. Journal of Materials Chemistry, 2001, 11(8): 2049-2060.

[27] (a) V. Pralong, A. Delahaye-Vidal, B. Beaudoin, B. Gérand, J M. Tarascon. Oxidation mechanism of cobalt hydroxide to cobalt oxyhydroxide [J]. Journal of Materials Chemistry, 1999, 9(4): 955-960. (b) Y. C. Zhu, H. L. Li, Y. Koltypin, A. Gedanken. Preparation of nanosized cobalt hydroxides and oxyhydroxide assisted by sonication [J]. Journal of Materials Chemistry, 2002, 12(3): 729-733. (c) J. C. Myers, R. L. Penn. Evolving surface reactivity of cobalt oxyhydroxide nanoparticles [J]. The Journal of Physical Chemistry C, 2007, 111(28): 10597-10602.

☆ X. H. Liu, R. Z. Ma, Y. Bando, T. Sasaki. Layered cobalt hydroxide nanocones: microwave-assisted synthesis, exfoliation and structural modification [J]. Angewandte Chemie International Edition, 2010, 49(44): 8253-8256.

A General Strategy to Layered Transition-Metal Hydroxide Nanocones: Tuning the Composition for High Electrochemical Performance

Abstract: A general and facile strategy for the synthesis of a large family of monometallic (Co, Ni) and bimetallic (Co − Ni, Co − Cu and Co − Zn) hydroxide NCs intercalated with DS ions is demonstrated. The basal spacing of the NCs can be varied by adjusting the intercalated DS amount. Especially, electrochemical characterizations reveal that bimetallic Co − Ni hydroxide NCs have a higher specific capacitance than their monometallic counterpart. These results suggest the importance of rational designing layered hydroxide NCs with tuned transition-metal composition for high-performance energy storage devices.

1 Introduction

Layered hydroxides, consisting of metal-hydroxyl host slabs and/or charge-balancing anions in the interlayer galleries, afford a large variety of functionality and hybrid possibility for potential applications as anion exchangers,[1] adsorbents,[2] catalysts,[3] active electrode materials,[4] drug delivery systems,[5] and photofunctional materials.[6] Especially, layered hydroxides can also be artificially exfoliated into unilamellar nanosheets.[7-9] The positively charged nanosheets can be employed as functional building blocks to fabricate diverse nanoarchitectures and novel nanodevices with unique functionalities.[10] A wide variety of layered transition-metal (Co, Ni, Zn) hydroxides, including layered double hydroxides (LDHs), and layered rare-earth hydroxides have recently received intense attention for the exploration of hydroxide nanosheets.[11-14] Generally, euhedral hydroxide crystals adopt a lamellar or plate-like morphology. Recently, we reported a unique synthetic process to prepare layered cobalt hydroxide nanocones (NCs) originating from the rolling-up of lamellae using hexamethylenetetramine (HMT, $C_6H_{12}N_4$) as an alkaline reagent, which then underwent exfoliation in formamide into unilamellar nanosheets.[15] If such a rolling-up protocol and peculiar morphological feature can be applied to other transition-metal hydroxides, it may offer a vast opportunity to rationally design NCs with controllable composition, further stimulating the research for functional nanomaterials derived from hydroxides.

In particular, layered transition-metal (Co, Ni, etc.) hydroxides with large interlayer spacing possess very desirable electrochemical activity derived from their redox character and better accessibility for the reaction species.[16] Furthermore, if Ni and Co are co-incorporated in the host layer, i.e., bimetallic Co – Ni hydroxides, an improved capacity and cycling stability may be expected in comparison with monometallic hydroxides,[16,17] which therefore offer an effective way for achieving high electrochemical performance.

Here we describe a facile and reliable synthetic strategy for the production of high-quality NCs with tunable transition-metal composition, variable basal spacing and high aspect ratio using urea [$CO(NH_2)_2$] as an alkaline reagent. Significantly, the current strategy is capable of synthesizing a large family of monometallic (Co, Ni) and bimetallic (Co – Ni, Co – Cu and Co – Zn) hydroxide NCs by tuning the molar ratios of metal sources. Furthermore, they can be exfoliated into unilamellar nanosheets with a homogeneous thickness of about one nanometer through soft chemical procedures. It was found that NCs with larger basal spacing could be more easily exfoliated possibly due to the advantage of an expanded interlayer gallery. Especially, electrochemical characterizations reveal that bimetallic Co – Ni hydroxide NCs exhibit very high specific capacitance and remarkable cycling stability, which may enable their practical applications as active electrode materials in supercapacitor and rechargeable battery.

2 Experimental Section

All chemicals used were of analytical grade and were purchased from Wako Chemical Reagents Company (Japan). They were used without further purification. Milli – Q water was used throughout the experiments.

2.1 Synthesis of layered monometallic hydroxides NCs. In a typical synthesis, 1 mmol $CoCl_2 \cdot 6H_2O$, 3 mmol urea, and 5 mmol SDS were charged into an autoclave of 100 mL capacity and dissolved in 50 mL Milli – Q water under stirring at room temperature. The autoclave was filled with water up to 80% of the total volume, sealed and microwave-heated at 100 – 140 ℃ for 1 – 3 h under magnetic stirring. The autoclave was allowed to cool down to room temperature naturally. The resulting green product was filtered, washed with Milli – Q water and absolute ethanol, and finally dried in air at 60 ℃ for 5 h.

2.2 Synthesis of layered bimetallic hydroxides NCs. In a typical synthesis, 0.5 mmol $CoCl_2 \cdot 6H_2O$, 3 mmol urea, and 5 mmol SDS were charged into an autoclave of 100 mL capacity and dissolved in 50 mL Milli − Q water under stirring at room temperature. The autoclave was filled with water up to 80% of the total volume, sealed and microwave-heated at 100 ℃ for 0.5 h under magnetic stirring. The autoclave was allowed to cool down to room temperature naturally. Subsequently, designed amount of $NiCl_2 \cdot 6H_2O$ (e.g., 0.5 mmol for $Co_{0.5}-Ni_{0.5}$) was added in the translucent red solution and stirred for 5 min. The autoclave was again sealed and microwave-heated at 100 − 140 ℃ for 1 − 3 h under magnetic stirring, and the system was allowed to cool down to room temperature naturally. The washing and collecting procedures were the same to those for monometallic hydroxide synthesis.

2.3 Anion exchange and exfoliation of layered cobalt hydroxides NCs. The NCs intercalated with various anions were prepared by an ethanol-assisted anion-exchange approach. For instance, as-prepared 0.2 g NCs was dispersed into the ethanol-water binary solution (200 mL, 1:1 v/v) of 1 M $NaNO_3$, 1 M NaCl, or 1 M $NaClO_4$ in a flask, respectively. The flask was tightly capped and mechanically shaken for 48 h at room temperature after purging with nitrogen gas. The product was filtered, washed with anhydrous ethanol and air-dried at 60 ℃ for 5 h.

To prepare hydroxides nanosheets, as-prepared NCs (0.05 g) were dispersed in 100 cm^3 of formamide in a flask. After purging with nitrogen gas and capped tightly, the flask was agitated in a mechanical shaker at a speed of 140 r/min for 12 − 48 h. The resulted translucent colloidal suspension was further treated by centrifugation at 6000 r/min for 20 min to remove possible unexfoliated NCs.

2.4 Characterizations. X-ray diffraction (XRD) data were recorded on a Rigaku Rint − 2200 diffractometer with a monochromatic Cu K_α radiation ($\lambda = 1.5405$ Å). The morphologies and dimensions of as-prepared products were examined with a JEOL JSM − 6700F field emission scanning electron microscope (FE − SEM). Transmission electron microscopy (TEM) characterizations were performed on a JEOL JEM − 3000F transmission microscope, and JEM − 3100F energy-filtering (Omega type) microscope equipped with an energy dispersive X-ray spectrometer (EDS) and elemental mapping/profiling capacity. A Seiko SPA 400 atomic force microscope (AFM) was used to examine the topography of the nanosheets deposited on Si wafers. A cleaned Si wafer was immersed in the colloidal formamide suspension for 20 min, which was followed by rinsing with a copious amount of water and drying under a N_2 stream. AFM images were acquired in tapping mode using a Si tip cantilever with a force constant of 20 N/m. Thermogravimetric differential thermal measurements (TG-DTA) were carried out using a Rigaku TGA – 8120 instrument in a temperature range of 25 – 1000 ℃ at a heating rate of 1 ℃/min under an air flow.

2.5 Electrochemical Measurement. Cyclic voltammetry (CV) and galvanic charge-discharge (CD) were carried out on a Solartron electrochemistry workstation using a three electrode cell with 1 M KOH as the electrolyte, Ag/AgCl electrode as reference electrode, platinum wire as counter electrode, respectively. The hydroxide NCs deposited on graphite substrates were used as working electrode. The working electrode was prepared as follows: 80 wt. % of as-prepared NCs was mixed with 10 wt. % of acetylene black (> 99.9%) and 10 wt. % of polytetrafluoroethylene (PTFE, > 99.9%) in an agate mortar until a homogeneous black powder was obtained. Then a few drops of ethanol were added. After briefly allowing the ethanol to evaporate, the suspension was dropped on a 1×1 cm^2 graphite substrate. The electrode was dried at 80 ℃ for 12 h in air and then was pressed under 40 MPa before electrochemical measurement. Specific capacitance could be calculated from the galvanostatic charge and discharge curves, using the following equation: $C = I\Delta t/m\Delta V$, where I is charge or discharge current, Δt is the time for a full charge or discharge, m indicates the mass of the active material, and ΔV represents the voltage change after a full charge or discharge.

3 Results and Discussion

Layered Co hydroxide NCs were successfully prepared via a microwave-assisted hydrothermal reaction of aqueous cobalt chloride and urea in the presence of sodium dodecyl sulfate (SDS). Figure 1(a) displays a typical scanning electron microscopy (SEM) image of the green-colored product obtained using 3 mmol urea and 5 mmol SDS at 100 ℃ for 1 h. Abundant cone-like objects with uniform morphology and almost 100% purity can be observed. The high-magnification SEM image shown in Figure 1(b) clearly reveals that the product possesses a regular conical shape with a hollow interior. Typical transmission electron microscopy (TEM) image [Figure 1(c)] also confirms the conical structures with hollow interiors. The NCs have an average bottom diameter of approximate 500 nm and tip diameter of about 50 nm, whereas a length up to 5 μm. The aspect ratio (length/diameter), > 10, is much higher than those previously obtained using HMT.[15] Urea, with a slower hydrolysis rate of releasing

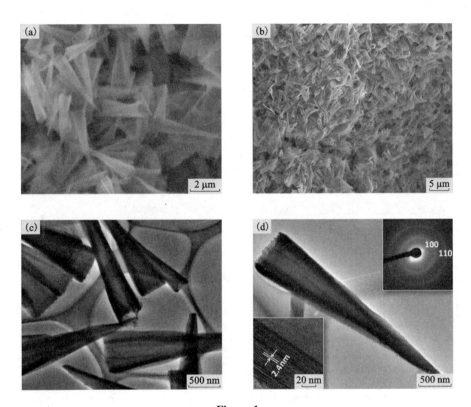

Figure 1
(a) Low- and (b) high-magnification SEM images of as-prepared NCs of layered Co hydroxide. (c) TEM image. (d) TEM and HRTEM (lower inset) images of an individual NC. The upper inset is an SAED pattern.

ammonia than that of HMT, provides more favorable conditions for the formation of hydroxide NCs. It is noteworthy that NCs could not be obtained in the absence of surfactant SDS, which can be indexed to cobalt hydroxide carbonate, namely $Co(OH)_{0.44}(CO_3)_{0.78} \cdot 0.29H_2O$. Figure 1(d) depicts a typical TEM image of an individual NC. The upper inset in Figure 1(d) shows the selected area electron diffraction (SAED) pattern, which can be indexed to [001] zone-axis diffraction of hexagonal cobalt hydroxide with an in-plane lattice constant of $a = 0.31$ nm. The "oval-shaped" diffraction pattern could result from the rolling-up and stacking of a nanosheet-like lamella.[18] The corresponding high-resolution transmission electron microscopy (HRTEM) image of the cone wall shown in lower inset reveals an interlayer spacing of about 2.4 nm.

As shown in Figure 2(a), the layered structure of as-prepared NCs was also identified by X-ray diffraction (XRD). A basal reflection series corresponding to an interlayer distance of 2.4 nm can be clearly discerned. The NCs, characteristic of a green color, might be classified as DS^- – intercalated α – type cobalt hydroxide with both octahedral and tetrahedral coordination.[19] The chemical formula of as-prepared NCs using 5 mmol SDS was estimated to be $Co(OH)_{1.75}DS_{0.25} \cdot 0.9H_2O$ based on thermogravimetric (TG) analysis. By increasing the amounts of SDS (e.g., using 20 mmol SDS), the interlayer distance of the NCs can be notably expanded to 3.7 nm.

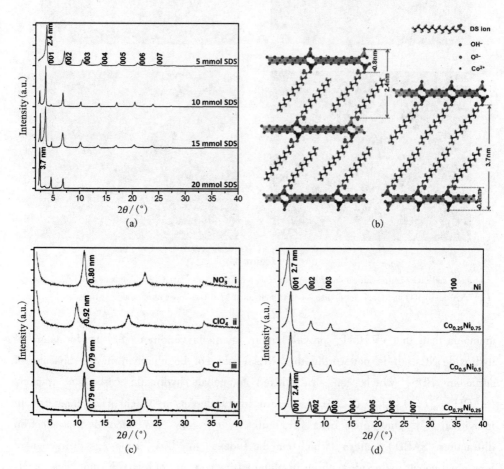

Figure 2

(a) XRD patterns of layered Co hydroxide NCs obtained using different amounts of SDS. (b) Plausible single-layer and double-layer models for NCs obtained using 5 mmol and 20 mmol SDS, respectively. (c) XRD patterns of NCs after anion-exchange: i)-iii) NCs obtained using 5 mmol SDS after exchange with NO_3^-, ClO_4^-, and Cl^-, respectively; iv) NCs obtained using 20 mmol SDS after exchange with Cl^-. (d) XRD patterns of layered Co – Ni hydroxide NCs at varied Co/Ni ratios.

The chemical formula was quantified as $Co(OH)_{1.60}DS_{0.40} \cdot H_2O$, which indicates a significant increase of DS contents in the gallery. The thickness of α – type hydroxide layers consisting of octahedral and tetrahedral coordination is about 0.80 nm,[20] whereas the length of DS ion is 1.78 nm.[21] The tetrahedral fourth apex pointing into the interlayer space may coordinate with the sulfate group of a DS ion. Therefore, the interlayer spacing of 2.4 nm and 3.7 nm might correspond to a single- and double-layer grafting of DS ions at a slant angle of about 53° between the host layers, i.e., $Co_3^{octa}Co^{tetra}(OH)_7DS$ and $Co_3^{octa}Co_2^{tetra}(OH)_8DS_2$ (Superscripts indicate octahedral and tetrahedral coordination), respectively. Figure 2(b) depicts possible structure models for the NCs obtained using 5 mmol and 20 mmol SDS, respectively. In other words, synthesis using a larger amount of SDS may cause an increase in tetrahedral sites. As a result, DS could not park in an interpenetrating form, leading to a double-layer intercalation.[21] In both cases, the interlayer DS could be readily exchanged with various anions by a conventional anion-exchange procedure at room temperature while the conical morphology is retained. As shown in Figure 2(c), the interlayer spacing, 2.4 nm for the single-layer DS^- – intercalated NCs, shifted to 0.8 nm for NO_3^-, 0.92 nm for ClO_4^-, and 0.79 nm for Cl^- – exchanged NCs, respectively. On the other hand, double-layer DS^- – intercalated NCs with a basal spacing of 3.7 nm was converted into Cl^- form with the same gallery height of 0.79 nm. The data validates that the expansion of interlayer gallery from 2.4 nm to 3.7 nm is originated from the increase of DS content, whereas the basic structural framework and thickness of host slabs seem identical.

More importantly, the current synthetic strategy could be extended to prepare layered hydroxide NCs with various transition-metal compositions. Typical XRD patterns of layered Co – Ni hydroxides at varied Co/Ni ratios are shown in Figure 2(d). The initial value of 2.4 nm for bimetallic Co – Ni hydroxides with a low Ni content is close to that of monometallic Co hydroxide NCs. However, the interlayer distance is slightly expanded to about 2.7 nm for Ni hydroxide NCs. The reason for the expansion might be related to some change in the coordination or orientation of DS in the gallery. SEM images of Ni, $Co_{0.25}-Ni_{0.75}$, $Co_{0.5}-Ni_{0.5}$, and $Co_{0.75}-Ni_{0.25}$ hydroxide NCs are given in Figure 3(a) – 3(d), respectively. With a higher ratio of cobalt, the conical feature appears more obvious and the aspect ratio is also larger. In addition, bimetallic Co – Cu and Co – Zn hydroxide NCs could also be successfully prepared in the same process, as shown in Figure 3(e) and 3(f), verifying the versatility of this synthetic approach.

The elemental maps of an individual DS^- – intercalated $Co_{0.5}-Ni_{0.5}$ hydroxide NC

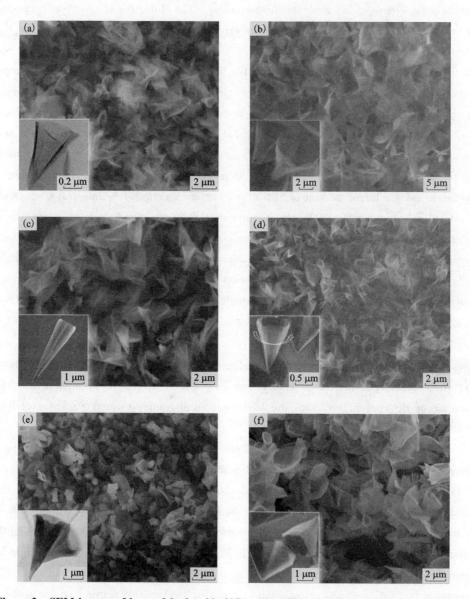

Figure 3 SEM images of layered hydroxide NCs with different transition-metal compositions
(a) Ni, (b) $Co_{0.25}$–$Ni_{0.75}$, (c) $Co_{0.5}$–$Ni_{0.5}$, (d) $Co_{0.75}$–$Ni_{0.25}$, (e) $Co_{0.5}$–$Cu_{0.5}$, and (f) $Co_{0.5}$–$Zn_{0.5}$. Insets depict the corresponding high-magnification images of individual NCs.

are displayed in Figure 4, which clearly demonstrates a homogeneous distribution of Co, Ni, and S elements. The line-scan analysis in Figure 4(a) reveals the profile of Co and

Ni atoms across the radial direction, indicating an apparent hollow feature. These results further confirm that Co and Ni were indeed incorporated into the host layer of individual NCs without segregation.

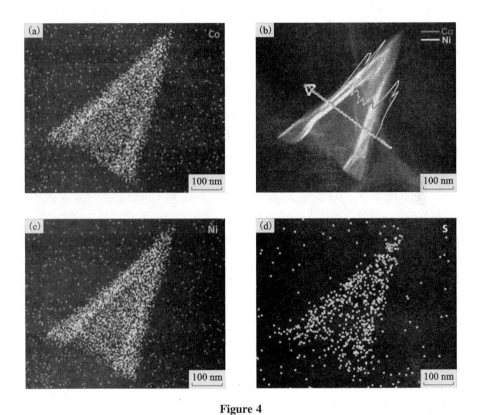

Figure 4

(a) TEM image of an individual DS$^-$ – intercalated Co$_{0.5}$ – Ni$_{0.5}$ hydroxide NC with line-scan profile across the radial direction. Elemental maps of (b) Co, (c) Ni and (d) S.

The evolution of the lamellar structure during the synthesis directly mirrors a possible rolling-up scenario for the formation of NCs. Figure 5(a) – 5(d) summarizes a series of morphological observations supposed to be in different stages of forming NCs. Firstly, layered hydroxide lamella intercalated with DS (e. g., thickness in the ranges of one to a few nanometers, width in the ranges of several hundred nanometers to several micrometers, etc.) was initially formed [Figure 5(a)]. The lamellar structure tends to form a conical angle ranging from 10° to 60° due to the curve of the edge [Figure 5(b)]. Subsequently, it gradually rolls up and naturally grows following the conical angle to form a cone-shape object [Figure 5(c)]. Finally, the lamella can fold up onto itself to form a perfect and regular NC, as shown in Figure 5(d).

Through dispersing the NCs in formamide and agitating in a mechanical shaker, a translucent colloidal suspension was yielded, suggesting the occurrence of exfoliation. Interestingly, NCs with an interlayer spacing of 3.7 nm could be more quickly exfoliated than those with a basal value of 2.4 nm, which might be attributed to the advantage of further separating the hydroxide layers and a more hydrophobic interlayer environment for the penetration of formamide. Atomic force microscopy (AFM) images of as-prepared nanosheets with different compositions are shown in Figure 5(e) – 5(h). The thickness of the nanosheets was measured to be unexceptionally about 1.0 nm, which is close to the crystallographic thickness of the host layer. All the results support that a conical structure may be formed by the rolling-up of a lamella, and it can also be unwrapped/exfoliated into unilamellar nanosheets. Such observations are fundamentally important for revealing the formation mechanism of NCs as well as the energy balance between nanocones/nanotubes and nanosheets.

The electrochemical behavior of layered $Co_{1-x}Ni_x$ ($x = 0, 0.25, 0.5, 0.75, 1.0$) hydroxide NCs were investigated by cyclic voltammetry (CV) and galvanic charge-discharge (CD) in a three-electrode cell with a Ag/AgCl reference electrode and 1 M KOH aqueous electrolyte. Figure 6(a) shows typical CV curves in a potential range of -0.2 V to 0.4 V at a scan rate of 10 mV/s. The redox current peaks were explicitly observed, which can be related to a combined effect of the following redox reactions derived from transition-metal hydroxides: $Co(OH)_2 + OH^- \rightleftharpoons CoOOH + H_2O + e^-$ and $Ni(OH)_2 + OH^- \rightleftharpoons NiOOH + H_2O + e^-$, respectively.[22-24] The background signal originated from graphite substrates was almost negligible. The oxidation (E_o) and reduction (E_r) peaks at -40 mV and -85 mV for monometallic Co hydroxide NCs shift to 260 mV and 145 mV for monometallic Ni hydroxide NCs, respectively, due to the higher redox potential of $Ni(OH)_2/NiOOH$ couple. More notably, the oxidation and reduction peaks are gradually enhanced and shifted to a higher potential with the increase of Ni content in the bimetallic $Co_{1-x}Ni_x$ ($x = 0.25, 0.5, 0.75$) host slabs. The highest redox current was observed for $Co_{0.5}-Ni_{0.5}$ hydroxide NCs, suggesting the feasibility of rationally tuning the transition-metal composition for optimum electrochemical performance.

Figure 6(b) shows the specific capacitances of $Co_{1-x}Ni_x$ hydroxide NCs at a galvanic current density of 10 A/g, measured from charge-discharge curves. The specific capacitances are approximate 490, 1100, 1580, 1400, and 740 F/g for Co, $Co_{0.75}-Ni_{0.25}$, $Co_{0.5}-Ni_{0.5}$, $Co_{0.25}-Ni_{0.75}$, and Ni hydroxides NCs, respectively.

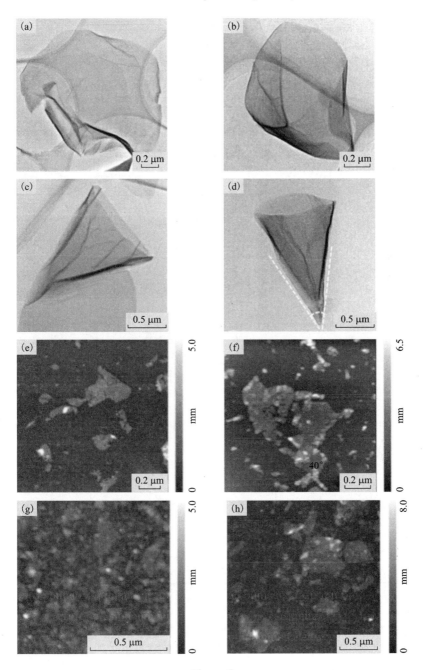

Figure 5

(a) – (d) TEM images exhibit a possible rolling-up process of lamellae into NCs. AFM images of the exfoliated nanosheets of (e) Co hydroxide NCs using 5 mmol SDS, (f) Co hydroxide NCs using 20 mmol SDS, (g) Ni hydroxide NCs, and (h) $Co_{0.5}$–$Ni_{0.5}$ hydroxide NCs.

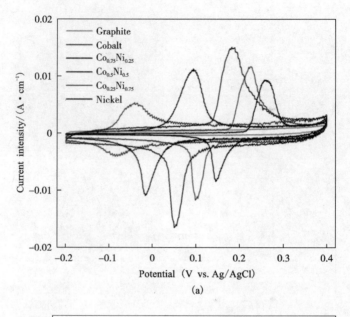

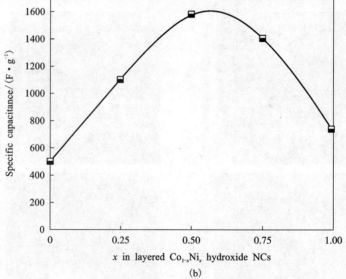

Figure 6

Electrochemical characterizations of layered hydroxide NCs with different composition of Co, $Co_{0.25}$–$Ni_{0.75}$, $Co_{0.5}$–$Ni_{0.5}$, $Co_{0.75}$–$Ni_{0.25}$, and Ni deposited on graphite substrates. (a) CV curvesat a scan rate of 10 mV/s. (b) Specific capacitance of NCs with different Co–Ni compositionat a charge and discharge current density of 10 A/g.

Interestingly, the specific capacitances of bimetallic hydroxides NCs are generally superior to those of monometallic hydroxides NCs. Consistent with the CV profiles, $Co_{0.5}$ – $Ni_{0.5}$ hydroxide NCs possess the highest specific capacitance, which may be attributed to the enhancement of the electro-active sites participated in the redox reaction due to possible valence interchange or charge hopping between Co and Ni cations.[25] The specific capacitance of layered $Co_{0.5}$ – $Ni_{0.5}$ hydroxides NCs is higher than that of reported $Ni(OH)_2$/graphene composites,[26] and is even comparable with that of noble metal oxide RuO_2.[27] In addition, the specific capacitance retains as high as 1560 F/g, 98.7% of the initial value, after 100 cycles. Such a high capacity and high rate capability may be particularly advantageous for potential uses in next-generation redox supercapacitors. Furthermore, in nickel-cadmium (Ni – Cd) and nickel-metal hydride (Ni – MH) alkaline rechargeable batteries, Co hydroxide is usually used as an additive to enhance the conductivity and stability of $Ni(OH)_2$ electrodes.[28] The present Co – Ni hydroxide NCs, taking advantage of their tunable bimetallic composition, peculiar hollow feature and large interlayer spacing, is more promising as an ideal active material for the development of high-performance energy storage applications.

4 Conclusion

In summary, a facile and general synthetic strategy was successfully developed for the production of a large family of layered transition-metal hydroxide NCs using urea as an alkaline reagent. The basal spacing of the layered hydroxide NCs can be adjusted by intercalating different amounts of DS, which in turn impacts the exfoliating property of resultant products into unilamellar nanosheets. These hydroxide nanocones and nanosheets with controllable transition-metal compositions may be very useful in exploring new magneto-optical or electrochemical devices. The electrochemical measurements revealed that transition-metal (Co, Ni) hydroxide NCs, exhibit a high specific capacitance and remarkable cycling stability, which is of great importance for their practical applications in high-performance supercapacitor and rechargeable battery. The general synthetic strategy reported here is expected to be applied to other systems (LDHs, layered rare-earth hydroxides, etc.), and is thus promising for achieving unique conical structure with hollow interior in various functional hydroxides for a wide range of potential applications.

References

[1] (a) D. L. Bish. Anion-exchange in takovite: applications to other hydroxide minerals [J]. Bull. Mineral, 1980, 103(1): 170 – 175. (b) P. J. Sideris, U. G. Nielsen, Z. Gan, C. P. Grey. Mg/Al ordering in layered double hydroxides revealed by multinuclear NMR spectroscopy [J]. Science, 2008, 321(5885): 113 – 117. (c) Y. Du, D. O'Hare. Observation of staging during intercalation in layered α – cobalt hydroxides: a synthetic and kinetic study [J]. Inorganic Chemistry, 2008, 47(24): 11839 – 11846. (d) B. Sylvia, K. P. Vishnu. Polytypism, disorder, and anion exchange properties of divalent ion (Zn, Co) containing bayerite-derived layered double hydroxides [J]. Inorganic Chemistry, 2010, 49(24): 11370 – 11377. (e) F. Geng, R. Ma, T. Sasaki. Anion-exchangeable layered materials based on rare-earth phosphors: unique combination of rare-earth host and exchangeable anions [J]. Accounts of Chemical Research, 2010, 43(9): 1177 – 1185. (f) R. Ma, J. Liang, K. Takada, T. Sasaki. Topochemical synthesis of Co – Fe layered double hydroxides at varied Fe/Co ratios: unique intercalation of triiodide and its profound effect [J]. Journal of the American Chemical Society, 2010, 133(3): 613 – 620.

[2] (a)S. Tezuka, R. Chitrakar, A. Sonoda, K. Ooi, T. Tomida. Studies on selective adsorbents for oxo-anions. Nitrate ion-exchange properties of layered double hydroxides with different metal atoms [J]. Green Chemistry, 2004, 6(2): 104 – 109. (b) X. Guo, F. Zhang, Q. Peng, S. Xu, X. Lei, D. G. Evans, X. Duan. Layered double hydroxide/eggshell membrane: An inorganic biocomposite membrane as an efficient adsorbent for Cr (VI) removal [J]. Chemical Engineering Journal, 2011, 166(1): 81 – 87.

[3] (a) B. Sels, D. De Vos, M. Buntinx, F. Pierard, A. Kirsch-De Mesmaeker, P. Jacobs. Layered double hydroxides exchanged with tungstate as biomimetic catalysts for mild oxidative bromination [J]. Nature, 1999, 400(6747): 855 – 857. (b)B. F. Sels, D. E. De Vos, P. A. Jacobs. Hydrotalcite-like anionic clays in catalytic organic reactions [J]. Catalysis Reviews, 2001, 43(4): 443 – 488. (c)B. M. Choudary, S. Madhi, N. S. Chowdari, M. L. Kantam, B. Sreedhar. Layered double hydroxide supported nanopalladium catalyst for Heck-, Suzuki-, Sonogashira-, and Stille-type coupling reactions of chloroarenes [J]. Journal of the American Chemical Society, 2002, 124(47): 14127 – 14136.

[4] T. Y. Wei, C. H. Chen, H. C. Chien, S. Y. Lu, C. C. Hu. A cost-effective supercapacitor material of ultrahigh specific capacitances: spinel nickel cobaltite aerogels from an epoxide-driven sol-gel process [J]. Advanced Materials, 2010, 22(3): 347 – 351.

[5] J. H. Yang, Y. S. Han, M. Park, T. Park, S. J. Hwang, J. H. Choy. New inorganic-based drug delivery system of indole-3-acetic acid-layered metal hydroxide nanohybrids with controlled release rate [J]. Chemistry of Materials, 2007, 19(10): 2679 – 2685.

[6] (a)M. Ogawa, K. Kuroda. Photofunctions of intercalation compounds [J]. Chemical Reviews,

1995, 95(2): 399 – 438. (b) K. Lang, P. Bezdička, J. L. Bourdelande, J. Hernando, I. Jirka, E. Káfuňková, F. Kovanda, P. Kubát, J. Mosinger, D. M. Wagnerová. Layered double hydroxides with intercalated porphyrins as photofunctional materials: subtle structural changes modify singlet oxygen production [J]. Chemistry of Materials, 2007, 19(15): 3822 – 3829.

[7] (a) R. Ma, Z. Liu, K. Takada, N. Iyi, Y. Bando, T. Sasaki. Synthesis and exfoliation of $Co^{2+} - Fe^{3+}$ layered double hydroxides: An innovative topochemical approach [J]. Journal of the American Chemical Society, 2007, 129 (16): 5257 – 5263. (b) R. Ma, K. Takada, K. Fukuda, N. Iyi, Y. Bando, T. Sasaki. Topochemical synthesis of monometallic ($Co^{2+} - Co^{3+}$) layered double hydroxide and its exfoliation into positively charged $Co(OH)_2$ nanosheets [J]. Angewandte Chemie International Edition, 2008, 47(1): 86 – 89. (c) J. Liang, R. Ma, N. Iyi, Y. Ebina, K. Takada, T. Sasaki. Topochemical synthesis, anion exchange, and exfoliation of Co – Ni layered double hydroxides: a route to positively charged Co – Ni hydroxide nanosheets with tunable composition [J]. Chemistry of Materials, 2009, 22(2): 371 – 378.

[8] S. Ida, Y. Sonoda, K. Ikeue, Y. Matsumoto. Drastic changes in photoluminescence properties of multilayer films composed of europium hydroxide and titanium oxide nanosheets [J]. Chemical Communications, 2010, 46: 877 – 879.

[9] (a) S. O'Leary, D. O'Hare, G. Seeley. Delamination of layered double hydroxides in polar monomers: new LDH-acrylate nanocomposites [J]. Chemical Communications, 2002, 14: 1506 – 1507. (b) T. Hibino, M. Kobayashi. Delamination of layered double hydroxides in water [J]. Journal of Materials Chemistry, 2005, 15(6): 653 – 656.

[10] R. Ma, T. Sasaki. Nanosheets of oxides and hydroxides: ultimate 2D charge-bearing functional crystallites [J]. Advanced Materials, 2010, 22(45): 5082 – 5104.

[11] (a) M. Adachi-Pagano, C. Forano, J. P. Besse, Delamination of layered double hydroxides by use of surfactants [J]. Chemical Communications, 2000, 1: 91 – 92; (b) T. Hibino. Delamination of layered double hydroxides containing amino acids [J]. Chemistry of Materials, 2004, 16(25): 5482 – 5488. (c) Q. Wu, A. Olafsen, Ø. B. Vistad, J. Roots, P. Norby. Delamination and restacking of a layered double hydroxide with nitrate as counter anion [J]. Journal of Materials Chemistry, 2005, 15(44): 4695 – 4700.

[12] (a) L. Li, R. Ma, Y. Ebina, N. Iyi, T. Sasaki. Positively charged nanosheets derived via total delamination of layered double hydroxides [J]. Chemistry of Materials, 2005, 17(17): 4386 – 4391. (b) R. Ma, Z. Liu, L. Li, N. Iyi, T. Sasaki. Exfoliating layered double hydroxides in formamide: a method to obtain positively charged nanosheets [J]. Journal of Materials Chemistry, 2006, 16(39): 3809 – 3813.

[13] (a) S. Ida, D. Shiga, M. Koinuma, Y. Matsumoto. Synthesis of hexagonal nickel hydroxide nanosheets by exfoliation of layered nickel hydroxide intercalated with dodecyl sulfate ions [J]. Journal of the American Chemical Society, 2008, 130 (43): 14038 – 14039. (b) O. Altuntasoglu, Y. Matsuda, S. Ida, Y. Matsumoto. Syntheses of zinc oxide and zinc hydroxide single nanosheets [J]. Chemistry of Materials, 2010, 22(10): 3158 – 3164.

[14] (a)L. Hu, R. Ma, T. C. Ozawa, T. Sasaki. Exfoliation of layered europium hydroxide into unilamellar nanosheets [J]. Chemistry-An Asian Journal, 2010, 5(2): 248-251. (b)B. I. Lee, S. Y. Lee, S. H. Byeon. Grafting of dodecylsulfate groups on gadolinium hydroxocation nanosheets for self-construction of a lamellar structure [J]. Journal of Materials Chemistry, 2011, 21(9): 2916-2923.

[15] X. H. Liu, R. Ma, Y. Bando, T. Sasaki. Layered cobalt hydroxide nanocones: microwave-assisted synthesis, exfoliation, and structural modification [J]. Angewandte Chemie International Edition, 2010, 49(44): 8253-8256.

[16] Z. A. Hu, Y. L. Xie, Y. X. Wang, H. Y. Wu, Y. Y. Yang, Z. Y. Zhang. Synthesis and electrochemical characterization of mesoporous Co_xNi_{1-x} layered double hydroxides as electrode materials for supercapacitors [J]. Electrochimica Acta, 2009, 54(10): 2737-2741.

[17] W. Chen, Y. Yang, H. Shao, J. Fan. Tunable electrochemical properties brought about by partial cation exchange in hydrotalcite-like Ni-Co/Co-Ni hydroxide nanosheets [J]. The Journal of Physical Chemistry C, 2008, 112(44): 17471-17477.

[18] T. Sasaki, Y. Ebina, Y. Kitami, M. Watanabe, T. Oikawa. Two-dimensional diffraction of molecular nanosheet crystallites of titanium oxide [J]. The Journal of Physical Chemistry B, 2001, 105(26): 6116-6121.

[19] R. Ma, Z. Liu, K. Takada, K. Fukuda, Y. Ebina, Y. Bando, T. Sasaki. Tetrahedral Co (II) coordination in α-type cobalt hydroxide: rietveld refinement and X-ray absorption spectroscopy [J]. Inorganic Chemistry, 2006, 45(10): 3964-3969.

[20] Q. Tao, J. Yuan, R. L. Frost, H. He, P. Yuan, J. Zhu. Effect of surfactant concentration on the stacking modes of organo-silylated layered double hydroxides [J]. Applied Clay Science, 2009, 45(4): 262-269.

[21] A. Clearfield, M. Kieke, J. Kwan, J. L. Colon, R. C. Wang. Intercalation of dodecyl sulfate into layered double hydroxides [J]. Journal of Inclusion Phenomena and Molecular Recognition in Chemistry, 1991, 11(4): 361-378.

[22] S. Chen, J. Zhu, X. Wang. One-step synthesis of graphene-cobalt hydroxide nanocomposites and their electrochemical properties [J]. The Journal of Physical Chemistry C, 2010, 114 (27): 11829-11834.

[23] T. Zhao, H. Jiang, J. Ma. Surfactant-assisted electrochemical deposition of α-cobalt hydroxide for supercapacitors [J]. Journal of Power Sources, 2011, 196(2): 860-864.

[24] J. W. Lee, J. M. Ko, J. D. Kim. Hierarchical microspheres based on α-Ni(OH)$_2$ nanosheets intercalated with different anions: synthesis, anion exchange, and effect of intercalated anions on electrochemical capacitance [J]. The Journal of Physical Chemistry C, 2011, 115(39): 19445-19454.

[25] V. Gupta, S. Gupta, N. Miura. Potentiostatically deposited nanostructured Co_xNi_{1-x} layered double hydroxides as electrode materials for redox-supercapacitors [J]. Journal of Power Sources, 2008, 175(1): 680-685.

[26] H. Wang, H. S. Casalongue, Y. Liang, H. Dai. Ni(OH)$_2$ nanoplates grown on graphene as advanced electrochemical pseudocapacitor materials [J]. Journal of the American Chemical Society, 2010, 132(21): 7472 - 7477.

[27] C. C. Hu, W. C. Chen, K. H. Chang. How to achieve maximum utilization of hydrous ruthenium oxide for supercapacitors [J]. Journal of the Electrochemical Society, 2004, 151(2): A281 - A290.

[28] (a) X. P. Gao, H. X. Yang. Multi-electron reaction materials for high energy density batteries [J]. Energy & Environmental Science, 2010, 3(2): 174 - 189. (b) X. Y. Zhao, L. Q. Ma, X. D. Shen, Co - based anode materials for alkaline rechargeable Ni/Co batteries: a review [J]. Journal of Materials Chemistry, 2012, 22(2): 277 - 285.

☆ X. H. Liu, R. H. Ma, Y. Bando, T. Sasaki. A general strategy to layered transition-metal hydroxide nanocones: tuning the composition for high electrochemical performance [J]. Advanced Materials, 2012, 24(16): 2148 - 2153.

High-Yield Preparation, Versatile Structural Modification, and Properties of Layered Cobalt Hydroxide Nanocones

Abstract: A low-cost oil bath synthetic route is presented to produce uniform and highly crystalline layered cobalt hydroxide nanocones (NCs) intercalated with dodecyl sulfate anions ($C_{12}H_{25}OSO_3^-$, DS^-). A new exfoliating procedure, by gradually unravelling/unzipping these NCs through heat treatment in formamide-water binary solution, is developed to prepare unilamellar nanosheets. Moreover, the NCs can be readily modified with various inorganic or organic anions via a conventional anion-exchange process at ambient temperature. The exchanged product, for example, NO_3^- – intercalated NCs, could be more easily and rapidly transformed into cobalt oxides (e.g., Co_3O_4 and CoO) than the original DS^- – intercalated form while retaining a conical feature. Both the cobalt hydroxide NCs and exfoliated nanosheets are electrochemically redoxable, exhibiting a Faradaic pseudocapacitive behavior. The magnetic measurements further reveal both antiferromagnetic behaviors for transformed Co_3O_4 and CoO NCs. Their Néel temperature values are lower than those of bulk oxides due to finite size and geometric confinement effect. The peculiar conical feature of NCs with a hollow interior and tunable layer spacing, as well as exfoliated unilamellar nanosheets with all surface area exposed, may show promise for potential applications in electrochemical energy storage and magnetic devices.

1 Introduction

During the past few decades, layered metal hydroxides have drawn immense attention due to their distinctive compositional flexibility and anion exchangeability, promising a wide variety of practical and potential applications, for example, anion exchangers, adsorbents, catalysts, active electrode materials, magnetic resonance imaging (MRI) contrast agents, photofunctional materials, drug delivery systems, and so forth.[1] Generally, layered hydroxides adopt a lamellar or platy morphology resulting from a hexagonal in-plane symmetry.[2] Recently, we have achieved a significant breakthrough in the preparation of unique layered transition-metal (Co, Ni, Co – Ni, etc.) hydroxide nanocones (NCs) under microwave-assisted hydrothermal conditions using either hexamethylenetetramine (HMT, $C_6H_{12}N_4$) or urea [$CO(NH_2)_2$] as hydrolysis agents and sodium dodecyl sulfate (SDS) as a surfactant and structure-directing agent.[3] The redoxable hydroxide NCs, possessing the advantages of a peculiar hollow features and large interlayer spacing, are very useful as active materials for the development of high-performance electrochemical energy storage devices.[3b] Nevertheless, the microwave-assisted synthetic protocol generally involves a special apparatus and a low yield, which is unfavorable to fulfilling the application prospects of these hydroxide NCs. Therefore, it remains an urgent research topic to develop a simple synthetic approach for producing transition-metal hydroxide NCs in large quantities and high uniformity.

On the other hand, layered metal hydroxides, including NCs, could be directly exfoliated/delaminated into unilamellar nanosheets via some soft chemical procedures, for example, mechanical shaking or ultrasonic dispersing in formamide.[3,4] The unilamellar nanosheets, typically with a thickness under one nanometer versus lateral size ranging from the submicrometer level to several micrometers, are ideal two-dimensional (2D) building blocks to fabricate multilayer nanofilms with tunable compositions and exciting functionalities based on layer-by-layer (LBL) and/or Langmuir-Blodgett (LB) molecular-scale assembling techniques.[5] Nevertheless, the exfoliating mechanism is still not well understood. The exfoliating/delaminating procedure needs to be further modified or improved to obtain well-defined hydroxide nanosheets with homogeneous thickness, as well as to provide a justifiable explanation for the exfoliation of layered hydroxides in general.

In current work, we have developed a convenient and cost-effective synthetic strategy, namely, oil bath synthesis, to prepare layered cobalt hydroxide NCs based on urea hydrolysis in the presence of anionic surfactant dodecyl sulfate (DS). The oil bath method may be regarded as one of the most facile and effective solution-based synthetic methods, which are widely used in many chemical laboratories. This synthesis results in a high yield and can be easily scaled-up, which may be of great importance for future industrial applications. The influence of reaction temperature, duration time, and amount of surfactant on the morphology and size of layered cobalt hydroxide NCs was carefully investigated in detail. In particular, it was found that layered cobalt hydroxide NCs could be gradually exfoliated into unilamellar nanosheets in a formamide-water binary solution via heat treatment. Observation of the exfoliation stages implies a plausible unravelling scenario for the hydroxide NCs. Furthermore, interlayer DS anions of layered cobalt hydroxide NCs were readily exchangeable for various anions (e. g. , Cl^-, ClO_4^-, CH_3COO^-, and NO_3^-, etc.) at ambient temperature, which offers an obvious advantage for structural and functional modifications. For example, NO_3^- - intercalated cobalt hydroxide NCs could be more readily transformed into cobalt oxides (e. g. , Co_3O_4 and CoO) retaining original morphological features in comparison with the original DS^- - intercalated form. Futhermore, electrochemical characterizations suggested that a suitable combination of interlayer spacing and intercalated anionic species are required for achieving optimum electrochemical performance. Meanwhile, the magnetic properties of cobalt oxide NCs have been examined, revealing a weak antiferromagnetic interaction in Co_3O_4 NCs while a strong one in CoO NCs. Based on these discoveries, layered hydroxide NCs and their derivatives may be regarded as a new class of advanced nanomaterials ideal for further fundamental research and technological applications in various fields.

2 Experimental Section

All chemicals of analytical grade were purchased from Wako Chemical Reagents Company (Japan). They were used without further purification. Milli - Q water was used throughout the experiments.

2.1 Synthesis of layered hydroxide NCs. In a typical synthesis, $CoCl_2 \cdot 6H_2O$ (99.5%, 5 mmol), urea (99.0%, 35 mmol), and SDS (98.0%, 0 - 25 mmol) were charged into a three-neck flask and dissolved in 500 mL Milli - Q water to give the final concentrations of 10, 70, and 0 - 50 mM, respectively. The pH of the initial mixture was about 7. The flask was then placed in an oil bath and heated at 100 - 120℃ for 2 - 12 h under continuous magnetic stirring and a nitrogen gas protection. The resulting product was filtered, washed with Milli - Q water and absolute ethanol, and finally dried in air at 60℃ for 4 h. The yield of NCs was estimated as about 90% based on the calculation of cobalt source.

2.2 Exfoliation of layered cobalt hydroxide NCs. Layered cobalt hydroxide NCs (0.2 g) were dispersed in a formamide-water binary solution (500 mL, 1∶4 v/v), which was then heated at 80℃ for 1 - 8 h under continuous magnetic stirring and a nitrogen gas protection. The resulting translucent colloidal suspension was further treated by centrifugation at 4000 r/min for 20 min to remove possible unexfoliated NCs.

2.3 Anion Exchange and calcination of layered cobalt hydroxide NCs. Layered cobalt hydroxide NCs intercalated with various anions were prepared by an ethanol-assisted anion-exchange approach. For instance, as-prepared DS^- – intercalated NCs (0.2 g) were dispersed into an ethanol-water binary solution (200 mL, 1:1 v/v) of 1 M $NaNO_3$, NaCl, $NaClO_4$ or $NaCOOCH_3$ in a flask, respectively. After being purged with nitrogen gas, the flask was tightly capped and mechanically shaken for 48 h at room temperature. The product was filtered, washed with anhydrous ethanol and finally dried in air at 60 ℃ for 4 h. DS^- – intercalated cobalt hydroxide NCs and corresponding NO_3^- – exchanged products were calcined in air or under nitrogen protection to prepare Co_3O_4 or CoO NCs at 400 ℃ or 800 ℃ for 2 h, respectively.

2.4 Characterizations. XRD data were recorded on a Rigaku Rint – 2200 diffractometer with a monochromatic Cu K_α radiation (λ = 0.15405 nm). The morphologies and dimensions of as-prepared products were examined with a Keyence VE8800 scanning electron microscope. TEM characterizations were performed on a JEOL JEM – 3000F transmission microscope. A Seiko SPA 400 atomic force microscope was used to examine the topography of the nanosheets deposited on Si wafers. A cleaned Si wafer was immersed in the colloidal formamide-water suspension for 20 min, which was followed by rinsing with a copious amount of water and drying under a N_2 stream. Atomic force microscope (AFM) images were acquired in tapping mode using a Si tip cantilever with a force constant of about 20 N/m. FT-IR spectra were recorded using the KBr pellet method on a Varian 7000e FT-IR spectrophotometer equipped with a liquid-nitrogen-cooled Mercury-Cadmium-Telluride (MCT) detector. Thermogravimetric and differential thermal analysis (TG-DTA) were carried out using a Rigaku TGA – 8120 instrument in a temperature range of 25 – 900 ℃ at a heating rate of 1 ℃/min under an air flow. Cyclic voltammetry and galvanic charge-discharge measurements were carried out on a Solartron electrochemistry workstation using a three electrode cell with 1 M KOH as the electrolyte, a Ag/AgCl electrode as reference electrode, and platinum wire as counter electrode, respectively. For electrochemical measurements, about 1 mg of hydroxide NCs or nanosheets deposited on graphite foils were used as working electrodes. The working electrodes were prepared as follows: 80 wt.% of as-prepared NCs or nanosheets was mixed with 10 wt.% of acetylene black (> 99.9%) and 10 wt.% of polytetrafluoroethylene (PTFE, > 99.9%) in an agate mortar until a homogeneous black powder was obtained. Then a few drops of ethanol were added. After briefly allowing the ethanol to evaporate, the suspension was dropped on a 1 cm × 1 cm graphite foil. The electrode was dried at 80 ℃ for 12 h in air and then was pressed under 40 MPa before electrochemical measurements. Magnetic measurements were conducted using a Quantum Design MPMS XP - 5 superconducting quantum interference device (SQUID).

3 Results and Discussion

Layered cobalt hydroxide NCs were synthesized via an oil bath method using $CoCl_2 \cdot 6H_2O$ as a cobalt source, urea as a hydrolysis reagent and SDS as a surfactant and structure-directing agent under a nitrogen gas atmosphere. The amounts of $CoCl_2 \cdot 6H_2O$ and urea were fixed at 5 mmol and 35 mmol, respectively. The amount of SDS was adjusted in a range of 0 - 25 mmol. In a typical synthesis, a green-colored product was obtained. Figure 1(a) depicts a representative scanning electron microscopy (SEM) image of the product prepared at 100 ℃ for 8 h using 12.5 mmol SDS. It is clearly seen that a large quantity of uniform conical structures were synthesized under such conditions. There was no apparent impurity observed other than the conical morphology. In the higher-magnification SEM image shown in Figure 1(b), hollow interiors of the regular conical structures were clearly identified. Such a conical feature is also notable in the transmission electron microscopy (TEM) image displayed in Figure 1(c). The NCs have an average bottom diameter of approximately 1 μm, tip diameter of about 200 nm, whereas a length up to 6 μm. The outer dimensions of as-prepared NCs under current oil bath conditions are similar, albeit somewhat thicker in wall thickness, to those previously obtained under microwave-assisted hydrothermal conditions also employing urea as a hydrolysis and alkaline agent.[3b] It is worth pointing out that both the outer dimensions, as well as wall thickness, are bigger than those of NCs obtained using HMT as an alkaline reagent.[3a] This might be derived from the different hydrolyzing rate of urea and HMT. Slower hydrolyzing rates of urea may be favorable for the formation of hydroxide NCs. The inset in Figure 1(c) represents a typical selected area electron diffraction (SAED) pattern, which can be indexed to in-plane or [001] zone-axis diffraction rings of hexagonal cobalt hydroxide with a lattice constant of $a = 0.310$ nm. The high-resolution TEM image of an individual cobalt hydroxide NC, as shown in Figure 1(d), demonstrates a layered structure with interlayer spacing measured at about 2.4 nm, in good agreement with our previous characterizations of α - type cobalt hydroxide intercalating DS anions in the gallery.[3]

The evolution in the morphology and structure of as-prepared products was monitored by adjusting the reaction time and temperature. Figure 2 displays X-ray powder diffraction (XRD) patterns of as-prepared products obtained at 100 - 120 ℃ for

Figure 1 Oil bath synthesis of layered cobalt hydroxide NCs obtained at 100 ℃ for 8 h (a), (b) SEM and (c) TEM images. The inset in (c) shows SAED pattern taken on an individual NC. (d) HRTEM image of the tip part of an individual NC.

different time duration from 2 h to 12 h. It is notable that the basal reflection peaks, corresponding to an interlayer spacing of about 2.4 nm, were gradually enhanced in intensity with the prolongation of the reaction time or elevation of temperature. The peaks also became sharp, implying that the crystallinity of the products was improved with a longer time or higher temperature. Such a trend was also verified in the microscopic observations. For example, when the reaction time was as short as 2 h, the product was mostly composed of small and short conical objects [Figure 3(a)]. The average sizes were estimated to be approximate 3 μm in length, 100 nm in tip diameter, and 500 nm in bottom diameter. At this stage, some lamellar objects were also observed. Extending the reaction time to 4 h, almost 100% conical structures with uniform morphology were obtained, as shown in Figure 3(b). The dimensions were correspondingly increased to 3.5 – 4.5 μm, 200 – 300 nm, and 600 – 800 nm,

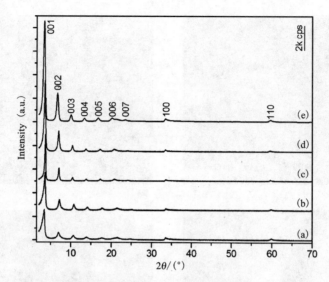

Figure 2 XRD patterns of layered cobalt hydroxide NCs obtained at 100 ℃ for varied time durations

(a) 2 h, (b) 4 h, (c) 8 h, (d) 12 h, (e) XRD pattern of NCs obtained at 120 ℃ for 8 h.

respectively. We speculate that the NCs appear to form through a rolling-up process of lamellar hydroxides, which is consistent with our previous reports.[3] The length was further elongated to 7.5 μm for NCs obtained from a reaction time of 12 h [Figure 3(c)]. This suggests both the elongation in length and the expansion in diameter with increasing reaction time. It was also discovered that the elevation of temperature was helpful to speed up the reaction. For example, the average length of NCs obtained at 120 ℃ for 8 h was estimated to be 7 μm [Figure 3(d)], comparable to that of NCs obtained at 100 ℃ for 12 h. Nevertheless, it is noteworthy that if the reaction time or temperature was further extended, some impurities might be produced, which appears to be related with the possible oxidation of Co(II) to Co(III).

Surfactant SDS seemed to play a crucial role in the formation of such a peculiar conical feature. Hexagonal cobalt hydroxide platelets, several micrometers in width and several tens of nanometers in thickness, were synthesized without using the surfactant SDS. The product appeared pink in color and was identified as brucite-like β-Co(OH)$_2$ with lattice constants of a = 0.318 nm and c = 0.465 nm [Figure 4(a)(i) and 4(b)]. On the other hand, if the amount of SDS was increased to 25 mmol, the product turned into agreen color same with NCs, also identified as DS$^-$ – intercalated

Figure 3 SEM images of layered cobalt hydroxide NCs obtained under different conditions
(a) 100℃, 2 h, (b) 100℃, 4 h, (c) 100℃, 12 h, (d) 120℃, 8 h.

α - type cobalt hydroxide. However, the majority of the product instead consisted of lamellar or platy objects. Only a small portion of NCs with average length of about 1.5 μm was discerned [Figure 4(a)(ii) and 4(c)]. The interlayer distance of the lamellar structure can be expanded by increasing the amount of SDS, which may affect the product morphology. Furthermore, inert gas protection was also found crucial in the formation of the NCs. Without a gas protection, DS^- - intercalated cobalt hydroxide product with irregular morphology was obtained [Figure 4(a)(iii) and 4(d)]. Based on above experimental observations, an appropriate amount of surfactant SDS and suitable inert gas protection may be considered as required conditions for the synthesis of NCs.

Layered metal hydroxides, including NCs, have been usually exfoliated via mechanical shaking in formamide.[3,4,6] In the current study, an alternative procedure, heat treating layered cobalt hydroxide NCs in a formamide-water binary solution, was

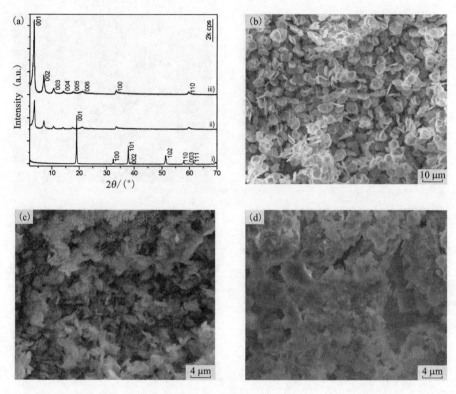

Figure 4
(a) XRD patterns of as-prepared products obtained at 100℃ for 2 h under different experimental conditions: i) 0 mmol SDS; ii) 25 mmol SDS; iii) without nitrogen gas protection. SEM images of as-prepared products (b), (c) and (d) correspond to (i), (ii) and (iii).

proposed and performed. The conditions are supposed to be milder than mechanical shaking and would bring new insights into their exfoliation behavior. Figure 5(a) – 5(e) depicts the morphological change of NCs during the treatment. After heat treatment for 2 h, the wall thickness generally became thinner than those of original NCs. In particular, some outer layers were observed to unravel from the NC as visualized by a typical TEM observation in Figure 5(a). The corresponding magnified part shown in Figure 5(b) clearly identifies such a detachment effect of some outer layers. Figure 5(c) depicts a high-resolution TEM (HRTEM) image of the tip part, indicating that the interlayer spacing of the unexfoliated part is kept at 2.4 nm. As displayed in Figure 5(d), prolonging the treatment time to 4 h made the unravelling and detaching effect more obvious. Ultimately, these NCs could becompletely

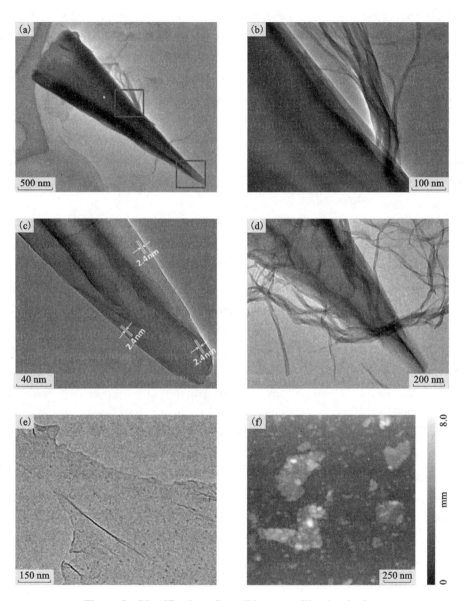

Figure 5 Identification of possible unravelling/unzipping stages of layered cobalt hydroxide NCs

(a) Typical TEM image of an individual NC after heat treatment for 2 h. (b), (c) High-magnification TEM images obtained from the different selected areas of (a) corresponding to outer wall part (red box) and tip part (green box), respectively. (d), (e) Typical TEM images of NC after heat treatment for 4 h and 8 h, respectively. (f) Tapping-mode AFM image of the exfoliated nanosheets deposited on Si substrate.

exfoliated/delaminated into unilamellar nanosheets after a treatment time of 8 h. A typical TEM image of the exfoliated nanosheets is given in Figure 5(e). A faint and homogeneous contrast indicates the ultrathin and uniform thickness. A tapping-mode atomic force microscope (AFM) image shown in Figure 5(f) also discerns ultrathin sheet-like objects with lateral size of several hundred nanometers on the Si wafer. The ultrathin nanosheets are irregular in morphology, indicating possible breakage or fracture of the NCs during the delamination process. The thickness of the nanosheets was measured to be approximately 1.0 nm, which is close to the crystallographic thickness of the host layer of α - type cobalt hydroxide, a clear proof supporting the unilamellar nature.[3,6]

The above characterizations validated that heat treatment in the formamide/water media was also effective in delaminating these NCs, providing an alternative route to the access of unilamellar hydroxide nanosheets. More intriguingly, the identification of such a novel unravelling/unzipping scenario of NCs for the first time is vital in understanding the transformation/evolution of a tubular structure, as well as providing new insights into the exfoliation mechanism of layered hydroxides in general.

The DS^- – intercalated α - type cobalt hydroxide NCs could be readily exchanged into other inorganic and organic anionic forms by a conventional ion-exchange procedure at room temperature. Figure 6 depicts XRD patterns of as-prepared products after exchange with Cl^-, ClO_4^-, CH_3COO^-, and NO_3^- anions. The basal diffraction peak, 2.4 nm for the original DS^- intercalated NCs, shifted to 0.78 nm for Cl^-, 0.91 nm for ClO_4^-, 0.92 nm for CH_3COO^-, and 0.79 nm for NO_3^- – exchanged NCs, respectively. The in-plane diffraction peaks of 100 and 110 retained their original positions, indicating that the host layers are basically intact. After anion-exchange, the conical features with hollow interiors are still apparent (Figure 7). On the other hand, the wall thickness of the exchanged products has been significantly reduced in contrast with that of DS^- – intercalated NCs, which is due to the decrease of the interlayer spacing, in good consistency with the XRD results.

As shown in Figure 8, the exchange of DS^- – intercalated cobalt hydroxide NCs into other anionic forms were further confirmed by Fourier transform infrared (FT-IR) spectroscopy. In all spectra, broad band at around 3500 cm^{-1} is attributable to the stretching vibration of O – H bonds, which indicates the presence of hydroxyl groups with hydrogen bonding and interlayer water molecules. The weak bands at around 1620 – 1635 cm^{-1} are well-known as the bending mode of the water molecules.[7] The absorption bands at low-frequency regions, below 635 cm^{-1}, are ascribed to Co – O

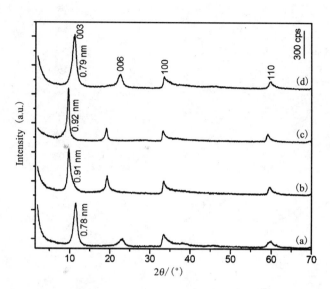

Figure 6 XRD patterns of layered cobalt hydroxide NCs exchanged with various anions
(a) Cl^-, (b) ClO_4^-, (c) CH_3COO^-, (d) NO_3^-.

stretching and Co – OH bending vibrations in the host layers.[8] Besides these common bands, in DS^-– intercalated NCs (spectrum a), the sharp bands at 2915 cm^{-1} and 2853 cm^{-1} are characteristic of the asymmetric and symmetric CH_2 stretching vibrations in the alkyl chain of dodecylsulfate, respectively, whereas the weak band at 2959 cm^{-1} is assigned to the stretching vibration of CH_3 group. The band at 1471 cm^{-1} corresponds to the CH_2 bending (scissor) mode whereas those in the range of 900 – 1300 cm^{-1} are associated with the stretching modes of sulfate group.[9] These features disappeared after the exchange into Cl^- form [spectrum (b)]. An additional weak band in spectrum (c) at 1126 cm^{-1} is derived from the exchanged perchlorate anions.[10] In spectrum (d), the bands at 1579 cm^{-1} and 1418 cm^{-1} are derived from the symmetric and asymmetric COO^- stretches, respectively, indicating the successful intercalation of acetate groups.[11] Similarly, a sharp and strong absorption band at 1394 cm^{-1} discerned in spectrum (e) is due to the $v3$ vibration mode of nitrate anions.[9,12] These evidences unambiguously confirmed that the cobalt hydroxide NCs possess a high anion exchangeability, desirable for versatile modification of interlayer contents as well as gallery spacing.

The intercalation of different anions might bring an impact to the phase transformation of the hydroxide NCs. The thermal decomposition behaviors of

Figure 7 SEM images of layered cobalt hydroxide NCs exchanged with various anions
(a) Cl^-, b) ClO_4^-, (c) CH_3COO^-, (d) NO_3^-

DS^- - and NO_3^- - intercalated NCs were investigated with thermogravimetric (TG) analysis in air in the temperature range of 25 – 900 ℃, as compared in Figure 9. In both curves, the gradual mass loss below 100 ℃ corresponds to the evaporation of the adsorbed water on the surface. The consecutive stages of weight loss in the range 100 – 170 ℃ can be attributed to the removal of interlayer water molecules. After that, the decomposition behavior becomes different. The weight-loss profile of DS^- - intercalated cobalt hydroxide NCs exhibits a major decrease between 170 ℃ and 800 ℃, which can be mainly attributed to the combustion of DS and dehydroxylation. The combustion of organic components in the DS group led to the formation of cobalt sulfate ($CoSO_4$) as a side product, which was further converted into Co_3O_4 at 800 ℃. Such a phase evolution was confirmed by XRD characterizations on calcined products detailed in Figure 10.

On the other hand, NO_3^- – intercalated cobalt hydroxide NCs was completely converted into Co_3O_4 at a temperature as low as 400 ℃ in air. Figure 11(a) shows a

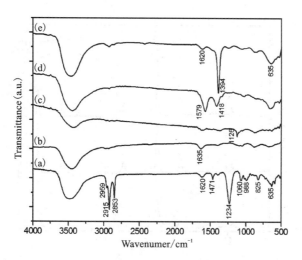

Figure 8 FT-IR spectra of layered cobalt hydroxide NCs exchanged with various anions
(a) DS^-, (b) Cl^-, (c) ClO_4^-, (d) CH_3COO^-, (e) NO_3^-

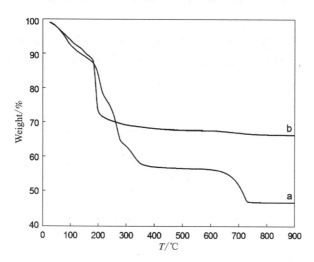

Figure 9 TG curves of layered cobalt hydroxide NCs intercalating (a) DS^- and (b) NO_3^- anions recorded at a heating rate of 1 ℃/min in air

typical XRD pattern of the product calcined in air. All of the reflections can be indexed as a pure face-centered cubic phase of spinel Co_3O_4 with lattice constant $a = 0.808$ nm (JCPDS 43 − 1003). That is to say, the exchange into an NO_3^- − intercalated form is more desirable for the phase transformation into cobalt oxides. In addition, CoO could be selectively obtained by calcination of NO_3^- − intercalated NCs at 400 ℃ for 2 h under nitrogen flow. The XRD pattern of the converted product is compared in Figure 11(b),

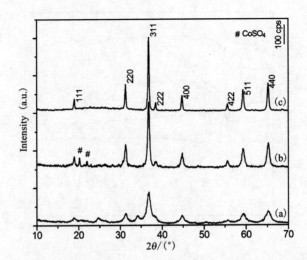

Figure 10 The evolution of XRD patterns of the products obtained by calcination of DS⁻ – intercalated cobalt hydroxide NCs at (a) 400℃, (b) 600℃ and (c) 800℃ for 2 h in air

which can be readily indexed as a cubic structure with lattice constant $a = 0.426$ nm, in good agreement with the standard data for CoO (JCPDS 65 – 2902). In both patterns, no impurity peak was recognized, indicating a complete conversion into corresponding oxides.

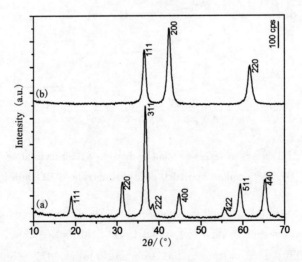

Figure 11 XRD patterns of the products obtained by calcination of NO_3^- – intercalated cobalt hydroxide NCs at 400℃ for 2 h (a) in air and (b) under nitrogen protection

More importantly, as-transformed cobalt oxides (Co_3O_4 and CoO) could retain the original conical framework with hollow interiors. Figure 12(a) and 12(b) depicts typical SEM images of the Co_3O_4 product obtained by calcination of NO_3^- - intercalated cobalt hydroxide NCs at 400 ℃ in air. It is clearly seen that the conical structures as well as hollow interiors are still obvious. Nevertheless, the surfaces of as-calcined Co_3O_4 NCs become wrinkled, which is different from the smooth surface feature of the starting hydroxide NCs. The average length of Co_3O_4 NCs was estimated to be about 4 μm, smaller than that of NO_3^- - intercalated NCs before calcination, implying the tendency to shrink during calcination. A typical TEM image of Co_3O_4 NCs is shown in Figure 12(c). In the inset, an SAED pattern was well indexed to spinel Co_3O_4, revealing a polycrystalline nature of the calcined NCs. This indicates the collapse of the layered structure and recrystallization into oxides during calcination. Lattice fringes resolved in the corresponding HRTEM image in Figure 12(d) were measured to be about 0.24 nm and 0.20 nm, which are consistent with the interplanar spacings between {311} and {400} in spinel Co_3O_4, respectively. In comparison, a typical SEM image of CoO NCs, obtained by calcination of NO_3^- - intercalated cobalt hydroxide NCs at 400 ℃ under nitrogen protection, is shown in Figure 12(e). Again, the calcined product retained the conical feature with a hollow interior. The hollow nature is also clearly revealed by TEM observation of an individual CoO NC in Figure 12(f). The upper inset displays the corresponding SAED pattern, which can be indexed to cubic CoO. Similar to the Co_3O_4 case, as-transformed CoO NCs are also polycrystalline. The lattice spacing in the HRTEM image shown in lower inset of Figure 12(f) was measured to be about 0.21 nm, agreeing well with the interplanar spacings of {200} for cubic CoO.

The electrochemical behavior of cobalt hydroxide NCs and exfoliated nanosheets in 1 M KOH aqueous electrolyte were investigated using cyclic voltammetry (CV) and galvanic charge-discharge cycling. Typical CV curves of layered cobalt hydroxide NCs intercalated with different anions in a potential range of -0.2 V to 0.4 V (vs. Ag/AgCl) at a scan rate of 10 mV/s are compared in Figure 13(a). Two pairs of redox peaks are commonly observed, revealing that there are two quasi-reversible electron-transfer processes, which are directly related with the following Faradaic reactions: $Co(OH)_2 + OH^- \rightleftharpoons CoOOH + H_2O + e^-$ and $CoOOH + OH^- \rightleftharpoons CoO_2 + H_2O + e^-$, respectively.[13] It is well known that the anodic peaks result from the oxidation of $Co(OH)_2$ to CoOOH (about 0.05 V) as well as CoOOH to CoO_2 (about 0.3 V), whereas the cathodic peaks correspond to the reverse reduction process. In

Figure 12 Characterizations of cobalt oxide NCs

(a), (b) SEM and (c) TEM images of Co_3O_4 NCs calcined at 400℃ in air. (d) HRTEM image with lattice fringes resolved. (e) SEM and (f) TEM images of CoO NCs calcined at 400℃ under nitrogen protection. The insets in (c), (f) show SAED patterns indexed to be Co_3O_4 and CoO, respectively.

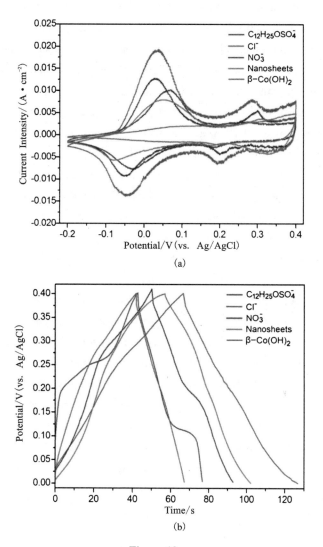

Figure 13

Electrochemical characterizations of (a) layered cobalt hydroxide NCs intercalated with various anions, unilamellar cobalt hydroxide nanosheets, and β-Co(OH)$_2$ hexagonal platelets. (b) Galvanostatic charge-discharge curves at a current density of 10 A/g.

comparison with DS$^-$-intercalated NCs, the redox current was enhanced to some extent after exchange into Cl$^-$ and NO$_3^-$. It is noteworthy that redox peaks of β-Co(OH)$_2$ hexagonal platelets are the weakest, revealing a poor electrochemical performance. Generally, the Faradaic pseudocapacitance is mainly based on redox reactions at the interface between the active material and the electrolyte. Based on the experimental results, we assume that the interlayer distance and the intercalated anions impart substantial effect on the electrochemical activity of layered transition-metal hydroxides, suggesting a suitable combination of interlayer spacing and anionic species may be required for achieving optimum electrochemical performance. Figure 13(b) exhibits the galvanostatic charge-discharge curves measured in the potential range between 0 V and 0.4 V at a current density of 10 A/g. The specific capacitance could be calculated from the galvanostatic charge-discharge curves using the equation: $C = I\Delta t/m\Delta V$, where I is charge-discharge current, Δt is the time for a full discharge, m indicates the mass of the active material, and ΔV represents the voltage change after a full discharge.[14] The specific capacitance of layered cobalt hydroxide NCs intercalated with DS^-, Cl^-, and NO_3^- anions was estimated to be approximately 340 F/g, 590 F/g, and 430 F/g, while that for cobalt hydroxide nanosheets and $\beta - Co(OH)_2$ hexagonal platelets was 450 F/g and 260 F/g, respectively. The capacitances are roughly close to our previously reported value (about 490 F/g) of layered cobalt hydroxide NCs synthesized under microwave-assisted conditions.[3b] It is generally assumed that the specific capacitances of layered hydroxides will be enhanced with the expansion of interlayer spacing for ion diffusion.[13b] Such an effect was verified by the result that the specific capacitance of $\beta - Co(OH)_2$ hexagonal platelets is the lowest among all the products. $\beta - Co(OH)_2$ is composed of neutral host layers without gallery space for anions, which may hinder ion diffusion of electrolyte. However, compared with DS^- - and NO_3^- - intercalated cobalt hydroxide NCs, the Cl^- - intercalated form exhibited the highest specific capacitance. The reason is not clear yet, but it probably originates from the exchange and transport behavior of interlayer OH^- associated with different anionic species, which has been reported for layered α - type cobalt or nickel hydroxides.[15]

The magnetic properties of as-transformed cobalt oxide NCs were measured on a superconducting quantum interference device (SQUID). The temperature dependences of the magnetic susceptibility (χ) for Co_3O_4 and CoO NCs under an applied field of 500 Oe are provided in Figure 14(a) and 14(b), respectively. Both products exhibited antiferromagnetic transitions at low temperature. For Co_3O_4, the transition occurs at about 29 K (Néel temperature, T_N), which is lower than that of bulk Co_3O_4 (40 K).[16] In addition to the antiferromagnetic transition, an upturn in the zero-field-cooled (ZFC) curve below T_N is also visible, possibly due to spin canting or other spin scattering effect. The antiferromagnetic transition for CoO NCs takes place at around 70 K, being far lower than the Néel temperature of bulk CoO that is known at about

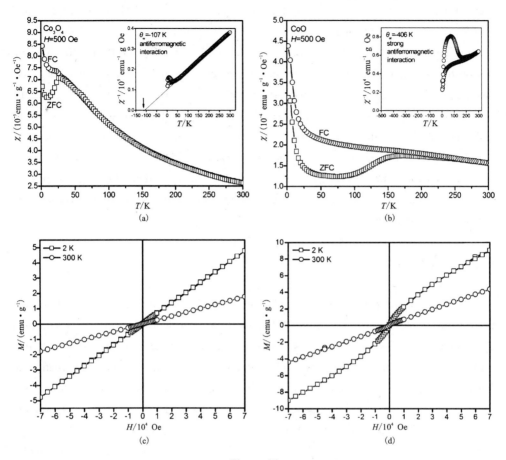

Figure 14

Temperature dependence of magnetic susceptibility (χ) for (a) Co_3O_4 and (b) CoO NCs measured under an applied field of 500 Oe. The insets in (a) and (b) illustrate the corresponding plots of inverse susceptibility (χ^{-1}) vs. temperature. Isothermal magnetization curves for (c) Co_3O_4 and (d) CoO NCs at 2 K and 300 K, respectively.

300 K. The lowering of Néel temperature possibly resulted from the size, shape and surface effects of cobalt oxide NCs, that is, the geometric confinement effect.[17] In order to further analyze the magnetic properties, the Curie-Weiss law was applied to the paramagnetic part, as illustrated by the corresponding insets. The analytical formula was $c(T) = N_A m_{eff}^2/3k_B(T-Q_W)$, where N_A is the Avogadro constant, m_{eff} is the effective magnetic moment, k_B is Boltzmann's constant and Q_W is the Weiss temperature. The plots yielded values of Q_W at -107 K and -406 K for Co_3O_4 and CoO NCs, respectively, suggesting weak antiferromagnetic interaction in Co_3O_4 while a

very strong one in CoO NCs. The hysteresis loops for Co_3O_4 and CoO NCs at 2 K and 300 K are depicted in Figure 14(c) and 14(d), respectively. The almost linear relation between the magnetization and magnetic field for Co_3O_4 at 2 K and 300 K again demonstrates that the product has an antiferromagnetic ground state. For CoO NCs, despite that the data at low temperature range reflect a slight curvature, the magnetization curves as a whole are nearly linear without forming clear hysteresis loops, also indicative of an antiferromagnetic ground state. Because of the peculiar morphological and structural characteristics, as well as finite size effect of cobalt oxide NCs, the magnetic properties are considerably different from those found in their bulk oxide counterparts, which might anticipate some prospective applications in magneto-optical or magnetic recording devices, etc.

4 Conclusion

In summary, a convenient and reliable synthetic strategy was presented for the preparation of DS^- – intercalated cobalt hydroxide NCs in large quantities based on urea hydrolysis in oil bath. The NCs could be gradually unwrapped/exfoliated into unilamellar nanosheets in the formamide-water binary solution through heat treatment. Meanwhile, DS^- – intercalated cobalt hydroxide NCs could be easily modified into various inorganic or organic anionic forms via an anion-exchange process at ambient temperature. The anion-exchanged products exhibited different oxidation behavior. Particularly, NO_3^- – intercalated NCs could be more readily calcined into cobalt oxides (e.g., Co_3O_4 and CoO) retaining original conical features in comparison with the DS^- – intercalated form. The electrochemical characterizations suggested that a suitable combination of interlayer gallery space and intercalated anions was required for achieving optimum electrochemical performance. The magnetic measurements revealed both antiferromagnetic properties of calcined Co_3O_4 and CoO NCs, but the Néel temperature values were lower than those of their bulk counterpart materials. The versatile structural modification of layered cobalt hydroxide NCs and their derivatives are expected to bring new opportunities for further fundamental research, as well as for technological applications in electrochemical energy storage or magneto-optical devices.

References

[1] (a) D. L. Bish. Anion-exchange in takovite: applications to other hydroxide minerals [J]. Bull. Mineral, 1980, 103: 170 – 175. (b) M. Ogawa, K. Kuroda. Photofunctions of intercalation compounds [J]. Chemical Reviews, 1995, 95(2): 399 – 438. (c) B. Sels, D. De Vos, M. Buntinx, F. Pierard, A. Kirsch-De Mesmaeker, P. Jacobs. Layered double hydroxides exchanged with tungstate as biomimetic catalysts for mild oxidative bromination [J]. Nature, 1999, 400(6747): 855 – 857. (d) B. F. Sels, D. E. De Vos, P. A. Jacobs. Hydrotalcite-like anionic clays in catalytic organic reactions [J]. Catalysis Reviews, 2001, 43 (4): 443 – 488. (e) B. M. Choudary, S. Madhi, N. S. Chowdari, M. L. Kantam, B. Sreedhar. Layered double hydroxide supported nanopalladium catalyst for Heck-, Suzuki-, Sonogashira-, and Stille-type coupling reactions of chloroarenes [J]. Journal of the American Chemical Society, 2002, 124(47): 14127 – 14136. (f) S. Tezuka, R. Chitrakar, A. Sonoda, K. Ooi, T. Tomida. Studies on selective adsorbents for oxo-anions. Nitrate ion-exchange properties of layered double hydroxides with different metal atoms [J]. Green Chemistry, 2004, 6(2): 104 – 109. (g) G. R. Williams, D. O'Hare, Towards understanding, control and application of layered double hydroxide chemistry [J]. Journal of Materials Chemistry, 2006, 16 (30): 3065 – 3074. (h) J. H. Yang, Y. S. Han, M. Park, T. Park, S. J. Hwang, J. H. Choy. New inorganic-based drug delivery system of indole – 3 – acetic acid-layered metal hydroxide nanohybrids with controlled release rate [J]. Chemistry of Materials, 2007, 19(10): 2679 – 2685. (i) P. J. Sideris, U. G. Nielsen, Z. Gan, C. P. Grey. Mg/Al ordering in layered double hydroxides revealed by multinuclear NMR spectroscopy [J]. Science, 2008, 321 (5885): 113 – 117. (j) T. Y. Wei, C. H. Chen, H. C. Chien, S. Y. Lu, C. C. Hu. A cost-effective supercapacitor material of ultrahigh specific capacitances: spinel nickel cobaltite aerogels from an epoxide-driven sol-gel process [J]. Advanced Materials, 2010, 22(3): 347 – 351.

[2] (a) Y. Zhao, F. Li, R. Zhang, D. G. Evans, X. Duan. Preparation of layered double-hydroxide nanomaterials with a uniform crystallite size using a new method involving separate nucleation and aging steps [J]. Chemistry of Materials, 2002, 14(10): 4286 – 4291. (b) Z. Liu, R. Ma, M. Osada, K. Takada, T. Sasaki. Selective and controlled synthesis of α – and β – cobalt hydroxides in highly developed hexagonal platelets [J]. Journal of the American Chemical Society, 2005, 127(40): 13869 – 13874. (c) Y. Han, Z. H. Liu, Z. Yang, Z. Wang, X. Tang, T. Wang, L. Fan, Kenta Ooi. Preparation of Ni^{2+}-Fe^{3+} layered double hydroxide material with high crystallinity and well-defined hexagonal shapes [J]. Chemistry of Materials, 2007, 20(2): 360 – 363. (d) S. Ida, D. Shiga, M. Koinuma, Y. Matsumoto. Synthesis of hexagonal nickel hydroxide nanosheets by exfoliation of layered nickel hydroxide intercalated with dodecyl sulfate ions [J]. Journal of the American Chemical Society, 2008, 130

(43): 14038 - 14039.

[3] (a) X. H. Liu, R. Ma, Y. Bando, T. Sasaki. Layered cobalt hydroxide nanocones: microwave-assisted synthesis, exfoliation, and structural modification [J]. Angewandte Chemie International Edition, 2010, 49(44): 8253 - 8256. (b) X. H. Liu, R. Ma, Y. Bando, T. Sasaki. A general strategy to layered transition-metal hydroxide nanocones: tuning the composition for high electrochemical performance [J]. Advanced Materials, 2012, 24(16): 2148 - 2153.

[4] (a) P. S. Braterman, Z. P. Xu, F. Yarberry. Layered double hydroxides (LDHs), handbook of Layered Materials [M]. (Eds: S. M. Auerbach, K. A. Carrado, P. K. Dutta) New Marcel Decker, Inc. , 2004: 373. (b) Q. Wang, D. O'Hare. Recent advances in the synthesis and application of layered double hydroxide (LDH) nanosheets [J]. Chemical Reviews, 2012, 112 (7): 4124 - 4155.

[5] (a) E. Coronado, C. Marti-Gastaldo, E. Navarro-Moratalla, A. Ribera, S. J. Blundell, P. J. Baker. Coexistence of superconductivity and magnetism by chemical design [J]. Nature Chemistry, 2010, 2(12): 1031 - 1036. (b) R. Ma, T. Sasaki, Nanosheets of oxides and hydroxides: ultimate 2D charge-bearing functional crystallites [J]. Advanced Materials, 2010, 22(45): 5082 - 5104. (c) D. P. Yan, J. Lu, J. Ma, M. Wei, D. G. Evans, X. Duan. Reversibly thermochromic, fluorescent ultrathin films with a supramolecular architecture [J]. Angewandte Chemie International Edition, 2011, 50(3): 720 - 723; (d) L. Wang, D. Wang, X. Y. Dong, Z. J. Zhang, X. F. Pei, X. J. Chen, B. Chen, J. Jin. Layered assembly of graphene oxide and Co - Al layered double hydroxide nanosheets as electrode materials for supercapacitors [J]. Chemical Communications, 2011, 47: 3556 - 3558.

[6] (a) Z. Liu, R. Ma, Y. Ebina, N. Iyi, K. Takada, T. Sasaki. General synthesis and delamination of highly crystalline transition-metal-bearing layered double hydroxides [J]. Langmuir, 2007, 23(2): 861 - 867. (b) R. Ma, K. Takada, K. Fukuda, N. Iyi, Y. Bando, T. Sasaki. Topochemical synthesis of monometallic (Co^{2+} - Co^{3+}) layered double hydroxide and its exfoliation into positively charged $Co(OH)_2$ nanosheets [J]. Angewandte Chemie International Edition, 2008, 47(1): 86 - 89.

[7] (a) Z. P. Xu, H. C. Zeng. Interconversion of brucite-like and hydrotalcite-like phases in cobalt hydroxide compounds [J]. Chemistry of Materials, 1999, 11(1): 67 - 74. (b) P. Jeevanandam, Y. Koltypin, Y. Mastai. Synthesis of α - cobalt (II) hydroxide using ultrasound radiation [J]. Journal of Materials Chemistry, 2000, 10(2): 511 - 514. (c) D. H. Sun, J. L. Zhang, H. J. Ren, Z. F. Cui, D. X. Sun. Influence of OH^- and SO_4^{2-} anions on morphologies of the nanosized nickel hydroxide [J]. The Journal of Physical Chemistry C, 2010, 114(28): 12110 - 12116.

[8] (a) Y. Hou, H. Kondoh, M. Shimojo, T. Kogure, T. Ohta. High-yield preparation of uniform cobalt hydroxide and oxide nanoplatelets and their characterization [J]. The Journal of Physical Chemistry B, 2005, 109(41): 19094 - 19098. (b) E. Lima, M. J. Martínez-Ortiz, R. I. G.

Reyes, M. Vera. Fluorinated hydrotalcites: the addition of highly electronegative species in layered double hydroxides to tune basicity [J]. Inorganic Chemistry, 2012, 51(14): 7774 – 7781.

[9] (a) F. X. Geng, H. Xin, Y. Matsushita, R. Ma, M. Tanaka, F. Izumi, N. Iyi, T. Sasaki. New layered rare-earth hydroxides with anion-exchange properties [J]. Chemistry-A European Journal, 2008, 14(30): 9255 – 9260. (b) T. Gao, B. P. Jelle. Paraotwayite-type $\alpha - Ni(OH)_2$ nanowires: structural, optical, and electrochemical properties [J]. The Journal of Physical Chemistry C, 2013, 117(33): 17294 – 17302.

[10] R. Ma, Z. Liu, K. Takada, N. Iyi, Y. Bando, T. Sasaki. Synthesis and exfoliation of Co^{2+}-Fe^{3+} layered double hydroxides: an innovative topochemical approach [J]. Journal of the American Chemical Society, 2007, 129(16): 5257 – 5263.

[11] (a) L. Poul, N. Jouini, F. Fievet. Layered hydroxide metal acetates (metal = zinc, cobalt, and nickel): elaboration via hydrolysis in polyol medium and comparative study [J]. Chemistry of Materials, 2000, 12(10): 3123 – 3132. (b) L. J. McIntyre, L. K. Jackson, A. M. Fogg. $Ln_2(OH)_5NO_3 \cdot xH_2O$ (Ln = Y, Gd – Lu): A novel family of anion exchange intercalation hosts [J]. Chemistry of Materials, 2007, 20(1): 335 – 340. (c) P. P. Wang, B. Bai, S. Hu, J. Zhuang, X. Wang. Family of multifunctional layered-lanthanum crystalline nanowires with hierarchical pores: hydrothermal synthesis and applications [J]. Journal of the American Chemical Society, 2009, 131(46): 16953 – 16960.

[12] M. Shao, M. Wei, D. G. Evans, X. Duan. Magnetic-field-assisted assembly of CoFe layered double hydroxide ultrathin films with enhanced electrochemical behavior and magnetic anisotropy [J]. Chemical Communications, 2011, 47: 3171 – 3173.

[13] (a) L. Cao, F. Xu, Y. Y. Liang, H. L. Li. Preparation of the novel nanocomposite $Co(OH)_2$/ultra-stable Y zeolite and its application as a supercapacitor with high energy density [J]. Advanced Materials, 2004, 16(20): 1853 – 1857. (b) L. Wang, Z. H. Dong, Z. G. Wang, F. X. Zhang, J. Jin. Layered $\alpha - Co(OH)_2$ nanocones as electrode materials for pseudocapacitors: understanding the effect of interlayer space on electrochemical activity [J]. Advanced Functional Materials, 2013, 23(21): 2758 – 2764.

[14] (a) H. Wang, H. S. Casalongue, Y. Liang, H. Dai. $Ni(OH)_2$ nanoplates grown on graphene as advanced electrochemical pseudocapacitor materials [J]. Journal of the American Chemical Society, 2010, 132(21): 7472 – 7477. (b) S. Chen, J. Zhu, X. Wang. One-step synthesis of grapheme-cobalt hydroxide nanocomposites and their electrochemical properties [J]. The Journal of Physical Chemistry C, 2010, 114(27): 11829 – 11834.

[15] (a) Z. A. Hu, Y. L. Xie, Y. X. Wang, L. J. Xie, G. R. Fu, X. Q. Jin, Z. Y. Zhang, Y. Y. Yang, H. Y. Wu. Synthesis of α – cobalt hydroxides with different intercalated anions and effects of intercalated anions on their morphology, basal plane spacing, and capacitive property [J]. The Journal of Physical Chemistry C, 2009, 113(28): 12502 – 12508. (b) J. W. Lee, J. M. Ko, J. D. Kim, Hierarchical microspheres based on $\alpha - Ni(OH)_2$ nanosheets

intercalated with different anions: synthesis, anion exchange, and effect of intercalated anions on electrochemical capacitance [J]. The Journal of Physical Chemistry C, 2011, 115(39): 19445-19454.

[16] (a) W. L. Roth. The magnetic structure of Co_3O_4 [J]. Journal of Physics and Chemistry of Solids, 1964, 25(1): 1-10. (b) L. He, C. P. Chen, N. Wang, W. Zhou, L. Guo. Finite size effect on Néel temperature with Co_3O_4 nanoparticles [J]. Journal of Applied Physics, 2007, 102(10): 103911.1-103911.4.

[17] M. Ghosh, E. V. Sampathkumaran, C. N. R. Rao. Synthesis and magnetic properties of CoO nanoparticles [J]. Chemistry of Materials, 2005, 17(9): 2348-2352.

☆ X. H. Liu, R. Z. Ma, Y. Bando, T. Sasaki. High-yield preparation, versatile structural modification and properties of layered cobalt hydroxide nanocones [J]. Advanced Functional Materials, 2014, 24(27): 4292-4302.

Layered Zinc Hydroxide Nanocones: Synthesis, Facile Morphological and Structural Modification, and Properties

Abstract: Layered zinc hydroxide nanocones intercalated with DS$^-$ have been synthesized for the first time via a convenient synthetic approach, using homogeneous precipitation in the presence of urea and sodium dodecyl sulfate (SDS). SDS plays a significant role in controlling the morphologies of as-synthesized samples. Conical samples intercalated with various anions were transformed through an anion-exchange route in ethanol solution, and the original conical structure was perfectly maintained. Additionally, these DS$^-$-inserted nanocones can be transformed into square-like nanoplates in aqueous solution at room temperature, fulfilling the need for different morphology-dependent properties. Corresponding ZnO nanocones and nanoplates have been further obtained through the thermal calcination of NO$_3^-$-intercalating zinc hydroxide nanocones/nanoplates. These ZnO nanostructures with different morphologies exhibit promising photocatalytic properties.

1 Introduction

Layered metal hydroxide salts with the general formula $M^{II}(OH)_{2-x}A^{n-}_{x/n} \cdot mH_2O$ [or $M^{II}_{1+x/2}(OH)_2A^{n-}_{x/n} \cdot mH_2O$], in which OH$^-$ deficiency on the host layers is compensated by coordinated/grafted guest anions (A^{n-}), are relatively rare anion-exchangeable hosts, in addition to well-known layered double hydroxides [LDH, $M^{II}_{1-x}M^{III}_x(OH)_2A^{n-}_{x/n} \cdot mH_2O$]. They have attracted extensive interests because of their rich interlayer chemistry, interacting with both inorganic/organic anions and showing potential for applications in many fields such as drug delivery, catalysis, or as supercapacitors.[1-7] In this category, layered zinc hydroxide salts [typical formula: $Zn_5(OH)_8A^{n-}_{2/n} \cdot mH_2O$, A = Cl$^-$, NO$_3^-$, etc.] are regarded as one of the most representative structures. The host layers are composed of both octahedral and tetrahedral coordination. Specifically, approximately a quarter of octahedral metal positions are vacant, and a pair of tetrahedral metal atoms, located above and below the missing octahedra, directly coordinate with anions in the interlayer gallery.[8-10] The anion exchangeability means they can be used as drug carriers with a controllable release rate, due to their low cytotoxicity in acidic environments.[11-16] In addition, layered zinc hydroxide is also widely used as a precursor for the preparation of ZnO, an important semiconductor with a large band gap (3 - 4.5 eV) and a high exciton binding energy (60 - 100 meV). ZnO transformed from layered zinc hydroxide via an environment-friendly thermal decomposition route has been thoroughly exploited, owing to its peculiar physical and chemical properties dependent on its size, morphology and orientation.[17-20] Moreover, ZnO was also revealed as a promising photocatalyst with high photosensitivity and a nontoxic nature, which might contribute to solving environmental problems due to its strong oxidizing power and solar energy conversion capabilities.[21-23] Layered zinc hydroxides, through a cost-efficient procedure, can also be used to synthesize mesoporous ZnO nanomaterials with superhigh specific surface area, on an industrial scale. Nevertheless, layered zinc hydroxide salts are known for their instability in aqueous solutions even at room temperature. In fact, there are very few studies dedicated to the morphological control of layered zinc hydroxide salts with various interlayer anions or resulting ZnO obtained through calcination.[17, 24-26]

On the other hand, cylindrical nanotubes and conical nanocones formed by rolling-up of layered lamellae or nanosheets, exhibiting high aspect ratios and hollow interiors, have been highly popular materials over the past decade due to their fascinating morphology-dependent properties and applications such as nanoelectronic and optoelectronic devices, solar cells, and for environmental protection, etc.[27-31] Nevertheless, apart from carbon and boron nitride, and cobalt or nickel hydroxide systems reported very recently, there are seldom successes in synthesizing large quantities of nanocones from other layered structures.[32-35]

In the current work, we demonstrate a synthetic strategy capable of producing unique zinc hydroxide nanocones in large quantities. Similarly to our previous reports on cobalt or nickel hydroxide systems,[33-35] layered zinc hydroxide nanocones could be synthesized by utilizing urea as a slowly-releasing alkaline source, together with anionic surfactants such as sodium dodecyl sulfate ($C_{12}H_{25}OSO_3Na$, SDS), via an oil-bath synthetic procedure. It was found that SDS not only directly determined the interlayer spacing through intercalation, but also affected the morphology of as-prepared layered zinc hydroxide products. DS^- – intercalated nanocones undergo an exchange reaction with different anions (Cl^-, NO_3^-, CH_3COO^-) in ethanol. However, the nanocone morphology would not be maintained if the same exchange reaction was carried out in aqueous solution at room temperature, instead transforming into square-like nanoplate morphology. This aspect reveals a unique feature of zinc hydroxide for facile morphology manipulation and thus provides a novel transformation protocol under mild conditions, e. g. , at room temperature. Furthermore, ZnO nanocones and nanoplates could be obtained via the calcination of corresponding layered zinc hydroxide precursors intercalated with nitrate anions. Under the irradiation of UV light, ZnO nanocones and nanoplates were revealed to exhibit remarkable photocatalytic properties for decomposing organic dyes.

2 Experimental Section

All reagents used were of analytical grade and were purchased from Wako Chemical Reagents Company, Japan. Milli – Q water was utilized throughout all the experiments.

2. 1 Synthesis of layered zinc hydroxide nanocones. In a typical synthetic procedure, $Zn(NO_3)_2 \cdot 6H_2O$, urea and SDS were dissolved in a 1000 mL two-neck flask with Milli – Q water under nitrogen gas, yielding the concentrations of 5 mM, 10 mM and 25 mM, respectively. Then, the mixed solution was heated at refluxing temperature for 5. 5 h under continuous magnetic stirring. The final product was recovered by washing with Milli – Q water and ethanol several times and was air-dried at room temperature.

2.2 Anion-exchange reaction. As-synthesized layered zinc hydroxide nanocones with DS^- were dispersed into 100 mL of ethanol containing 25 mmol $NaNO_3$, NaCl and CH_3COONa, respectively. The vessels were tightly capped and mechanically shaken for several hours at room temperature. The products were filtered and air-dried at room temperature.

2.3 Transformation of zinc hydroxide nanocones to nanoplates. The layered zinc hydroxide nanocones were dispersed into 200 mL of 1 M NaCl aqueous solution without any agitation for 24 h at room temperature. The samples were collected by filtering, washing, and air-dried at room temperature.

2.4 Synthesis of ZnO nanocones and nanoplates. The as-prepared layered zinc hydroxide nanocones and nanoplates in NO_3^- form were calcined at a heating rate of 1℃/min in air to 200℃ and 800℃.

2.5 Evaluation of photocatalytic performance. 100 mg of ZnO nanocones or nanoplates were dispersed in 50 mL aqueous methylene blue (MB) with a concentration of 15 mg/L. The adsorption of MB on the surface of the photocatalysts was facilitated under magnetic stirring in the dark for 1 h. A 500 W Xe lamp (XEF - 501S San-ei Electric) was used for UV irradiation of the samples. Samples were placed about 10 cm from the light source, corresponding to a UV intensity of approximately 1 mW/cm^2 at $\lambda < 300$ nm, as determined with a spectroradiometer (USR - 30 Ushio Electric). The UV - Vis absorption spectrum of the aqueous solution of MB was recorded using a Hitachi U - 4100 UV - Vis spectrophotometer.

2.6 Charaterization. Phase identification of as-prepared samples was conducted by X-ray diffraction (XRD, with a Rigaku Ultima IV diffractometer operated at 40 kV/40 mA using Cu - K_α radiation). The morphology and dimensions of as-prepared products were characterized using scanning electron microscopy (SEM, JSM - 6010LA). High-resolution transmission electron microscopy (HRTEM) images and selected area electron diffraction (SAED) patterns were obtained on a JEM - 3000F transmission electron microscope. Metal contents were determined by inductively coupled plasma (ICP) atomic emission spectroscopy (SPS1700HVR, Seiko Instruments) after dissolving a weighed amount of the sample in an aqueous HCl solution. Thermogravimetric analysis (TGA) measurements were carried out using a Thermo Plus 2 TG8120 instrument in a temperature range of room temperature - 1000 ℃ at a heating rate of 1 ℃/min under air flow.

3 Results and Discussion

Figure 1(a) and 1(b) show typical SEM images, at different magnifications, of as-prepared zinc hydroxide nanocones obtained through oil-bath precipitation of zinc nitrate (5 mM) in the presence of urea (10 mM) and SDS (25 mM). It can be seen that a large quantity of uniform conical objects were synthesized. The length of as-obtained zinc hydroxide nanocones is several micrometers, whereas the average tip and bottom diameters are approximately 0.2 μm and 2 μm, respectively. The hollow interior structure is confirmed by the transmission electron microscopy (TEM) image shown in Figure 1(c). A difference in contrast is discernable between the dark periphery and greyish intermediate portion. The wall thickness of these nanocones are several hundreds of nanometres. A selected area electron diffraction (SAED) pattern of an individual nanocone is given in the top-left inset of Figure 1(c), which can be readily indexed to the in-plane or [001] zone-axis diffraction of hexagonal zinc hydroxide with a lattice constant of $a = 0.32$ nm. The diffraction rings with d-spacings of 0.27 nm and 0.16 nm correspond to [100] and [110], respectively, indicating that the rolling-up of the hydroxide layers lacks crystallographic interlayer registry. The layered stacking with an interlayer spacing of approximately 3.2 nm is visualized by the high-resolution transmission electron microscopy (HRTEM) image given in the bottom-right inset of Figure 1(c).

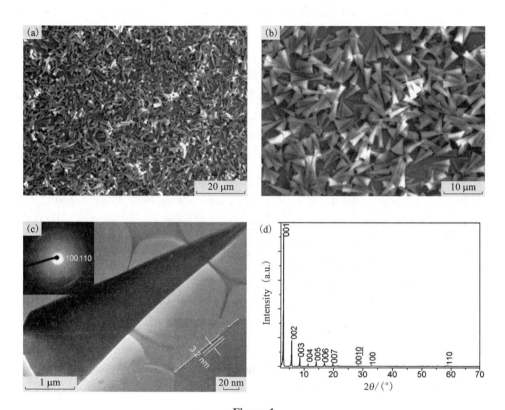

Figure 1

(a) Low-magnification and (b) high-magnification SEM images of as-prepared layered zinc hydroxide nanocones. (c) TEM image of resulting nanocone. The insets in (c) show an SAED pattern and a HRTEM image taken of an individual nanocone. (d) XRD pattern of as-synthesized zinc hydroxide nanocones intercalated with DS⁻ anions (inset) magnified profile in the region of 30 – 35° and 55 – 65°.

A typical XRD pattern of as-prepared nanocones intercalated with DS⁻ anions is displayed in Figure 1(d). All the diffraction peaks can be indexed in a hexagonal cell of a = 0.3150(2) nm, c = 3.1532(8) nm, consistent with the layered zinc hydroxides reported in the literature.[36-38] Sharp (001) basal reflections at the low angular range of the XRD pattern reveal the high crystallinity of the sample along the layer stacking direction (c axis). The in-plane diffraction peaks around 33° and 59° in the magnified XRD pattern in the top-right inset of Figure 1(d) indicate that layered zinc hydroxide was synthesized by this convenient route. The basal spacing close to 3.2 nm indicates that the DS$^-$ anions are likely to arrange in bilayers, similar to previous reports.[1,39] The basal spacing (d_L) for DS$^-$ in the hydroxide gallery may be expressed as: $d_L = (0.73 + 2 \times 0.15 + 2 \times 0.197) + 2 \times 0.127 \times 12 \cdot \sin\theta$, where the thickness of the host layer is taken as 0.73 nm, and the carbon chain length of dodecyl sulphate is around $0.127n$ nm ($n = 12$), with a sulphate head of 0.197 nm and an end tail of 0.15 nm, respectively.[1,40] As a result, the tilting angle of the bistratal DS$^-$ chains can be estimated as approximate 35°. According to the results of wet chemical analysis as well as thermogravimetric (TG) analysis, the TG curve of layered zinc hydroxide nanocones under ambient condition in the temperature range of room temperature (RT) to 1000 ℃ with a heating rate of 1 ℃/min. The first weight loss of 0.64% below 55 ℃ was ascribed to the evaporation of adsorbed water molecules on the zinc hydroxide surface. With the increasing of temperature from 55 ℃ to 130 ℃, the weight of 2.76% was attributed to the removal of interlayer water molecules. The interlayer spacing of layered zinc hydroxide is about 3.0 nm. The decrease of interlayer spacing from original 3.2 nm is consistent with the loss of water molecules. With the increasing of temperature to 200 ℃, partial ZnO was synthesized, then ZnO and $Zn_3O(SO_4)_2$ (PDF 31-1469) were obtained at about 600 ℃. When the temperature was raised to 1000 ℃, the loss weight of 59.06% is due to the removal of dodecyl sulfate (DS$^-$) and hydroxyl (OH$^-$). XRD pattern clearly demonstrates that a pure ZnO phase can be obtained via calcination of layered zinc hydroxide nanocones intercalated with DS$^-$ anions. The chemical composition of as-prepared nanocones was estimated as $Zn(OH)_{1.55}DS_{0.45} \cdot 0.3H_2O$ (Anal. calcd: Zn, 30.2%; S, 6.7%; ignition loss, 62.4%. Found: Zn, 28.9%; S, 6.31%; ignition loss, 61.8%). The data indicate that the nanocones are layered zinc hydroxide salts with a composition very close to the theoretical formula of $Zn_5(OH)_8DS_2 \cdot 2H_2O$.

The amount or concentration of SDS surfactant has a great influence on the morphology and phase of as-prepared products. Without using a surfactant, the products evolved into ZnO nanorods. Figure 2(a) depicts a typical SEM image of uniform nanorods with a length of approximate 5 μm. The XRD pattern shown in Figure 2(d)(i) reveals a hexagonal ZnO structure with lattice constants of $a = 0.3259(1)$ nm and $c = 0.5224(4)$ nm. When 12.5 mmol of SDS was used, the majority of the product appeared as hexagonal nanoplates, although a small fraction of nanocones was also observed [Figure 2(b)]. Upon increasing the amount of SDS to 25

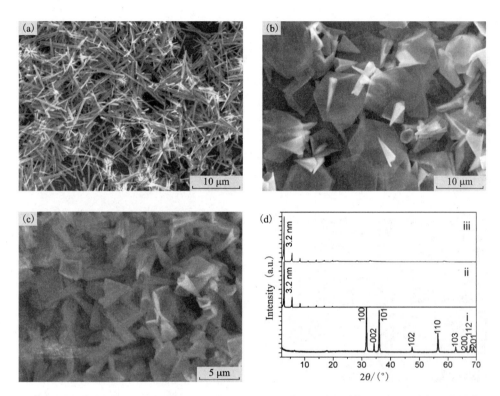

Figure 2 SEM images of as-prepared samples obtained with different quantities of SDS
(a) 0 mM, (b) 12.5 mM, (c) 50 mM; (d) corresponding XRD patterns: i) 0 mM, ii) 12.5 mM, iii) 50 mM.

mM, high-purity nanocones were synthesized [Figure 1(a)]. When the amount was further increased to 50 mM, nanocones were still obtained as the main product, as presented in Figure 2(c). Figure 2(d) compares the XRD patterns of as-obtained samples under different SDS concentrations, which indicates the formation of layered zinc hydroxide salts with similar interlayer spacing under varied amounts of SDS. We speculate that SDS works as both a surfactant and a structure-directing agent for the formation of peculiar nanocones of layered zinc hydroxides, the same as earlier reports on Co – (Ni) hydroxides.[34] A relatively low energy barrier with a tubular structure is responsible for the rolling-up of a thin sheet/layer along a conical angle into a nanocone under the template role of SDS.[41]

Figure 3(a) – 3(c) shows SEM images of layered zinc hydroxide nanocones exchanged with NO_3^-, Cl^- and CH_3COO^-, using ethanol as the solvent at room temperature. Figure 3(d) displays the XRD patterns of zinc hydroxide nanocones

Figure 3

SEM images of zinc hydroxide nanocones exchanged with different anions in ethanol (a) NO_3^-, (b) Cl^- and (c) CH_3COO^-. (d) XRD patterns of zinc hydroxide nanocones exchanged with different anions.

intercalated with Cl^-, NO_3^- and CH_3COO^- anions. The interlayer spacing shifted to 0.90 nm for NO_3^-, 1.14 nm for Cl^- and 1.44 nm for CH_3COO^-, from the original 3.15 nm for DS^- anions. Compared with the previous report on layered Co – (Ni) hydroxide nanocones intercalating intercalated with NO_3^- (0.79 nm), Cl^- (0.78 nm) and CH_3COO^- (0.91 nm), the interlayer spacing of the current zinc hydroxide nanocones is noticeably larger. The reason for this is not yet clear. We speculate that exchange in different solvents, e.g., in ethanol or in water, might contribute to different anion arrangements and consequently interlayer spacings.

In fact, the interlayer spacing was reduced to 0.79 nm for the exchange with Cl^- if the anion-exchange process was carried out by using 1M NaCl in ethanol-water binary solvents (1:1 v/v). However, the nanocone morphology could not be well retained when water was added to the solvents. It was found that the nanocones deteriorated into

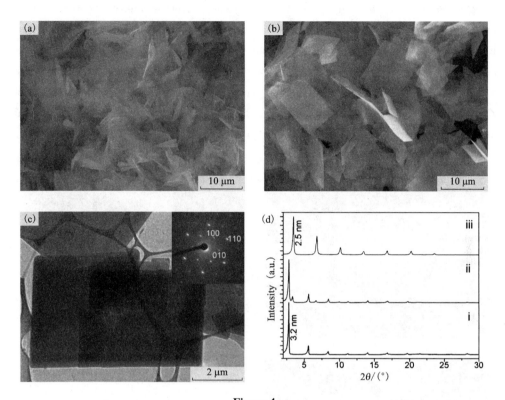

Figure 4

SEM images of zinc hydroxide in contact with aqueous NaCl (1 M) solution for different durations: (a) 6 h and (b) 24 h, (c) TEM image of as-transformed product at 24 h, (Inset) SAED pattern. (d) XRD patterns of transformed products in NaCl solution for different durations: i) 0 h, ii) 6 h, iii) 24 h.

irregular nanoplates in the deionized water. In Figure 4(a) and 4(b), the conical structure was gradually transformed into monodisperse square-like nanoplates in aqueous 1 M NaCl solution. The lateral length of the square-like plates is approximately 4 – 5 μm, while the very thin and uniform thickness is estimated to be a few tens of nm. The same morphological change was also observed in aqueous 1 M NaNO$_3$ solution. The SAED pattern collected on an individual nanoplate shown in the inset of Figure 4(c) confirmed the single-crystal nature with hexagonally arranged diffraction spots. The slight deviation from an ideal hexagonal symmetry, e. g., into an orthorhombic one, which might be attributed to an occupancy fluctuation or position shift of the Zn^{2+} in the hydroxide layer, is considered to be responsible for the formation of rectangular platelets. The XRD patterns in Figure 4(d) show that the interlayer spacing was reduced to 2.5 nm from original 3.2 nm, revealing a partial deintercalation of DS⁻ and the possible transformation from a bilayer to a monolayer arrangement in the gallery. The lower DS content was supported by a smaller weight loss in TG analysis. In other words, the anion-exchange process was not fully facilitated in aqueous solutions for zinc hydroxide nanocones, which is significantly different from the previous reports on Co – (Ni) hydroxide nanocones.[33-35] Such a unique feature of Zn based hydroxide nanostructures may be effectively used to apply facile morphological and structural modification at room temperature.

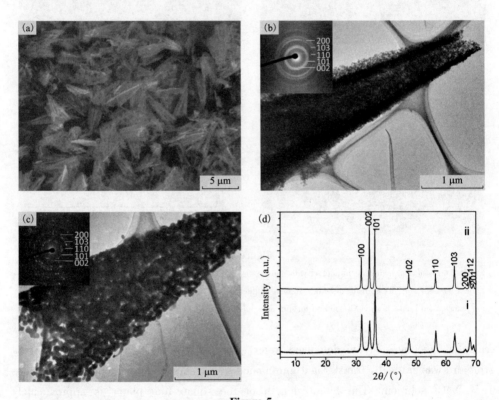

Figure 5
(a) SEM and (b) TEM images of ZnO nanocones obtained at 200 ℃ (Inset) SAED pattern. (c) TEM image of ZnO nanocones obtained at 800 ℃, (Inset) SAED pattern. (d) XRD patterns of ZnO nanocones obtained at different temperature: i) 200 ℃ and ii) 800 ℃.

The layered zinc hydroxide nanocones could be transformed into ZnO nanocones via a simple calcination treatment. In particular, inorganic anions (e. g., NO_3^-) intercalated hydroxide nanocones show lower pyrolysis temperatures and release less gas content during calcination, as confirmed by the TG analysis. They are more suitable for thermal transformation into ZnO nanocones. To preserve the original nanocone structure well, a slow heating rate was also found to be favourable. Figure 5 shows the SEM, TEM images and XRD patterns of ZnO nanocones obtained by calcination at different temperatures. The average diameter of the ZnO nanocones is smaller than the starting hydroxide nanocones owing to dehydration and removal of anionic species in the galleries through pyrolysis. The hollow interior nature of nanocones was identified by TEM characterization, as shown in Figure 5(b) and 5(c). In general, the oxide nanocones are polycrystalline, being composed of many tiny particles. The interstices between them may be formed due to the release of foamed water and gas contents. With the increase in calcination temperature to 800 ℃, the polycrystalline nanoparticle assembly into a conical framework is obvious [Figure 5(c)]. The XRD patterns of calcined nanocones obtained at different temperatures given in Figure 5(d) can both be indexed to be hexagonal ZnO with almost the same lattice constants [$a = 0.32466(4)$ nm and $c = 0.5197(4)$ nm for 200 ℃, $a = 0.32517(7)$ and $c = 0.5205(2)$ nm for 800 ℃].

The photocatalytic performances of the ZnO nanocones and nanoplates were evaluated by examining the photodegradation of methylene blue (MB) as a representative pollutant under irradiation by ultraviolet (UV) light. The time-dependences of UV – Vis absorption spectra of the aqueous solution of MB with 100 mg of ZnO nanocones and nanoplates were compared in Figure 6. As can be seen from Figure 6(a) and 6(b), well-defined absorption peaks of MB positioned at 610 nm and 660 nm rapidly decreased with prolonged time duration. MB was almost completely degraded in an hour, verified by both UV – Vis absorption spectra and photographs shown in the insets of Figure 6(a) and 6(b). Compared with a blank experiment without a photocatalyst, the remarkable efficiency of photodegradation within one hour using ZnO nanocones and nanoplates is evident [Figure 6(c)]. Furthermore, the photocatalytic efficiency of ZnO nanocones appears higher than that of the ZnO nanoplates. As the band structures of these ZnO nanomaterials should not differ too much, the reason might be derived from their morphological differences, e.g., specific surface area, exposed crystal facets, etc. It seems that ZnO nanocones exhibit a higher specific surface area than that of nanoplates due to a porous assembly of the nanoparticles. Furthermore, it is known that the efficiency of photocatalysis is significantly affected by the specific exposed crystal facets. It has been reported that the (002) polar face of ZnO possesses higher activity than that of (100). Consequently, the catalytic activity of hexagonal ZnO would be enhanced with more exposed (002) facets. Based on the XRD patterns for ZnO nanocones in Figure 5(d)(ii) and ZnO

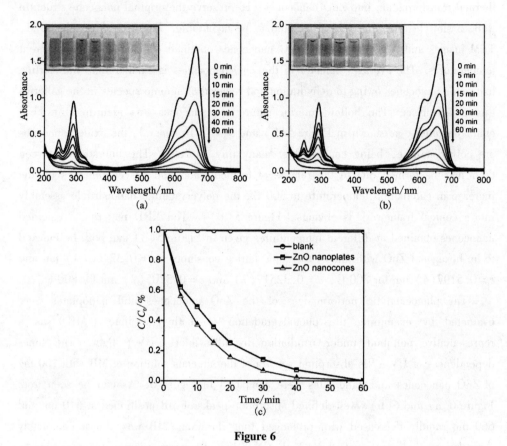

Figure 6

Evolution of the absorption spectra of methylene blue (15 mg/L, 50 mL) with different photocatalysts: (a) ZnO nanocones and (b) ZnO nanoplates. Insets in both panels show photographs of the methylene blue suspensions corresponding to different UV illumination times. (c) Time-dependent photodegradation of methylene blue.

nanoplates, the peak intensity ratios were tabulated. It can be seen that the peak intensity ratio of (100) to (002) is higher in ZnO nanocones than in ZnO nanoplates. In other words, the nanoplates hold a preferred growth tendency oriented along the c axis, which may diminish the proportion of (002) polar faces, namely the exposed (002) facets.[42] The above factors might be among the reasons for ZnO nanocones to show a higher photocatalytic activity than that of ZnO nanoplates. This indicates the high potential of altering morphological characteristics for functional optimization, as well as employing ZnO with this unique nanocone structure in photochemical and optoelectronic applications.

4 Conclusion

In summary, layered zinc hydroxide nanocones intercalated with DS⁻ anions have been synthesized via a convenient and reliable synthetic approach, using homogeneous precipitation in the presence of urea and SDS, in which SDS imparts a significant influence on the synthesis as both a surfactant and a structure-directing agent. These nanocones could be transformed into square-like nanoplates in aqueous solution even at room temperature. Corresponding ZnO nanocones and nanoplates were synthesized through further calcination, which provide convenient and effective routes to design peculiar conical and plate-like functional oxides under mild conditions. The fine control of the morphological features, e. g., nanocone vs. nanoplate, may be employed to tune their properties, e. g., photocatalytic performance. The conical structure with a hollow interior as well as its nanoplate counterpart, is promising for various applications such as photocatalysis, optoelectronics and biomedicine, etc.

References

[1] M. Meyn, K. Beneke, G. Lagaly. Anion-exchange reactions of hydroxy double salts [J]. Inorganic Chemistry, 1993, 32(7): 1209 – 1215.

[2] G. R. Williams, J. Crowder, J. C. Burley, A. M. Fogg. The selective intercalation of organic carboxylates and sulfonates into hydroxy double salts [J]. Journal of Materials Chemistry, 2012, 22(27): 13600 – 13611.

[3] V. Laget, C. Hornick, P. Rabu, M. Drillon, R. Ziessel. Molecular magnets: Hybrid organic-inorganic layered compounds with very long-range ferromagnetism [J]. Coordination Chemistry Reviews, 1998, 178 – 180(2): 1533 – 1553.

[4] R. Rojas, C. Barriga, M. A. Ulibarri, P. Malet, V. Rives. Layered Ni(Ⅱ)-Zn(Ⅱ) hydroxyacetates. Anion exchange and thermal decomposition of the hydroxysalts obtained [J]. Journal of Materials Chemistry, 2002, 12(4): 1071 – 1078.

[5] L. Xu, Y. S. Ding, C. H. Chen, L. Zhao, C. Rimkus, R. Joesten, S. L. Suib. 3D flowerlike α – nickel hydroxide with enhanced electrochemical activity synthesized by microwave-assisted hydrothermal method [J]. Chemistry of Materials, 2007, 20(1): 308 – 316.

[6] W. K. Hu, D. Noréus. Alpha nickel hydroxides as lightweight nickel electrode materials for alkaline rechargeable cells [J]. Chemistry of Materials, 2003, 15(4): 974 – 978.

[7] V. Gupta, T. Kusahara, H. Toyama, S. Gupta, N. Miura. Potentiostatically deposited nanostructured α – Co(OH)$_2$: A high performance electrode material for redox-capacitors [J].

Electrochemistry Communications, 2007, 9(9): 2315 - 2319.

[8] W. Stahlin, H. R. Oswald. The crystal structure of zinc hydroxide nitrate, $Zn_5(OH)_8(NO_3)_2 \cdot 2H_2O$ [J]. Acta Crystallographica Section B: Structural Crystallography and Crystal Chemistry, 1970, 26(6): 860 - 863.

[9] R. Ma, Z. P. Liu, K. Takada, K. Fukuda, Y. Ebina, Y. Bando, T. Sasaki. Tetrahedral Co (II) coordination in α - type cobalt hydroxide: Rietveld refinement and X-ray absorption spectroscopy [J]. Inorganic Chemistry, 2006, 45(10): 3964 - 3969.

[10] A. Moezzi, A. McDonagh, A. Dowd, M. Cortie. Zinc hydroxyacetate and its transformation to nanocrystalline zinc oxide [J]. Inorganic Chemistry, 2012, 52(1): 95 - 102.

[11] S. Inoue, S. Fujihara. Liquid-liquid biphasic synthesis of layered zinc hydroxides intercalated with long-chain carboxylate ions and their conversion into ZnO nanostructures [J]. Inorganic Chemistry, 2011, 50(8): 3605 - 3612.

[12] L. Poul, N. Jouini, F. Fiévet. Layered hydroxide metal acetates (metal = zinc, cobalt, and nickel): elaboration via hydrolysis in polyol medium and comparative study [J]. Chemistry of Materials, 2000, 12(10): 3123 - 3132.

[13] H. Morioka, H. Tagaya, M. Karasu, J. Kadokawa, K. Chiba. Effects of zinc on the new preparation method of hydroxy double salts [J]. Inorganic Chemistry, 1999, 38(19): 4211 - 4216.

[14] J. H. Yang, Y. S. Han, M. Park, T. Park, S. J. Hwang, J. H. Choy. New inorganic-based drug delivery system of indole - 3 - acetic acid-layered metal hydroxide nanohybrids with controlled release rate [J]. Chemistry of Materials, 2007, 19(10): 2679 - 2685.

[15] M. Z. Hussein, S. H. Al Ali, Z. Zainal, M. N. Hakim. Development of antiproliferative nanohybrid compound with controlled release property using ellagic acid as the active agent [J]. Internationa Journal of Nanomedicine, 2011, 6: 1373 - 1383.

[16] S. H. H. Al Ali, M. A. Qubaisi, M. Z. Hussein, Z. Zainal, M. N. Hakim. Preparation of hippurate-zinc layered hydroxide nanohybrid and its synergistic effect with tamoxifen on HepG2 cell lines [J]. Internationa Journal of Nanomedicine, 2011, 6: 3099 - 3111.

[17] A. Tsukazaki, A. Ohtomo, T. Onuma, M. Ohtani, T. Makino, M. Sumiya, K. Ohtani, S. F. Chichibu, S. Fuke, Y. Segawa, H. Ohno, H. Koinuma, M. Kawasaki. Repeated temperature modulation epitaxy for p-type doping and light-emitting diode based on ZnO [J]. Nature Materials, 2005, 4(1): 42 - 46.

[18] K. Hümmer. Interband magnetoreflection of ZnO [J]. Physica Status Solidi (b), 1973, 56 (1): 249 - 260.

[19] X. D. Wang, C. J. Summers, Z. L. Wang. Large-scale hexagonal-patterned growth of aligned ZnO nanorods for nano-optoelectronics and nanosensor arrays [J]. Nano Letters, 2004, 4(3): 423 - 426.

[20] A. B. Djurisic, Y. H. Leung. Optical properties of ZnO nanostructures [J]. Small, 2006, 2 (8-9): 944 - 961.

[21] L. Q, Jing, Y. C. Qu, B. Q. Wang, S. D. Li, B. J. Jiang, L. B. Yang, W. Fu, H. G. Fu, J. Z. Sun. Review of photoluminescence performance of nano-sized semiconductor materials and its relationships with photocatalytic activity [J]. Solar Energy Materials and Solar Cells, 2006, 90(12): 1773 – 1787.

[22] P. Li, Z. Wei, T. Wu, Q. Peng, Y. Li. Au – ZnO hybrid nanopyramids and their photocatalytic properties [J]. Journal of the American Chemical Society, 2011, 133(15): 5660 – 5663.

[23] C. L. Carnes, K. J. Klabunde. The catalytic methanol synthesis over nanoparticle metal oxide catalysts [J]. Journal of Molecular Catalysis A: Chemical, 2003, 194(1): 227 – 236.

[24] M. H. Huang, Y. Y. Wu, H. Feick, N. Tran, E. Weber, P. D. Yang. Catalytic growth of zinc oxide nanowires by vapor transport [J]. Advanced Materials, 2001, 13(2): 113 – 116.

[25] M. Law, L. E. Greene, J. C. Johnson, R. Saykally, P. D. Yang. Nanowire dye-sensitized solar cells [J]. Nature Materials, 2005, 4(6): 455 – 459.

[26] E. Kandare, J. M. Hossenlopp. Thermal degradation of acetate-intercalated hydroxy double and layered hydroxy salts [J]. Inorganic Chemistry, 2006, 45(9): 3766 – 3773.

[27] K. Tibbetts, R. Doe, G. Ceder. Polygonal model for layered inorganic nanotubes [J]. Physical Review B, 2009, 80(1): 014102.

[28] S. Berber, Y. K. Kwon, D. Tománek. Electronic and structural properties of carbon nanohorns [J]. Physical Review B, 2000, 62(4): R2291.

[29] A. Krishnan, E. Dujardin, M. M. J. Treacy, J. Hugdahl, S. Lynum, T. W. Ebbesen. Graphitic cones and the nucleation of curved carbon surfaces [J]. Nature, 1997, 388(6641): 451 – 454.

[30] D. Golberg, Y. Bando, Y. Huang, T. Terao, M. Mitome, C. C. Tang, C. Y. Zhi. Boron nitride nanotubes and nanosheets [J]. ACS Nano, 2010, 4(6): 2979 – 2993.

[31] K. Li, J. Ju, Z. Xue, J. Ma, L. Feng, S. Gao, L. Jiang. Structured cone arrays for continuous and effective collection of micron-sized oil droplets from water [J]. Nature Communications, 2013, 4: 2276 – 2282.

[32] L. Bourgeois, Y. Bando, W. Q. Han, T. Sato. Structure of boron nitride nanoscale cones: ordered stacking of 240 and 300 disclinations [J]. Physical Review B, 2000, 61(11): 7686.

[33] R. Ma, Y. Bando, T. Sasaki. Directly rolling nanosheets into nanotubes [J]. The Journal of Physical Chemistry B, 2004, 108(7): 2115 – 2119.

[34] X. H. Liu, R. Ma, Y. Bando, T. Sasaki. Layered cobalt hydroxide nanocones: microwave-assisted synthesis, exfoliation, and structural modification [J]. Angewandte Chemie International Edition, 2010, 49(44): 8253 – 8256.

[35] X. H. Liu, R. Ma, Y. Bando, T. Sasaki. A general strategy to layered transition-metal hydroxide nanocones: tuning the composition for high electrochemical performance [J]. Advanced Materials, 2012, 24(16): 2148 – 2153.

[36] M. Vucelic, G. D. Moggridge, W. Jones. Thermal properties of terephthalate- and benzoate-

intercalated LDH [J]. The Journal of Physical Chemistry, 1995, 99(20): 8328 – 8337.

[37] F. Millange, R. I. Walton, D. O'Hare. Time-resolved in situ X-ray diffraction study of the liquid-phase reconstruction of Mg – Al – carbonate hydrotalcite-like compounds [J]. Journal of Materials Chemistry, 2000, 10(7): 1713 – 1720.

[38] Y. Zhao, F. Li, R. Zhang, D. G. Evans, X. Duan. Preparation of layered double-hydroxide nanomaterials with a uniform crystallite size using a new method involving separate nucleation and aging steps [J]. Chemistry of Materials, 2002, 14(10): 4286 – 4291.

[39] V. V. Naik, R. Chalasani, S. Vasudevan. Composition driven monolayer to bilayer transformation in a surfactant intercalated Mg – Al layered double hydroxide [J]. Langmuir, 2011, 27(6): 2308 – 2316.

[40] C. H. Liang, Y. Shimizu, M. Masuda, T. Sasaki, N. Koshizaki. Preparation of layered zinc hydroxide/surfactant nanocomposite by pulsed-laser ablation in a liquid medium [J]. Chemistry of Materials, 2004, 16(6): 963 – 965.

[41] P. C. Tsai, T. H. Fang. A molecular dynamics study of the nucleation, thermal stability and nanomechanics of carbon nanocones [J]. Nanotechnology, 2007, 18(10): 105702.

[42] H. Lin, L. Li, M. Zhao, X. Huang, X. Chen, G. Li, R. Yu. Synthesis of high-quality brookite TiO_2 single-crystalline nanosheets with specific facets exposed: tuning catalysts from inert to highly reactive [J]. Journal of the American Chemical Society, 2012, 134(20): 8328 – 8331.

☆ W. Ma, R. Z. Ma, J. B. Liang, C. X. Wang, X. H. Liu, K. C. Zhou, T. Sasaki. Layered zinc hydroxide nanocones: synthesis, facile morphological and structural modification, and properties [J]. Nanoscale, 2015, 6(22): 13870 – 14004.

Chapter II

Inorganic Nanotubes

Selective and Controlled Synthesis of Single-Crystalline Yttrium Hydroxide/Oxide Nanosheets and Nanotubes

Abstract: Single-crystalline yttrium hydroxide nanotubes could be successfully synthesized in large quantities via a metastable nanosheet precursor reacted with sodium hydroxide under hydrothermal conditions. The nanosheet precursors were obtained through a facile hydrothermal synthetic method using soluble yttrium nitrate as the yttrium source and triethylamine as both an alkaline and complexing reagent. The influences of reaction time and concentration of sodium hydroxide on the formation of yttrium hydroxide nanotubes were investigated. Yttrium oxide and europium-doped yttrium oxide nanosheets and nanotubes could also be selectively obtained via a thermal decomposition method using corresponding hydroxides as precursor. The phase structures, morphologies and properties of as-prepared products were investigated in detail by X-ray powder diffraction (XRD), transmission electron microscopy (TEM), scanning electron microscopy (SEM), selected area electron diffraction (SAED), high-resolution transmission electron microscopy (HRTEM), and photoluminescence spectroscopies. The formation mechanisms of yttrium hydroxide nanotubes were discussed based on the experimental results. These low-dimensional nanostructures could be expected to bring new opportunities in the vast research and application areas.

1 Introduction

The design and controlled synthesis of inorganic materials with desired sizes and shapes has drawn considerable attention for many years due to their unique size-dependent and shape-dependent physical properties and their importance in basic scientific research and potential technology applications.[1-6] In particular, one-dimensional (1D) nanotubes have attracted continuous research attention during the past decades based on their distinctive structural features and fascinating properties, such as superior photocatalytic behavior, enhanced luminescence efficiency and lower lasing threshold, which have been demonstrated to be of considerable applications in,

e. g. , electronic, optical, sensors, energy, and catalysts fields, etc.[7-15] It is generally believed that two-dimensional (2D) nanosheets could be rolled into the 1D nanotubes under appropriate experimental conditions. The well-known examples could be found in the preparation of Bi, MnO_2, TiO_2, WS_2, $Ca_2Nb_3O_{10}$, $H_2La_2Ti_3O_{10}$ nanotubes, etc.[16-22] These systems imply a possible "rolling mechanism" from 2D nanosheets to 1D nanotubes, which inspires us to explore the fabrication of nanotubes through rolled-up nanosheets.

Rare earth compounds have stimulated great interest because of their unique properties arising from the electron transitions within the 4f shell and now have been widely applications in various fields, such as high-performance luminescence, up-conversion materials, catalysts, biochemical probes, medical diagnostics, and so forth.[23-24] As an important member of rare earth compounds, low-dimensional yttrium compounds have attracted significant research interest due to their fascinating properties and potential applications over the past several years. Considerable investigations have been conducted for preparation of low-dimensional yttrium compounds, which offer a wide range of structure dimensions and sizes for tuning of relevant properties. In recent years, a variety of novel shapes of low-dimensional yttrium compounds have been reported.[25-32] However, to the best of our knowledge, the formation mechanisms of yttrium compounds nanotubes were still not well understood. Furthermore, the production of square Y_2O_3 and Y_2O_3/Eu^{3+} nanosheets have not been realized until now. Herein, we demonstrated that yttrium hydroxide nanotubes could be successfully synthesized via a rolled-up metastable nanosheet precursor under hydrothermal conditions. Y_2O_3 and Y_2O_3/Eu^{3+} nanosheets or nanotubes could also be selectively obtained by calcining the corresponding hydroxide precursor. The photoluminescence properties of as-synthesized Y_2O_3/Eu^{3+} nanosheets and nanotubes were also investigated in detail.

2 Experimental Section

All chemicals in this work, such as aqueous yttrium nitrate [$Y(NO_3)_3 \cdot 6H_2O$], europium oxide (Eu_2O_3, purity, 99.99%), sodium hydroxide (NaOH) and triethylamine (TEA), were analytical grade from the Beijing Chemical Factory, China. They were used without further purification. Deionized water was used throughout.

2.1 Synthesis of square nanosheets of yttrium hydroxide precursor. In a typical synthesis, 0.001 mol $Y(NO_3)_3 \cdot 6H_2O$ was put into a Teflon-lined autoclave of 50 mL capacity and dissolved in 20 mL deionized water under stirring at room temperature. And then 3 mL of TEA solution was added dropwise in the solution. The autoclave was filled with deionized water up to 80% of the total volume; after 10 min stirring, it was sealed and maintained at 160℃ for 24 h without shaking or stirring. The autoclave was allowed to cool down to room temperature naturally after heat treatment. The resulting products were filtered, washed with deionized water and absolute ethanol, and finally dried in air at 60℃ for 5 h.

Synthesis of yttrium hydroxide nanotubes. The synthesis of yttrium hydroxide nanotubes was based on the preparation of yttrium hydroxide precursor and the subsequent hydrothermal treatment under alkaline conditions. In a typical synthesis, 0.1 g yttrium hydroxide nanosheet precursor was put into a Teflon-lined autoclave of 50 mL capacity and dispersed in 40 mL 0.125 - 0.5 M NaOH solutions. After stirring for 10 min, the autoclave was sealed and heated at 160℃ for a period of 2 - 24 h, and then the system was allowed to cool to room temperature naturally after heat treatment. The washing and collecting of the procedures were the same as those for yttrium hydroxide precursor.

2.2 Synthesis of yttrium oxide square nanosheets and nanotubes. Y_2O_3 nanotubes and square nanosheets were obtained by calcining the yttrium hydroxide nanotubes and square nanosheets precursor in a furnace for 2 h at 600℃, respectively. The preparation procedure of europium-activated yttrium oxide (Y_2O_3/Eu^{3+}) nanotubes and square nanosheets was similar to that of Y_2O_3 except that Eu^{3+} solution was added in the $Y(NO_3)_3 \cdot 6H_2O$ solution. Eu^{3+} solution was prepared as follows: 0.0001 mol Eu_2O_3 was dissolved in 10% diluted nitric acid to form a clear aqueous solution that was subsequently heated to vaporize the redundant nitric acid to form europium nitrate, and then the resulting product was redissolved in 10 mL of deionized water.

2.3 Characterization. The obtained specimens were characterized on a D/max2550 VB + X-ray powder diffraction (XRD) instrument with Cu K_α radiation ($\lambda = 1.54178$ Å). The operation voltage and current were kept at 40 kV and 40 mA, respectively. The size and morphology of as-synthesized products were determined at 10 kV by a XL30 S - FEG scanning electron microscope (SEM) and at 160 kV by a JEM - 200CX transmission electron microscope (TEM) and a JEOL JEM - 2010F high-resolution transmission electron microscope (HRTEM). Energy-dispersive spectrometer (EDS) images were recorded on a XL30 S-FEG Microscope. Selected area electron diffraction (SAED) was further performed to identify the crystallinity. Photoluminescence experiments were conducted on an F - 4500 fluorescence spectrophotometer with a Xe lamp at room temperature.

3 Results and Discussion

Figure 1(a) shows the typical XRD pattern of as-synthesized product prepared using 3 mL TEA as complexing agent at 160 ℃ for 24 h. However, all reflection peaks of the pattern cannot be finely indexed to a pure phase consistent with the reported data in JCPDS cards or any other reported literatures, which can be regarded as the precursor of $Y(OH)_3$. Furthermore, the Fourier transform infrared spectrometer (FT-IR) spectrum have shown the obvious presence of Y—O, O—H, C—H_3, and C—N bonds in precursor of $Y(OH)_3$, so the precursor may be the result of TEA [$(C_2H_5)_3N$] molecules inserting layers of layered $Y(OH)_3$, which form a complex compound regarded as inorganic-surfactant $Y(OH)_3 - (C_2H_5)_3N$ mesostructured material. Figure 2(a) shows the typical TEM of as-prepared $Y(OH)_3$ precursor and indicates that square nanosheets with orthogonal corners were achieved in large quantity using this approach. These square nanosheets had an edge length range from 180 nm to 250 nm. Furthermore, there are few quasi-circle nanosheets which may convert into square nanosheets with the elongation of reaction time.

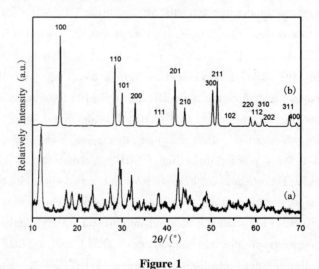

Figure 1

XRD patterns (a) from $Y(OH)_3$ precursor and (b) from $Y(OH)_3$ nanotubes prepared using $Y(OH)_3$ precursor reacted with 0.5 M NaOH solutions at 160 ℃ for 24 h.

Figure 2

(a) TEM images of as-prepared Y(OH)$_3$ precursor. (b) SEM image of as-prepared Y(OH)$_3$ nanotubes. The inset is a higher-magnified SEM image of Y(OH)$_3$ nanotubes. (c) TEM image of an individual Y(OH)$_3$ nanotube. The inset shows an SAED image taken on the individual Y(OH)$_3$ nanotube under [010] zone axis. (d) HRTEM image obtained from a selected top area of the individual Y(OH)$_3$ nanotube. The inset is a further magnified images to show the crystal lattice.

Nanosheets and broad nanobelts have the trend of curling into nanorods and nanotubes.[20, 22, 33] We get enlightenment from this phenomenon: Do square nanosheets precursor curl into nanotubes? We think the as-prepared precursor is inorganic-surfactant lamellar intercalations: (i) Soluble yttrium nitrate will firstly dissociate in water into Y^{3+}, which then reacts with OH$^-$ hydrolyzed by TEA to form Y(OH)$_3$. (ii) (C$_2$H$_5$)$_3$N molecules inserted into layers of layered Y(OH)$_3$ to form inorganic-surfactant Y(OH)$_3$ - (C$_2$H$_5$)$_3$N lamellar intercalations based on relatively weak hydrogen-bond and van der Waals force interactions. The inserted (C$_2$H$_5$)$_3$N molecules could be extracted from interlayers under proper conditions; the square nanosheet precursor with vacant interplanars would more easily roll into tubular nanostructures.[16] Therefore, transformation of these nanosheets into $Y(OH)_3$ nanotubes should be expected through removal of the $(C_2H_5)_3N$ molecules. Some researchers have stated that various nanotubes could be synthesized through rolling nanosheets into nanotubes by alkali treatment;[20,22] the similar process could be employed in the synthesis of $Y(OH)_3$ nanotubes.

Figure 1(b) shows the typical XRD pattern of the product prepared using 0.1 g of $Y(OH)_3$ precursor reacted with 0.5 M NaOH solutions at 160℃ for 24 h. All peaks of the pattern can be finely indexed to a pure hexagonal phase of $Y(OH)_3$ with the cell constants of $a = 6.261$ Å and $c = 3.544$ Å, consistent with the reported data in JCPDS cards (JCPDS 83 – 2042). No characteristic peaks of impurities can be detected, indicating that pure phase of $Y(OH)_3$ is successfully obtained under the current synthetic conditions. The morphologies and structures of as-prepared $Y(OH)_3$ sample were investigated by SEM, which can provide the information on the quantity and quality of $Y(OH)_3$ nanotubes. The SEM images indicate that a large quantity of $Y(OH)_3$ nanotubes was achieved using this approach, as seen in Figure 2(b). The size of the $Y(OH)_3$ nanotubes is about 400 nm in diameter and up to 15 μm in length. The inset in Figure 2(b) is a higher-magnified SEM image. Herein, the hollow interior structure of $Y(OH)_3$ nanotubes with wall thickness ranging from several to several hundred nanometers can be clearly observed. SEM observations also indicated that almost 100% of as-prepared products are uniform $Y(OH)_3$ nanotubes. HRTEM provided further insight into the structure of as-prepared $Y(OH)_3$ nanotubes. Figure 2(c) shows a typical image of an individual $Y(OH)_3$ nanotube with a mean diameter of about 40 nm and wall thicknesses about 6 nm. The inset in Figure 2(c) is the SAED image, perpendicular to axis of the individual nanotube, which can be indexed to the diffraction pattern of the [010] zone axis, and indicates that the individual $Y(OH)_3$ nanotube is a single crystal and grew along the [100] direction. Figure 2(d) shows the HRTEM image obtained from a selected top area of the individual $Y(OH)_3$ nanotube, from which the hollow structure features can be clearly seen with inner diameter about 24 nm and outer diameter about 40 nm. The inset in Figure 2(d) taken from the area marked by the square shows the nanotube is structurally uniform with an interplanar spacing in the wall about 1.77 Å, which corresponds to the (002) lattice plane of the hexagonal $Y(OH)_3$.

Further studies indicated that the phase structures and morphologies of final products strongly depended on reaction conditions such as reaction time and concentration of NaOH solution. Figure 3(a) – 3(d) shows the evolution of the XRD patterns of as-prepared products obtained using 0.5 M NaOH solutions at 160 ℃ for 2 h, 4 h, 8 h and 12 h, respectively. When the reaction time was reduced to 2 h, the main reflection peaks of the pattern are similar to the results of the Y(OH)$_3$ precursor, as shown in Figure 3(a). With careful observation, the slender reflection peaks from the hexagonal Y(OH)$_3$ can also be found, implying few Y(OH)$_3$ precursor transformed to hexagonal Y(OH)$_3$. Prolonging the reaction time to 4 h and 8 h, the reflection peaks from the hexagonal Y(OH)$_3$ are continuously improved. When the reaction time was

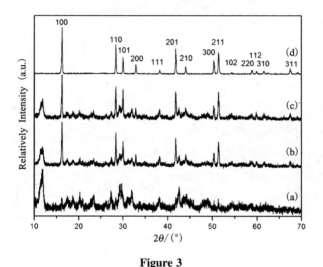

Figure 3

Evolution of XRD patterns of the products obtained using Y(OH)$_3$ precursor reacted with 0.5 M NaOH solutions at 160 ℃ for different reaction time: (a) 2 h, (b) 4 h, (c) 8 h, and (d) 12 h.

prolonged to 12 h, the XRD pattern of the product is shown in Figure 3(d). All of the peaks can be identified as the pure hexagonal phase of Y(OH)$_3$, which agrees well with the XRD results of the above-mentioned sample prepared at 160 ℃ for 24 h. This indicates the precursor has completely transformed at 160 ℃ for 12 h and sodium hydroxide is helpful for the formation of hexagonal Y(OH)$_3$. Furthermore, with the elongation of reaction time, the square nanosheets are gradually transformed into tubular nanostructures under hydrothermal conditions. The as-prepared product obtained at 160 ℃ for 2 h is mainly the square nanosheets with an edge length range from 180 nm to 250 nm besides the short nanotubes which seemed to be inherited from the precursor, as shown in Figure 4(a). When the reaction time is lifted to 4 h and 8 h, the products were a mixture of square nanosheets and nanotubes, which can be clearly seen in

Figure 4

TEM images of as-prepared products prepared using 0.5 M NaOH solutions at 160℃ for (a) 2 h, (b) 4 h, (c) 8 h, and (d) 12 h show the transformation of nanotubes from nanosheets.

Figure 4, parts B and C. Further elongating the reaction time to 12 h, most of the products were hexagonal Y(OH)$_3$ nanotubes [in Figure 4(d)]. If the reaction time was further extended, the products completely comprising nanotubes were obtained, as shown in Figure 2(b). The concentrations of sodium hydroxides are of great importance in the formation of the hexagonal Y(OH)$_3$ nanotubes. When the product is prepared by using 0 mL NaOH solution and reacting for 24 h, the morphology has not visibly changed, compared with precursor of Y(OH)$_3$ nanosheets. Figure 5(a) is the typical TEM image of the product prepared using 0.125 M NaOH solutions at 160℃ for 24 h. Herein, it is presented that the large quantity of square nanosheets still remain besides the short nanotubes. When the concentration of NaOH solution was increased to 0.25 M, tubular nanostructures were mainly formed [Figure 5(b)]. This result suggested that the square nanosheet precursor could be easily transformed into Y(OH)$_3$

Figure 5

TEM and SEM images of as-prepared products obtained using (a) 0.125 M and (b) 0.25 M NaOH solutions at 160 ℃ for 24 h.

nanotubes, and the growth process of Y(OH)$_3$ nanotubes could be controlled step by step by adjusting reaction time and concentration of NaOH solution.

There are some hypotheses that tubular structures might also form through the rolling process of lamellar structures, which have attracted many researchers devoting themselves into the fabrication of nanotubes through the rolling mechanism. Iijima thought that the carbon nanotubes were formed by curling and rolling of graphite nanosheets.[34] Viculis et al. have reported the exfoliated graphite nanosheets curl onto themselves.[35] The "rolling mechanism" inspired us to investigate the formation of Y(OH)$_3$ nanotubes. We have mentioned that the Y(OH)$_3$ precursor may be the result of (C$_2$H$_5$)$_3$N molecules inserted to form an inorganic-surfactant Y(OH)$_3$ - (C$_2$H$_5$)$_3$N lamellar mesostructured materials based on relatively weak hydrogen-bond and van der Waals force interactions, as shown in Figure 6(a). The addition of NaOH solution formed a strong alkaline environment that cause the (C$_2$H$_5$)$_3$N molecules to be extracted from precursor due to the influence of NaOH.[20] Then, Na ions were intercalated into the intersheet gallery and pinned the adjacent layers; the extraction of (C$_2$H$_5$)$_3$N molecules and insertion of Na ions cause the Y(OH)$_3$ interplanar to be vacant and the variance in bonding character between adjacent layers. Furthermore, the embedded Na ions were easily to be removed through washing. As a result, the defective nanosheets curled or overlapped themselves under hydrothermal conditions, as described in Figure 6(b), which shows the nanosheets start to roll up. When the concentration of sodium hydroxide and reaction time reached some extent, the rolling

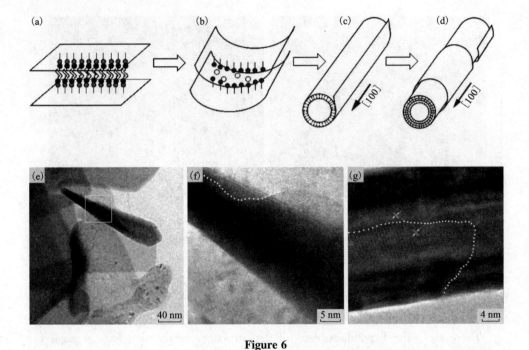

Figure 6

(a) - (d) Schematic illustration of the formation of Y(OH)$_3$ nanotubes by the conversion of Y(OH)$_3$ nanosheet precursor. (e) TEM image of a curling or rolling nanosheet. (f) HRTEM of the area marked in (e) exhibits the curling and overlapping part of the nanosheet in detail. (g) HRTEM of an individual nanotube, the area marked by broken line showing evidently an individual attached nanosheet.

nanosheets formed short nanotubes [Figure 6(c)]. With the elongation of reaction time, there were lots of nanosheets attached on the surface of nanotubes, causing nanotubes to grow longer and thicker [Figure 6(d)], and the attachment of nanosheets made tubes grow along a specific [100] zone axis. To further confirm the process, TEM and HRTEM were employed to testify the typical formation process of the nanotubes. Figure 6(e) indicates a typical higher-magnified TEM image of a rolling nanosheet. Figure 6(f) is an HRTEM image obtained from a selected area marked by the square in Figure 6(e), which exhibits the curling and overlapping part of the nanosheet in detail. The broken line marked is the interface between the curling part and the rest of nanosheet, which clearly confirmed that the nanotubes could be successfully synthesized from the rolling of nanosheets precursor. Figure 6(g) indicates the typical HRTEM image of an individual nanotube, and the area marked by broken line shows evidently an individual attached nanosheet. The crystal face features of the attached nanosheet are identical with those of nanotubes, consistent with the SAED result presented above. However, cubic phase Y_2O_3 nanotubes could not be obtained by curing Y_2O_3 nanosheets because cubic phase Y_2O_3 nanosheets are stable and cannot easily curl into nanotubes, different from metastable precursor of $Y(OH)_3$ nanosheets.

Single-crystalline Y_2O_3 nanosheets and nanotubes could be selectively obtained by calcining the corresponding nanosheet precursor and hexagonal $Y(OH)_3$ nanotubes obtained by using 0.5 M NaOH solutions at 160℃ for 24 h in air at 600℃ for 2 h, respectively. The calcination temperature was confirmed based on the results of differential scanning calorimetric analysis (DSC) and thermogravimetric analysis (TG). Parts A and B of Figure 7 are the XRD patterns of as-prepared Y_2O_3 nanosheets and nanotubes. All the reflections of the XRD patterns can be finely indexed to a pure body-centered cubic phase [space group: Ia3 (206)] of Y_2O_3 with lattice constants $a = 10.60$ Å (JCPDS 79-1257). No impurity peaks were observed, which indicates that as-prepared nanosheet precursor and hexagonal $Y(OH)_3$ nanotubes have been converted completely into fluorite structure Y_2O_3 by calcination in air at 600℃ for 2 h.

Figure 8(a) shows the typical TEM image of Y_2O_3 square nanosheets obtained by calcination of the corresponding nanosheet precursor. The edge lengths of square nanosheets range from 150 nm to 200 nm with orthogonal corners. Although there are

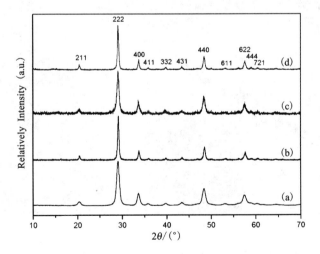

Figure 7

XRD patterns of Y_2O_3 (a) nanosheets and (b) nanotubes obtained by calcining corresponding $Y(OH)_3$ nanosheet precursor and $Y(OH)_3$ nanotubes. XRD patterns of (c) Y_2O_3/Eu^{3+} nanosheets and (d) Y_2O_3/Eu^{3+} nanotubes.

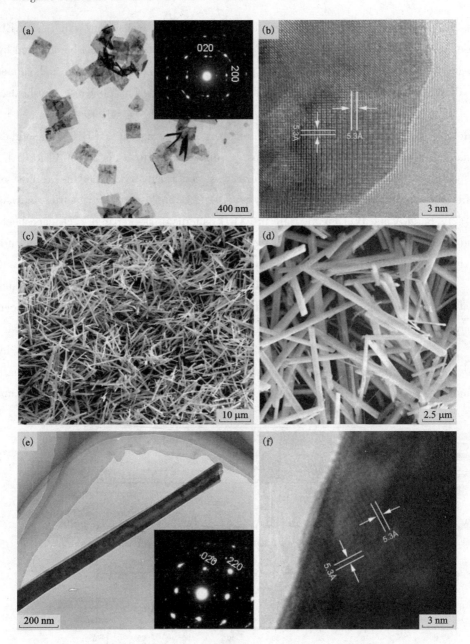

Figure 8

(a) TEM image of as-prepared Y_2O_3 nanosheets; the inset shows the corresponding SAED image indexed to the diffraction pattern of the [001] zone axis. (b) HRTEM image of an individual Y_2O_3 nanosheet. (c) Low- and (d) higher-magnified SEM images of as-prepared Y_2O_3 nanotubes. (e) TEM image of an individual Y_2O_3 nanotube; the inset shows the SAED image under [001] zone axis. (f) HRTEM image obtained from the wall under [001] zone axis.

few curled sheets caused by calcination, the structural features of the mainly square nanosteets are still retained well after calcination. The inset shows the corresponding SAED pattern of a single nanosheet under [001] zone axis, which exhibits each Y_2O_3 nanosheet is well crystallized. Figure 8(b) shows the HRTEM image of an individual Y_2O_3 nanosheet, and the interplanar distances are calculated as 5.3 Å, corresponding to {002} crystal planes of body-centered cubic Y_2O_3. Figure 8(c) is the SEM image of Y_2O_3 nanotubes obtained by calcining corresponding $Y(OH)_3$ nanotubes, which indicates the large quantity of uniform Y_2O_3 nanotubes with length up to about 15 μm were achieved using this approach. Figure 8(d) shows a higher-magnified SEM image of as-prepared Y_2O_3 nanotubes, which provides further information of the nanotubes with hollow interior. Figure 8(e) indicates a TEM image of an individual Y_2O_3 nanotube with a mean diameter of about 100 nm and wall thicknesses about 20 nm. The inset is the corresponding SAED pattern index to the diffraction pattern of the [001] zone axis, which shows that each Y_2O_3 nanotube is a single crystal and crystallized well. Figure 8(f) shows the HRTEM image of tube wall under [001] zone axis; the spacing of 5.3 Å between two adjacent lattice planes corresponds to {002} crystal planes.

Parts (c) and (d) of Figure 7 are the XRD patterns of Y_2O_3/Eu^{3+} nanosheets and nanotubes prepared by calcining the corresponding europium-doped nanosheet precursor and $Y(OH)_3$ nanotubes in air at 600℃ for 2 h, respectively. Because the content of europium is relatively low and the atom radii of yttrium and europium are nearly equal, the phase structures of Y_2O_3/Eu^{3+} are identical with those of Y_2O_3, as shown in Figure 7, parts (c) and (d). The images of Y_2O_3/Eu^{3+} nanosheets and nanotubes are, respectively, shown in Figure 9, parts (a) and (b), which indicate the morphologies of Y_2O_3/Eu^{3+} nanosheets and nanotubes are maintained perfectly as that Y_2O_3 nanosheets and nanotubes. Figure 9, parts (c) and (d), shows the EDS measurements on Y_2O_3/Eu^{3+} nanosheets and nanotubes, respectively. Only the characteristic peaks Y, O, Eu and C are observed, and no other contamination elements were detected (the element C is caused by the conducting resin), confirming the formation of Y_2O_3/Eu^{3+} nanosheets and nanotubes.

The Y_2O_3/Eu^{3+} nanosheets and nanotubes also displayed interesting optical properties. The room-temperature photoluminescence excitation and emission spectra of as-synthesized Y_2O_3/Eu^{3+} nanosheets and nanotubes are presented in Figure 10 which exhibit characteristic optical properties of Eu^{3+} ions in the body-centered cubic Y_2O_3 host structure. The inset is the excitation spectrum of the Y_2O_3/Eu^{3+} nanosheets,

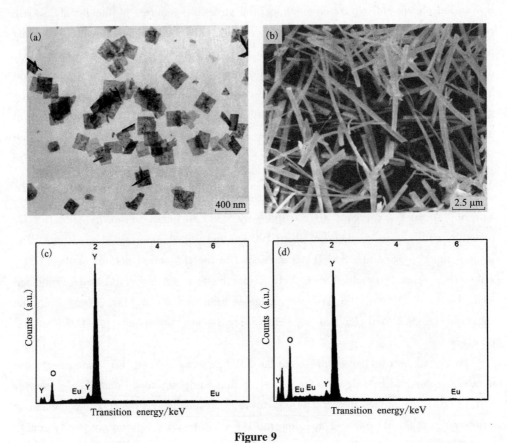

Figure 9

(a) TEM image of Y_2O_3/Eu^{3+} nanosheets and (b) SEM image of Y_2O_3/Eu^{3+} nanotubes; EDS of Y_2O_3/Eu^{3+} (c) nanosheets and (d) nanotubes showing only the characteristic peaks Y, O, and Eu are observed (the element C is caused by conducting resin).

which has only one broad band centered at around 248 nm. This is attributed to charge transfer between the Eu^{3+} cations and the surrounding anions.[36-37] The dashed lines and solid lines are the emission spectra of the Y_2O_3/Eu^{3+} nanosheets and nanotubes, respectively. In the Y_2O_3 lattice cell, there are two Y^{3+} sites in body-centered cubic Y_2O_3, which means 75% of these sites are noncentrosymmetric with C_2 symmetry, and the remaining 25% are centrosymmetric with S_6 symmetry.[26,38] The main peaks are observed from 610 nm to 630 nm, which are electronic dipole allowed and correspond to the $^5D_0-^7F_2$ transition, due to the Eu^{3+} on a C_2 site. The emission band from 580 nm to 589 nm corresponds to the $^5D_0-^7F_0$ transition, and emission band from 591 nm to 600 nm corresponds to the $^5D_0-^7F_1$ transition. These bands are magnetic dipole transitions and hardly vary with crystal field strength around Eu^{3+}, expected to arise from both Eu^{3+} C_2 and S_6 sites. Furthermore, bands from 648 nm to 668 nm and 706 nm to 718 nm are also observed, which correspond to the $^5D_0 - {}^7F_3$ and $^5D_0 - {}^7F_4$ transitions, respectively.[39-40] The photoluminescence intensity of nanotubes is higher than that of nanosheets, as shown in Figure 10. These results may exhibit the correlations between the physical properties of products and their size or shape.

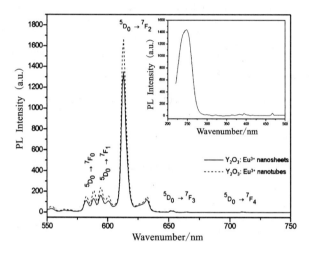

Figure 10

Emission spectrum of Y_2O_3/Eu^{3+} nanosheets (solid lines) and Y_2O_3/Eu^{3+} nanotubes (dashed lines) at excitation of 248 nm. The inset is the excited spectrum of Y_2O_3/Eu^{3+} nanosheets.

4 Conclusion

In summary, single-crystalline yttrium hydroxide nanotubes could be successfully synthesized in large quantities via a rolled-up nanosheets precursor obtained by using soluble yttrium nitrate as the yttrium source and TEA as both an alkaline and complexing reagent under proper hydrothermal conditions. The influences of reaction time and concentration of sodium hydroxide on the formation of yttrium hydroxide nanotubes were investigated. Single-crystalline yttrium oxides and europium-doped yttrium oxides square nanosheets and nanotubes could be selectively obtained by calcination of the corresponding square nanosheet precursor and hexagonal yttrium hydroxide nanotubes, respectively. Optical properties of as-prepared europium-doped yttrium oxide nanosheets and nanotubes were investigated. On the basis of our experimental results, the formation of yttrium hydroxide nanotubes was regarded as follows: nanosheet precursor curled or rolled to form nanotubes which then grew longer and thicker drived by the attached nanosheets. This method will have a good prospect in the future large-scale application due to its high yields, simple reaction apparatus, and low reaction temperature. Furthermore, this novel method can be surely carried out to synthesize other high-quality metal hydroxide and corresponding oxide nanosheets and nanotubes, and we find some rare earth hydroxides and transition metal hydroxide nanosheets exhibited similar "curling" phenomenon, which will be reported later. These low-dimensional nanostructures can also be expected to bring new opportunities in the vast research and applications in optical devices, catalysts, magnets, and other fields of functional materials.

References

[1] J. T. Hu, M. Ouyang, C. M. Lieber. Controlled growth and electrical properties of heterojunctions of carbon nanotubes and silicon nanowires [J]. Nature, 1999, 399(6731): 48 – 51.

[2] S. J. Tans, A. R. M. Verschueren, C. Dekker. Room-temperature transistor based on a single carbon nanotube [J]. Nature, 1998, 393(6680): 49 – 52.

[3] C. Zhou, J. Kong; E. Yenilmez, H. Dai. Modulated chemical doping of individual carbon nanotubes [J]. Science, 2000, 290(5496): 1552 – 1555.

[4] Z. Yao, H. W. Ch. Postma, L. Balents, C. Dekker. Carbon nanotube intramolecular junctions [J]. Nature, 1999, 402(6759): 273 – 276.

[5] R. D. Antonov, A. T. Johnson. Subband population in a single-wall carbon nanotube diode [J]. Physical Review Letters, 1999, 83(16): 3274 – 3276.

[6] J. Elias, R. Tena-Zaera, C. Levy-Clement. Effect of the chemical nature of the anions on the electrodeposition of ZnO nanowire arrays [J]. The Journal of Physical Chemistry C, 2008, 112 (15): 5736 – 5741.

[7] J. Goldberger, R. He, Y. Zhang, S. Lee, H. Yan, H. J. Choi, P. Yang. Single-crystal gallium nitride nanotubes [J]. Nature, 2003, 422(6932): 599 – 602.

[8] J. C. Hulteen, C. R. Martin. A general template-based method for the preparation of nanomaterials [J]. Journal of Materials Chemistry, 1997, 7(7): 1075 – 1087.

[9] Z. Deng, D. Chen, F. Tang, X. Meng, J. Ren, L. Zhang. Orientated attachment assisted self-

assembly of Sb_2O_3 nanorods and nanowires: end-to-end versus side-by-side [J]. The Journal of Physical Chemistry C, 2007, 111(14): 5325-5330.

[10] L. D. Hicks, M. S. Dresselhaus. Thermoelectric figure of merit of a one-dimensional conductor [J]. Physical Review B, 1993, 47(24): 16631.

[11] G. R. Patzke, F. Krumeich, R. Nesper. Oxidic nanotubes and nanorods-anisotropic modules for a future nanotechnology [J]. Angewandte Chemie International Edition, 2002, 41(14): 2446-2461.

[12] R. Fan, Y. Y. Wu, D. Y. Li, M. Yue, A. Majumdar, P. D. Yang. Fabrication of silica nanotube arrays from vertical silicon nanowire templates [J]. Journal of the American Chemical Society, 2003, 125(18): 5254-5255..

[13] Y. Sun, B. Mayers, Y. Xia. Metal nanostructures with hollow interiors [J]. Advanced Materials, 2003, 15(7-8): 641-646.

[14] B. Mayers, Y. Xia. Formation of tellurium nanotubes through concentration depletion at the surfaces of seeds [J]. Advanced Materials, 2002, 14(4): 279-282.

[15] S. Xiong, B. Xi, C. Wang, G. Zou, L. Fei, W. Wang, Y. Qian. Shape-controlled synthesis of 3D and 1D structures of CdS in a binary solution with L-cysteine's assistance [J]. Chemistry-A European Journal, 2007, 13(11): 3076-3081.

[16] Y. Li, X. Li, R. He, J. Zhu, Z. Deng. Artificial lamellar mesostructures to WS_2 nanotubes [J]. Journal of the American Chemical Society, 2002, 124(7): 1411-1416.

[17] S. Amelinckx, D. Bernaerts, X. B. Zhang, G. Van Tendeloo, J. Van Landuyt. A structure model and growth mechanism for multishell carbon nanotubes [J]. Science, 1995, 267(5202): 1334-1338.

[18] S. F. Braga, V. R. Coluci, S. B. Legoas, R. Giro, D. S. Galvão, R. H. Baughman. Structure and dynamics of carbon nanoscrolls [J]. Nano Letters, 2004, 4(5): 881-884.

[19] Y. Li, J. Wang, Z. Deng, Y. Wu, X. Sun, D. Yu, P. Yang. Bismuth nanotubes: a rational low-temperature synthetic route [J]. Journal of the American Chemical Society, 2001, 123(40): 9904-9905.

[20] Q. Chen, W. Zhou, G. Du, L. Peng. Trititanate nanotubes made via a single alkali treatment [J]. Advanced Materials, 2002, 14(17): 1208-1211.

[21] R. E. Schaak, T. E. Mallouk. Prying apart ruddlesden-popper phases: exfoliation into sheets and nanotubes for assembly of perovskite thin films [J]. Chemistry of Materials, 2000, 12(11): 3427-3434.

[22] R. Ma, Y. Bando, T. Sasaki, Directly rolling nanosheets into nanotubes [J]. The Journal of Physical Chemistry B, 2004, 108(7): 2115-2119.

[23] G. Y. Adachi, N. Imanaka. The binary rare earth oxides [J]. Chemical Reviews, 1998, 98

[24] R. Si, Y. Zhang, L. You, C. Yan. Rare-earth oxide nanopolyhedra, nanoplates, and nanodisks [J]. Angewandte Chemie International Edition, 2005, 117(21): 3320-3324.

[25] X. Wu, Y. Tao, F. Gao, L. Dong, Z. Hu. Preparation and photoluminescence of yttrium hydroxide and yttrium oxide doped with europium nanowires [J]. Journal of Crystal Growth, 2005, 277(1): 643-649.

[26] X. Li, Q. Li, Z. Xia, L. Wang, W. Yan, J. Wang, R. I. Boughton. Growth and characterization of single-crystal Y_2O_3: Eu nanobelts prepared with a simple technique [J]. Crystal Growth & Design, 2006, 6(10): 2193-2196.

[27] X. Wang, X. Sun, D. Yu, B. Zou, Y. Li. Rare earth compound nanotubes [J]. Advanced Materials, 2003, 15(17): 1442-1445.

[28] X. Wang, Y. Li. Rare-earth-compound nanowires, nanotubes, and fullerene-like nanoparticles: synthesis, characterization, and properties [J]. Chemistry-A European Journal, 2003, 9(22): 5627-5635.

[29] Q. Tang, Z. Liu, S. Li, S. Zhang, X. Liu, Y. Qian. Synthesis of yttrium hydroxide and oxide nanotubes [J]. Journal of Crystal Growth, 2003, 259(1): 208-214.

[30] X. Bai, H. Song, G. Pan, Z. Liu, S. Lu, W. Di, X. Ren, Y. Lei, Q. Dai, L. Fan. Luminescent enhancement in europium-doped yttria nanotubes coated with yttria [J]. Applied Physics Letters, 2006, 88(14): 143104-143106.

[31] X. Bai, H. Song, L. Yu, L. Yang, Z. Liu, G. Pan, S. Lu, X. Ren, Y. Lei, L. Fan. Luminescent properties of pure cubic phase Y_2O_3/Eu^{3+} nanotubes/nanowires prepared by a hydrothermal method [J]. The Journal of Physical Chemistry B, 2005, 109(32): 15236-15242.

[32] Y. P. Fang, A. Xu, L. P. You, R. Q. Song, J. C. Yu, H. X. Zhang, Q. Li, H. Q. Liu. Hydrothermal synthesis of rare earth (Tb, Y) hydroxide and oxide nanotubes [J]. Advanced Functional Materials, 2003, 13(12): 955-960.

[33] O. Zhou, R. M. Fleming, D. W. Murphy, C. H. Chen, R. C. Haddon, A. P. Ramirez, S. H. Glarum. Defects in carbon nanostructures [J]. Science, 1994, 263(5154): 1744-1747.

[34] S. Iijima. Helical microtubules of graphitic carbon [J]. Nature, 1991, 354(6348): 56-58.

[35] L. M. Viculis, J. J. Mack, R. B. Kaner. A chemical route to carbon nanoscrolls [J]. Science, 2003, 299(5611): 1361-1361.

[36] B. R. Judd. Optical absorption intensities of rare-earth ions [J]. Physical Review, 1962, 127(3): 750-761.

[37] Z. Lu, D. Qian, Y. Tang. Facile synthesis and characterization of sheet-like Y_2O_3: Eu^{3+} microcrystals [J]. Journal of Crystal Growth, 2005, 276(3): 513-518.

[38] J. Silver, M. I. Martunez, T. G. Ireland, R. J. Withnall. Yttrium oxide upconverting phosphors. Part 2: Temperature dependent upconversion luminescence properties of erbium in yttrium oxide [J]. The Journal of Physical Chemistry B, 2001, 105(30): 7200 - 7204.

[39] S. Ray, P. Pramanik, A. Singha, A. Roy. Optical properties of nanocrystalline Y_2O_3: Eu^{3+} [J]. Journal of Applied Physics, 2005, 97(9): 094312.

[40] M. A. Flores-Gonzalez, K. Lebbou, R. Bazzi, C. Louis, P. Perriat, O. Tillemen. Eu^{3+} addition effect on the stability and crystallinity of fiber single crystal and nano-structured Y_2O_3 oxide [J]. Journal of Crystal Growth, 2005, 277(1): 502 - 508.

☆ N. Zhang, X. H. Liu, R. Yi, R. R. Shi, G. H. Gao, G. Z. Qiu. Selective and controlled synthesis of single-crystalline yttrium hydroxide/oxide nanosheets and nanotubes [J]. Journal of Physical Chemistry C, 2008, 112(46): 17788 - 17795.

Rational Synthetic Strategy: From ZnO Nanorods to ZnS Nanotubes

Abstract: We demonstrate here that ZnS nanotubes can be successfully synthesized via a facile conversion process from ZnO nanorods precursors. During the conversion process, ZnO nanorods are first prepared as sacrificial templates and then converted into tubular ZnO/ZnS core/shell naonocomposites through a hydrothermal sulfidation treatment by using thioacetamide (TAA) as sulfur source. ZnS nanotubes are finally obtained through the removal of ZnO cores of tubular ZnO/ZnS core/shell nanocomposites by KOH treatment. The photoluminescence (PL) characterization of as-prepared products shows much enhanced PL emission of tubular ZnO/ZnS core/shell nanocomposites compared with their component counterparts. The probable mechanism of conversion process is also proposed based on the experimental results.

1 Introduction

During the past decades, wide band gap semiconductor materials have attracted considerable attention due to their size-dependent properties and important technological applications. In particular, ZnS is one of the important semiconductor materials with a wide band gap of 3.7 eV and received a wide range of research interest because of its potential applications in various fields such as nonlinear optical devices, displays, sensors, infrared windows and lasers.[1-3] Various synthetic routes, for example, solvothermal procedure, hydrothermal method, solid-vapor deposition technique, pulsed laser vaporization,[4-6] have been employed to prepared ZnS with different morphologies like nanobelts, nanorods, nanowires, nanoribbons, hollow nanospheres, and other hierarchical nanostructures.[7-12] Recently, a new approach to the preparation of ZnS and ZnS nanocomposites with various morphologies based on corresponding ZnO precursors has been a focus of materials research. Hexagonal ZnS nanotubes have been obtained via a conversion and etching process of ZnO nanocolumns.[13] ZnO/ZnS nanocable and ZnS nanotube arrays have been successfully synthesized by a thioglycolic acid-assisted solution route with removal of ZnO cores[14] and by the conversion from

ZnO nanorods due to the Kirkendall effect under simple ultrasonic irradiation,[15] respectively. ZnO/ZnS core/shell microspheres have been prepared with ZnO microspheres as reactive templates and they exhibit a distinct enhancement of photoluminescence compared with the uncoated ZnO microspheres.[16] ZnS hollow nanospheres with nanoporous shell were successfully synthesized through the evolvement of ZnO nanospheres.[12] Heterostructured ZnO/ZnS core/shell nanotube arrays were synthesized by a conversion process from ZnO nanorods arrays grown by atomic-layer deposition.[17] The examples above are generally two-step routes while there are still other reports on one-pot process in which ZnO precursors were not point out by the authors but the conversion from ZnO to ZnS was most likely one of the growth mechanisms. For example, hollow ZnS microspheres,[18] bifunctional ZnO/ZnS nanoribbons decorated by $\gamma - Fe_2O_3$ clusters,[19] and flowerlike ZnS-ZnO heterogeneous microstructures built up by ZnS-particle-strewn ZnO microrods.[20] However, there are few reports of tubular ZnO/ZnS core/shell nanocomposites and ZnS nanotubes using ZnO nanorods precursors via a hydrothermal route.

In this paper, tubular ZnO/ZnS core/shell nanocomposites could be successfully obtained via a sulfidation conversion by using ZnO nanorods as sacrificial templates in which thioacetamide (TAA) was used as as sulfur source. ZnS nanotubes could also be successfully obtained via an alkali treatment process of tubular ZnO/ZnS core/shell nanocomposites. Compared with ZnO nanorods, as-prepared ZnO/ZnS core/shell nanocomposites and ZnS nanotubes exhibit much different room-temperature photoluminescence (PL) properties. The possible formation mechanism of tubular ZnO/ZnS core/shell nanocomposites was also proposed. This strategy may provide more opportunities for the preparation of core/shell structured and hollow metal chalcogenides.

2 Experimental Section

All chemicals were of analytical grade from Shanghai Chemical Reagents Co., and used as starting materials without further purification.

2.1 Preparation of ZnO nanorods. In a typical procedure, 0.297 g of $Zn(NO_3)_2 \cdot 6H_2O$ and 0.142 g of Na_2SO_4 were loaded into a Teflon-lined stainless steel autoclave of 50 mL capacity and dissolved in 30 mL deionized water. Then 5.0 mL of hydrazine monohydrate (80% v/v) solution was added dropwise during vigorous stirring. Next, the autoclave was filled with deionized water up to 80% of the total volume, after 10 min stirring, sealed and maintained at 180 ℃ for 24 h. Subsequently, the system was allowed to cool to room temperature naturally. The resulting precipitate was collected by filtration and washed with absolute ethanol and distilled water in sequence for several times. The final product was dried in a vacuum box at 50 ℃ for 4 h.

2.2 Preparation of tubular ZnO/ZnS core/shell nanocomposties. ZnO nanorods (0.0081 g) and thioacetamide (TAA, 0.075 g and 0.75 g, respectively) were put into a Teflon-lined stainless steel autoclave of 50 mL capacity and dissolved in 30 mL deionized water. The solution was stirred vigorously for 10 min and sealed and maintained at 100 ℃ for 6 h, and then cooled down to room temperature. The resulting precipitate was collected by filtration and washed with absolute ethanol and distilled water in sequence for several times. The final product was dried in a vacuum box at 50 ℃ for 4 h.

2.3 Preparation of ZnS nanotubes. Tubular ZnO/ZnS core/shell nanocomposites (0.01 g) were soaked in KOH solution (4 M) under constant stirring for 1 h at room temperature. Finally, the obtained products were collected by centrifugation and washed with distilled water repeatedly and then dried in vacuum box at 50 ℃ for 4 h.

2.4 Characterization. The obtained products were characterized on a D/max2550 VB + X-ray powder diffractometer (XRD) with CuK_α radiation ($\lambda = 1.54178$ Å). The operation voltage and current were kept at 40 kV and 40 mA, respectively. The size and morphology of as-synthesized products were determined at 20 kV by a XL30 S – FEG scanning electron microscope (SEM) and at 160 kV by a JEM – 200CX transmission electron microscope (TEM) and a JEOL JEM – 2010F high-resolution transmission electron microscope (HRTEM). Energy-dispersive X-ray spectroscopy (EDS) was taken on the SEM. The room-temperature photoluminescence (PL) measurement was carried out on an F – 2500 spectrophotometer using the 325 nm excitation line of Xe light. ZnO nanorods, ZnO/ZnS core/shell nanocomposties and ZnS nanotubes powder were used as the standards. UV – Vis absorption spectra (UV – Vis) were performed on a Perkin-Elmer Lambda – 20 spectrometer at room temperature. For UV spectroscopic analysis samples were dispersed thoroughly in distilled water.

3 Results and discussion

The crystal structure and crystallinity of as-prepared products were investigated by XRD. Figure 1(a) shows a typical XRD pattern of ZnO precursors, in which all the diffraction peaks can be well indexed to hexagonal ZnO with lattice constants $a = 3.249$ Å and $c = 5.208$ Å, in good agreement with the standard PDF data (JCPDS 36 – 1451). The XRD patterns of tubular ZnO/ZnS core/shell nanocomposite prepared with different molar ratio of ZnO to TAA were shown in Figure 1(b) and 1(c), respectively. Besides those of ZnO, characteristic diffraction peaks corresponding to face-centered cubic ZnS can be observed after the sulfidation treatment

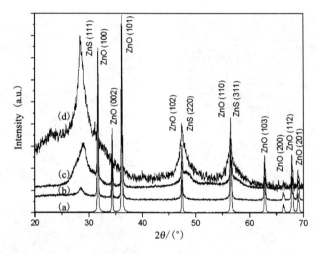

Figure 1

XRD patterns of as-prepared ZnO nanorods (a), tubular ZnO/ZnS core/shell nanocomposite obtained with different molar ratio of ZnO to TAA: (b) 1:1; (c), 1:10, and ZnS nanotubes(d).

of ZnO precursors, indicating the formation of ZnO/ZnS nanocomposite. The intensity of main diffraction peaks of ZnS in Figure 1(c) is much stronger than that in Figure 1(b), showing more ZnS was formed after sulfidation treatment due to the larger molar ratio of ZnO to TAA. Figure 1(d) shows the XRD patterns of ZnS nanotubes obtained via an alkali treatment process of tubular ZnO/ZnS core/shell nanocomposites [Figure 1(c)]. The main diffraction peaks in the pattern can be indexed as the face-centered cubic ZnS with lattice constants $a = 5.406$ Å [space group: $F\bar{4}3m(216)$], which are consistent with the values in the literature (JCPDS 05 – 0566). The average nanocrystallite size of ZnS products is estimated to be about 2.8 nm by employing the well-known Debye-Scherrer formula, close to exciton Bohr radius of ZnS (about 2.5 nm).[21]

The morphology, structure, and size of ZnO nanorods were characterized with SEM, TEM and HRTEM. Figure 2(a) and 2(b) shows typical low and high magnification SEM images of ZnO precursors, respectively, in which large quantities of ZnO nanorods with diameter in the range of 100 – 800 nm and length up to 10 μm can be clearly observed. EDS analysis exhibits the products only consist of Zn and O in addition to C which is caused by the C substrate, consistent with XRD characterization. Figure 2(c) is a representative TEM image of ZnO nanorods, which shows the morphology of several nanorods is relatively straight. Detailed information regarding to

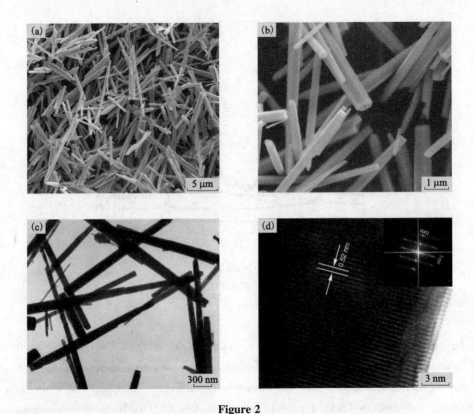

Figure 2
(a) Low- and (b) high-magnification SEM images, (c) TEM images, and (d) HRTEM image of as-prepared ZnO nanorods. Inset of (d) fast Fourier transform (FFT) image of the HRTEM result.

the crystal structure of final products has been investigated by HRTEM. Figure 2(d) shows a typical HRTEM image of an individual ZnO nanorod. The lattice spacing was calculated to be 0.52 nm, corresponding to the d-spacing of (0001) crystal plane of the wurtzite ZnO, which suggests that those nanorods are single crystalline and grow along the [0001] direction. The inset of Figure 2(d) is the corresponding Fast Fourier Transform (FFT) image of HRTEM result, showing that the nanorod is of single-crystalline structure with preferential [0001] growth direction.

The conversion from ZnO to ZnO/ZnS core/shellnanocomposites can be realized due to the much different solubility between them (solubility product constant K_{sp}: 6.8×10^{-17} for ZnO and 1.2×10^{-24} for ZnS) at the same temperature.[18] Figure 3(a) depicts a typical low magnification SEM image of ZnO/ZnS core/shell nanocomposites prepared with molar ratio of ZnO to TAA as 1:10, in which many tube-like products with average length about 5 μm can be found. Compared with ZnO nanorods precursors, those products are shorter in length while the diameter is nearly same, indicating ZnO nanorods have broken into smaller sections during the conversion process. Figure 3(b) is a high magnification SEM image selected from Figure 3(a). A tube-like product with irregular open tip can be clearly observed. EDS analysis shows that S was found besides Zn and O, agreeing with the XRD pattern. The as-prepared products have also been investigated by TEM and the result is shown in Figure 3(c).

Figure 3

(a) Low- and (b) high-magnification SEM images, (c) TEM images, and (d) SAED patterns of as-prepared ZnO/ZnS core/shell nanocomposites obtained with molar ratio of ZnO to TAA as 1:10.

The distinct hollow interior spaces reveal the core/shell structured nature of as-prepared products. In contrast to the smooth inner surface, the outer surface of ZnS shell is relatively rough. Parts of ZnO cores connected with shells can be found with careful observation, as marked by a white rectangle. Two kinds of electron diffraction patterns can be clearly observed [Figure 3(d)], of which one is indexed to single-crystalline ZnO core and the other refers to ZnS shell with polycrystalline structure, further confirming the core/shell structure of products. A comparative experiment has been carried out to study the influence of the amount of TAA on the morphology of final products and it is found that less TAA leads to the blurred boundaries between cores and shells which are difficult to distinguish.

Since ZnO has an amphoteric character and ZnS exhibits a good resistivity to alkaline solution,[18] strong base such as KOH can be employed to remove the ZnO cores while leave the ZnS shells intact. Figure 4(a) is a typical SEM image of ZnS product obtained with the removal of ZnO cores by KOH treatment. Herein, ZnS nanotubes with open tips can be clearly observed. In addition, the walls of some nanotubes are so thin that they look almost transparent, further confirming the hollow nature of products. There also exist some fragments, which can be attributed to the inevitable collapse of ZnS nanotubes during the removal process due to their relatively thin walls, similar to the collapse of final products in other chemical conversion process such as galvanic replacement in our previous study.[22] Figure 4(b) shows a higher magnification SEM image exhibiting clearly the open tips of nanotubes. The rough outer surface of nanotubes can also be observed, suggesting the morphology has been well inherited from the shells of ZnO/ZnS core/shell nanocomposite. EDS spectrum of ZnS nanotubes presents prominent Zn and S peaks. Figure 4(c) is a representative TEM image of as-prepared product, in which ZnS nanotubes with lengths ranging from 500 nm to 2 μm and average diameter of 300 nm can be found. The average thickness of the wall of those nanotubes is estimated to be approximately 50 nm. Moreover, the surface of those nanotubes is rough and some fragments exist, consistent well with the SEM observation. The inset of Figure 4(c) is the corresponding SAED pattern taken on a single nanotube, revealing the polycrystalline nature of the product. The diffraction rings can be outwards indexed to (111), (220), and (311) crystal planes of cubic ZnS, respectively. The inset of Figure 4(d) is a high magnification TEM image which exhibits that a single ZnS nanotube with a closed tip is built up of numerous nanoparticles and some particles loosely stack on the wall of the tube. HRTEM analysis provides further insight into the structures of those nanoparticles. Figure 4(d) shows a

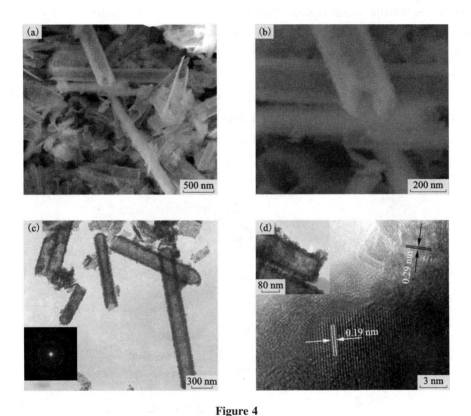

Figure 4

(a) Low- and (b) high-magnification SEM images, (c) TEM image, and (d) SAED patterns of as-prepared ZnS nanotubes. Inset of (c) SAED pattern and inset of (d) high magnification TEM image of the corresponding product.

HRTEM image taken on the edge of an individual ZnS nanotube, in which many tiny nanocrystallites can be observed and lattice spacings corresponding to (101) and (110) crystal planes of cubic ZnS are distinguished, providing further evidence for the polycrystalline nature of ZnS nanotubes, agreeing well with the SAED patterns. The nanocrystallites size of ZnS is clearly shown in another HRTEM image and it comes out to be 3 nm, consistent with the estimation from the Debye-Scherrer formula.

The conversion process from ZnO nanorods to ZnS nanotubes which involves sulfidation and etching steps is shown in Figure 5(a). The conversion from solid templates to core/shell or hollow nanostructures is interesting and several formation mechanisms have been proposed to elucidate the process, including the well known Ostwald ripening,[23] classic and newly prevalent Kirkendall effect,[24] and inequivalent exchange of metallic atoms regarding the galvanic replacements reaction, and so on.[25] We believe that a diffusion and consumption process could explain well the formation of ZnO/ZnS core/shell nanocomposites in our case. Figure 5(b) is the schematic illustration of formation mechanism of ZnO/ZnS core/shell nanocomposites, which presents more details. In the initial stage, TAA react with water when it is introduced at certain temperature to produce H_2S, and thus ions exchange happens as S^{2-} react with Zn^{2+} slowly dissolved from the surface of ZnO nanorods to form initial ZnS shells under the driving force caused by the fact that ZnS is more thermodynamically stable due to its

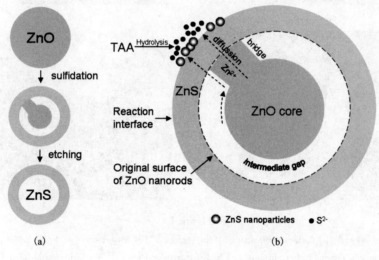

Figure 5

Schematic illustration of (a) the conversion process from ZnO nanorods to ZnS nanotubes and (b) formation mechanism of ZnO/ZnS core/shell nanocomposites.

lower solubility.[16] Small gaps will form between the ZnO cores and initial ZnS shells as ZnO on the surface is gradually converted into ZnS. Meanwhile, part of ZnO in contact with ZnS shells stochastically remains [Figure 3(c)] and serves as convenient transportation channels, similar to bridges, for Zn ions to reach the reaction interface via surface and bulk diffusion process. Then zinc ions continuously diffuse from inside of ZnO cores to the outer surface of the ZnS shells along certain ZnO bridges—it is more easy for Zn^{2+} to penetrate through the ZnS shells than S^{2-} does due to the smaller size of Zn^{2+} to react with S^{2-} in solution and thus ZnS shells become thicker as more and more ZnS nanoparticles pile up on the initial ZnS shells. Note some ZnS nanoparticles which do not act as building units of ZnS shells but exist separately in the solution

[Figure 4(d)]. With the proceeding of the conversion reaction, the intermediate gap is enlarged due to the continuous consumption of ZnO cores and ZnO/ZnS core/shell nanocomposites are obtained correspondingly. When the core/shell nanocomposites are etched by KOH, the cores are removed and ZnS nanotubes finally form.

Figure 6 presents the UV–Vis absorption spectra of tubular ZnO/ZnS core/shell nanocomposite prepared with different molar ratios of ZnO to TAA and ZnS nanotubes. The peak at 317 nm can be indexed to ZnS absorption peak, showing an obvious blue shift from 345 nm for bulk ZnS, which is due to the quantum size effects. The peak attributed to ZnO at 378 nm can be found in Figure 6(a) and 6(b) which corresponds to ZnO cores in core/shell nanocomposites, and no such a peak but the peak at 378 nm is observed in Figure 6(c), suggesting the removal of ZnO cores and the existence of ZnS shells.

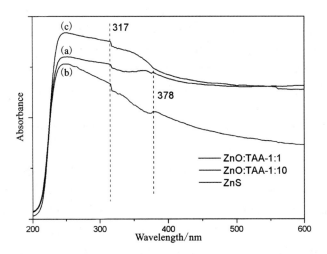

Figure 6　UV–Vis absorption spectra of ZnO/ZnS core/shell nanocomposites prepared with different molar ratio of ZnO to TAA
(a) 1∶1, (b) 1∶10, and (c) ZnS nanotubes.

The room-temperature photoluminescence (PL) under the excitation wavelength of 325 nm has been performed to investigate the optical properties of as-prepared products. Figure 7(a) presents the PL spectrum of ZnO nanorods, in which a broad ultraviolet emission with peak at 387 nm, a weak blue band, and a negligible green band can be observed. The ultraviolet emission originates from the excitonic recombination corresponding to the near-band edge emission of ZnO and the green luminescence can be attributed to the recombination of an electron and a photogenerated hole caused by

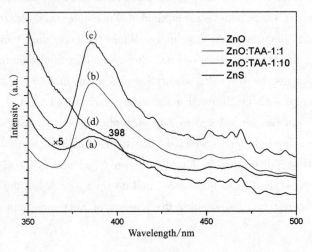

Figure 7

Photoluminescence spectra of ZnO nanorods (a), ZnO/ZnS core/shell nanocomposites prepared with different molar ratio of ZnO to TAA: (b) 1:1, (c)1:10, and ZnS nanotubes (d) under the photon excitation of 325 nm at room temperature.

surface defects and oxygen vacancies[26, 27] while the exact mechanism in regard to blue emission is not yet clear.[28] Compared with ZnO nanorods, both of ZnO/ZnS core/shell nanocomposites prepared with different molar ratio of ZnO to TAA exhibit much enhanced ultraviolet and visible emissions, similar to the previous report.[29] In addition, no obvious red-shift or blue-shift is observed [ultraviolet emission peaks at 386 and 385 for Figure 7(b) and 7(c), correspondingly] compared with Figure 7(a). It is also found that all of the emission peaks in Figure 7(c) are stronger than those in Figure 7(b). The portion of ZnS shells with a wider band gap closely coated on ZnO cores provides an efficient passivation of the surface trap states, giving rise to a strongly enhanced fluorescence quantum yield and thus resulting in the obvious enhancement in the ultraviolet emission of core/shell ZnO/ZnS nanocomposites.[16, 30, 31] The improved PL emission has also been reported in the study of CdS/ZnS wires and CdSe/CdS/ZnCdS/ZnS nanostructures.[32, 33] Moreover, the PL properties of ZnS nanotubes have been investigated and the result is shown in Figure 7(d). Several blue emission bands similar to those in Figure 7(c) can be seen, except for the weaker intensity. Moreover, a broad emission peak centered at 398 nm was observed in PL spectrum of ZnS nanotubes, showing a remarkable blue shift compared with emission position of bulk

ZnS (440 – 500 nm),[34] which could be attributed to the quantum size effects of ZnS nanocrystallines. The comparison among those spectra suggests that the PL properties of ZnO/ZnS core/shell nanocomposites are superior to their component counterparts.

4 Conclusion

In summary, ZnS nanotubes could be successfully prepared through a facile conversion strategy. ZnO/ZnS core/shell nanocomposites were obtained via a sulfidation process by using ZnO nanorods as sacrificial templates. A subsequent removal process of ZnO cores with KOH treatment finally converts ZnO/ZnS core/shell nanocomposites to ZnS nanotubes. PL study reveals that ZnO/ZnS core/shell nanocomposites exhibit enhanced UV and visible emissions compared with ZnO nanorods and ZnS nanotubes, showing their unique and superior optical properties. A diffusion and consumption mechanism has been proposed for the formation of ZnO/ZnS core/shell nanocomposites. This strategy may be extended to prepare other hollow and core/shell structured metal chalcogenides by using corresponding metal oxides with specific morphologies as precursors.

References

[1] D. Moore, Z. L. Wang. Growth of anisotropic one-dimensional ZnS nanostructures [J]. Journal of Materials Chemistry, 2006, 16(40): 3898 – 3905.
[2] P. Hu, Y. Liu, L. Fu, L. Cao, D. Zhu. Self-assembled growth of ZnS nanobelt networks [J]. The Journal of Physical Chemistry B, 2004, 108(3): 936 – 938.
[3] G. Z. Shen, Y. Bando, D. Golberg, Carbon-coated single-crystalline zinc sulfide nanowires [J]. The Journal of Physical Chemistry B, 2006, 110(42): 20777 – 20780.
[4] S. Yu, M. Yoshimura. Shape and phase control of ZnS nanocrystals: template fabrication of wurtzite ZnS single-crystal nanosheets and ZnO flake-like dendrites from a lamellar molecular precursor ZnS($NH_2CH_2CH_2NH_2$)$_{0.5}$[J]. Advanced Materials, 2002, 14(4): 296 – 300.
[5] Y. Zhang, Q. Peng, X. Wang, Y. D. Li. Synthesis and characterization of monodisperse ZnS nanospheres [J]. Chemistry Letters, 2004, 33(10): 1320 – 1321.
[6] Q. H. Xiong, G. Chen, J. D. Acord, X. Liu, J. J. Zengel, H. R. Gutierrez, J. M. Redwing, L. C. Lew Yan Voon, B. Lassen, P. C. Eklund. Optical properties of rectangular cross-sectional ZnS nanowires [J]. Nano Letters, 2004, 4(9): 1663 – 1668.
[7] M. V. Limaye, S. Gokhale, S. A. Acharya, S. K. Kulkarni. Template-free ZnS nanorod synthesis by microwave irradiation [J]. Nanotechnology, 2008, 19(41): 415602 – 415606.

[8] C. Ma, M. Moore, J. Li, Z. L. Wang. Nanobelts, nanocombs, and nanowindmills of wurtzite ZnS [J]. Advanced Materials, 2003, 15(3): 228-231.

[9] X. S. Fang, L. D. Zhang. One-dimensional (1D) ZnS nanomaterials and nanostructures [J]. Journal of Materials Science & Technology, 2006, 22(6), 721-736.

[10] S. Kar, S. Chaudhuri. Controlled synthesis and photoluminescence properties of ZnS nanowires and nanoribbons [J]. The Journal of Physical Chemistry B, 2005, 109(8): 3298-3302.

[11] C. H. Liang, Y. Shimizu, T. Sasaki, H. Umehara, N. Koshizaki. Au-mediated growth of wurtzite ZnS nanobelts, nanosheets, and nanorods via thermal evaporation [J]. The Journal of Physical Chemistry B, 2004, 108(28): 9728-9733.

[12] H. F. Shao, X. F. Qian, Z. K. Zhu. The synthesis of ZnS hollow nanospheres with nanoporous shell [J]. Journal of Solid State Chemistry, 2005, 178(11): 3522-3528.

[13] L. Dloczik, R. Engelhardt, K. Ernst, S. Fiechter, I. Sieber, R. Könenkamp. Hexagonal nanotubes of ZnS by chemical conversion of monocrystalline ZnO columns [J]. Applied Physics Letters, 2001, 78(23): 3687-3689.

[14] C. L. Yan, D. F. Xue. Conversion of ZnO nanorod arrays into ZnO/ZnS nanocable and ZnS nanotube arrays via an in situ chemistry strategy [J]. The Journal of Physical Chemistry B, 2006, 110(51): 25850-25855.

[15] H. F. Shao, X. F. Qian, B. C. Huang. Fabrication of single-crystal ZnO nanorods and ZnS nanotubes through a simple ultrasonic chemical solution method [J]. Materials Letters, 2007, 61(17): 3639-3643.

[16] Y. F. Zhu, D. H. Fan, W. Z. Shen. A general chemical conversion route to synthesize various ZnO-based core/shell structures [J]. The Journal of Physical Chemistry C, 2008, 112(28): 10402-10406.

[17] H. C. Liao, P. C. Kuo, C. C. Lin, S. Y. Chen. Synthesis and optical properties of ZnO-ZnS core-shell nanotube arrays [J]. Journal of Vacuum Science & Technology B, 2006, 24(5): 2198-2201.

[18] C. L. Yan, D. F. Xue. Room temperature fabrication of hollow ZnS and ZnO architectures by a sacrificial template route [J]. The Journal of Physical Chemistry B, 2006, 110(14): 7102-7106.

[19] X. B. Cao, X. M. Lan, Y. Guo, C. Zhao. S. M. Han, J. Wang, Q. R. Zhao, Preparation and characterization of bifunctional ZnO/ZnS nanoribbons decorated by $\gamma - Fe_2O_3$ clusters [J]. The Journal of Physical Chemistry C, 2007, 111(51): 18958-18964.

[20] Y. H. Ni, S. Yang. J. M. Hong, L. Zhang, W. L. Wu, Z. S. Yang, Fabrication, characterization and properties of flowerlike ZnS-ZnO heterogeneous microstructures built up by ZnS-particle-strewn ZnO microrods [J]. The Journal of Physical Chemistry C, 2008, 112(22): 8200-8205.

[21] Z. Chen, Q. M. Gao, M. L. Ruan. Electronic coupling one-dimensional Ag/ZnS nanocomposites in a nanoporous nickel phosphate host [J]. Nanotechnology, 2007, 18(25):

255607-255611.

[22] R. Yi, R. R. Shi, G. H. Gao, N. Zhang, X. M. Cui, Y. H. He, X. H. Liu. Hollow metallic microspheres: fabrication and characterization [J]. The Journal of Physical Chemistry C, 2009, 113(4): 1222-1226.

[23] Y. Chang, J. J. Teo, H. C. Zeng. Formation of colloidal CuO nanocrystallites and their spherical aggregation and reductive transformation to hollow Cu_2O nanospheres [J]. Langmuir, 2005, 21(3): 1074-1079.

[24] H. J. Fan, U. Gösele, M. Zacharias. Formation of nanotubes and hollow nanoparticles based on Kirkendall and diffusion processes: a review [J]. Small, 2007, 3(10): 1660-1671.

[25] L. Au, X. M. Lu, Y. N. Xia. A comparative study of galvanic replacement reactions involving Ag nanocubes and $AuCl_2^-$ or $AuCl_4^-$ [J]. Advanced Materials, 2008, 20(13): 2517-2522.

[26] J. Liang, J. Liu, Q. Xie, S. Bai, W. Yu, Y. Qian. Hydrothermal growth and optical properties of doughnut-shaped ZnO microparticles [J]. The Journal of Physical Chemistry B, 2005, 109(19): 9463-9467.

[27] J. X. Duan, X. T. Huang, E. Wang, H. H. Ai. Synthesis of hollow ZnO microspheres by an integrated autoclave and pyrolysis process [J]. Nanotechnology, 2006, 17(6): 1786-1790.

[28] L. Dai, X. L. Chen, W. J. Wang, T. Zhou, B. Q. Hu. Growth and luminescence characterization of large-scale zinc oxide nanowires [J]. Journal of Physics: Condensed Matter, 2003, 15(13): 2221-2226.

[29] F. Li, Y. Jiang, L. Hu, L. Y. Liu, Z. Li, X. T. Huang. Structural and luminescent properties of ZnO nanorods and ZnO/ZnS nanocomposites [J]. Journal of Alloys and Compounds, 2009, 474(1): 531-535.

[30] L. Yu, X. F. Yu, Y. F. Qiu, Y. J. Chen, S. H. Yang. Nonlinear photoluminescence of ZnO/ZnS nanotetrapods [J]. Chemical Physics Letters, 2008, 465(4): 272-274.

[31] P. Reiss, M. Protière, L. Li. Core/shell semiconductor nanocrystals [J]. Small, 2009, 5(2): 154-168.

[32] A. Datta, S. K. Panda, S. Chaudhuri. Synthesis and optical and electrical properties of CdS/ZnS core/shell nanorods [J]. The Journal of Physical Chemistry C, 2007, 111(46): 17260-17264.

[33] R. G. Xie, U. Kolb, J. X. Li, T. Basche, A. Mews. Synthesis and characterization of highly luminescent CdSe-core $CdS/Zn_{0.5}Cd_{0.5}S/ZnS$ multishell nanocrystals [J]. Journal of the American Chemical Society, 2005, 127(20): 7480-7488.

[34] C. L. Jiang, W. Q. Zhang, G. F. Zou, W. C. Yu, Y. T. Qian. Hydrothermal synthesis and characterization of ZnS microspheres and hollow nanospheres [J]. Materials Chemistry and Physics, 2007, 103(1): 24-27.

☆ R. Yi, G. Z. Qiu, X. H. Liu. Rational synthetic strategy: from ZnO nanorods to ZnS nanotubes, Journal of Solid State Chemistry, 2009, 182(10): 2791-2795.

Selective Synthesis and Magnetic Properties of Uniform CoTe and CoTe$_2$ Nanotubes

Abstract: Uniform CoTe and CoTe$_2$ nanotubes have been selectively synthesized in large quantities via a facile hydrothermal co-reduction route based on the precipitate slow-release controlled process under mild conditions, in which H$_2$TeO$_3$ were employed to supply Te source and sodium hydroxide and aqueous hydrazine was used as alkaline and reducing agent. CoTe and CoTe$_2$ nanotubes exhibit ferromagnetic and paramagnetic behaviour, respectively. These low-dimensional nanostructures could be expected to bring new opportunities in the vast research and application areas.

1 Introduction

The design and controlled synthesis of inorganic materials with novel morphologies and desired compositions have drawn immense attention for many years due to their unique morphology- and composition-dependent physical properties and their importance in basic scientific research and potential technology applications.[1,2] In particular, semiconducting tellurides have attracted enormous attention due to their distinctive physicochemical properties and wide variety of practical and potential applications in, e. g., optical recording materials, solar cell, sensor and laser materials, thermoelectric materials, and conductivity fields, etc.[3-8] The properties of tellurides sensitively depend on their morphologies and compositions. Thus it is a significant challenges to fabricate tellurides with novel morphologies and different compositions. Traditionally, tellurides have been synthesized by solid-state reactions, sonochemical synthesis, and self-propagating high-temperature synthesis.[9-12] Most of these approaches are based on high-temperature processes that generally require intricate processing and the use of high temperature for initiating the reactions. Furthermore the solution phase synthesis of tellurides is often involved relatively dangerous and highly toxic because of the use of H$_2$Te as the Te source. Recently, one-dimensional (1D) telluride nanomaterials have become the focus due to their brilliant prospects.[13-6] Herein we demonstrated that uniform CoTe and CoTe$_2$ nanotubes have been selectively synthesized via a facile hydrothermal co-reduction route based on the precipitate slow-release controlled process under mild conditions.[17]

2 Experimental Section

All chemicals were of analytical grade, and were used without further purification.

2.1 Synthesis. In a typical procedure for fabrication of CoTe nanotubes, 1 mmol $Co(NO_3)_2 \cdot 6H_2O$ was put into a Teflon-lined autoclave with 50 mL capacity and dissolved in 20 mL deionized water. Then 1 mmol H_2TeO_3 and 20 mL $N_2H_4 \cdot H_2O$ (80 wt.%) were added into the autoclave. After stirring for 5 min, the autoclave was sealed and heated at 140℃ for 24 h. The autoclave was allowed to cool down to room temperature naturally after heat treatment. The resulting product was obtained and collected by filtration, washed with deionized water, absolute ethanol and then dried in a vacuum box at 60℃ for 4 h. To prepare $CoTe_2$ nanotubes, 1 mmol $Co(NO_3)_2 \cdot 6H_2O$ was put into an autoclave and dissolved in 10 mL deionized water. Subsequently 10 mL 0.5 M NaOH, 2 mmol H_2TeO_3 and 1 mmol sodium dodecyl sulfate (SDS) were added into the autoclave. The other procedures were the same as those for the CoTe nanotubes except elongating the reaction time to 48 h.

2.2 Characterization. The obtained specimens were characterized on a D/max 2550 VB + X-ray powder diffraction (XRD) with Cu K_α radiation ($\lambda = 1.54178$ Å). The operation voltage and current were kept at 40 kV and 40 mA, respectively. The size and morphology of as-synthesized products were determined at 20 kV by a XL30 S-FEG scanning electron microscope (SEM) and at 160 kV by a JEM-200CX transmission electron microscope (TEM) and at 300 kV by a Tecnai FEG30 high-resolution transmission electron microscope (HRTEM). Electron energy loss spectroscopy (EELS) was used to examine the chemical composition of individual nanotubes. Selected area electron diffraction (SAED) was further performed to identify the crystallinity. Magnetic properties of the samples were measured with the magnetic property measurement (MPMS XP-5, Quantum Design SQUID magnetometer).

3 Results and Discussion

Figure 1(a) shows a typical SEM image of CoTe nanotubes and indicates large quantities of nanotubes were achieved using this approach. The morphology of the nanotubes is relatively straight and uniform with outer diameters of about 50 nm and lengths up to 1 μm. A typical TEM image of CoTe nanotubes is shown in Figure 1(b), from which it can be seen that the mean diameter of CoTe nanotubes is about 50 nm. The inset shows the select area electron diffraction (SAED) pattern of CoTe nanotubes taken on a mass of CoTe nanotubes reveals the satisfactory crystallinity of as-prepared product, which can be indexed to the hexagonal phase of CoTe. Figure 1(c) shows a typical HRTEM image of an individual CoTe nanotube. The inset shows a partly further magnified image. The interlayer spacing is about 2.10 Å, which corresponds to the interlayer distance of the (102) crystal plane in CoTe. The crystal structures of as-prepared products were characterized by X-ray powder diffraction (XRD). All of the reflections of the XRD pattern of as-prepared nanotubes in Figure 1(d) can be readily indexed to a hexagonal phase [space group: P63/mmc (194)] CoTe with NiAs

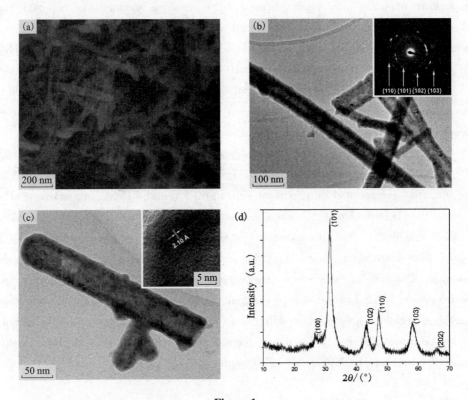

Figure 1

(a) SEM and (b) TEM images of CoTe nanotubes. The inset shows a SAED pattern of CoTe nanotubes. (c) A high-magnification TEM image of an individual CoTe nanotube. The inset is further magnified image to show the crystal lattice. (d) XRD pattern of as-synthesized CoTe nanotubes.

structure (JCPDS 34 − 0420; $a = 3.892$ Å and $c = 5.374$ Å). Figure 2(a) showed the TEM image of CoTe nanotubes obtained by using H_2TeO_3 as Te source and with the surfactant SDS (1 mmol) processed at 140 ℃ for 24 h, which indicated that the product was tubular structures with floccule surface and length 2 − 3 μm. Further studies indicated that H_2TeO_3 is a dominant factor for formation of CoTe nanotubes. Figure 2(b) − 2(d) shows the overviews of the SEM images of CoTe nanorods and indicates large quantities of nanorods with the mean diameter of about 120 nm were achieved using Na_2TeO_3 as Te source. Compared to soluble Na_2TeO_3, H_2TeO_3 can dissociate in water and form TeO_3^{2-} slowly. When the reaction temperature was elevated to 140 ℃, dissociative TeO_3^{2-} and Co^{2+} can be reduced by hydrazine to form Te and Co monomers, which will react with each other to form CoTe product under hydrothermal conditions.

Figure 3(a) shows the typical XRD patterns of $CoTe_2$ nanotubes prepared by using 10 mL 0.5 M NaOH solution as precipitator and molar ratio of 1∶2 for cobalt nitrate∶ tellurous acid with the surfactant SDS process at 140 ℃ for 48 h. The main reflections in this pattern can be indexed as the orthorhombic $CoTe_2$ [space group: Pnnm (58); JCPDS 11 − 0533; $a = 5.329$ Å, $b = 6.322$ Å and $c = 3.908$ Å]. The peaks marked with asterisks are from unreacted Te. Figure 3(b) shows a typical SEM image of as-synthesized $CoTe_2$ nanotubes with the mean diameter of about 50 nm and length up to 2 μm. A representative TEM image of as-prepared products shown in Figure 3(c) indicates clearly $CoTe_2$ nanotubes with hollow interior and uniform diameter. The inset of Figure 3(c) is the corresponding SAED pattern of $CoTe_2$ nanotubes, which can be indexed to the orthorhombic phase of $CoTe_2$. Figure 3(d) shows a typical HRTEM image of $CoTe_2$ nanotube, revealing that the $CoTe_2$ nanotube is structurally uniform with interlayer spacing about 2.82 Å, which corresponds to the value of the (111) lattice plane of the orthorhombic $CoTe_2$. The $CoTe_2$ nanotubes was further characterized by electron energy loss spectroscopy (EELS), the Te M edge and Co L edge were acquired at 577 eV and 779 eV respectively and there were no other element peaks detected, which confirmed that the nanotubes were only composed of Co and Te.

Furthermore, further studies suggested that surfactants have influences on the size and morphologies of $CoTe_2$ products. Figure 4(a) shows the typical SEM image of $CoTe_2$ nanorods with diameter in the range of 40 − 60 nm and length up to about 1 μm obtained in the absence of surfactants. Figure 4(b) is the SEM image of $CoTe_2$ nanorods synthesized by using sodium oleate (1 mmol) as surfactant, suggesting the rodlike structure nature with the mean diameter of about 30 nm and length up to

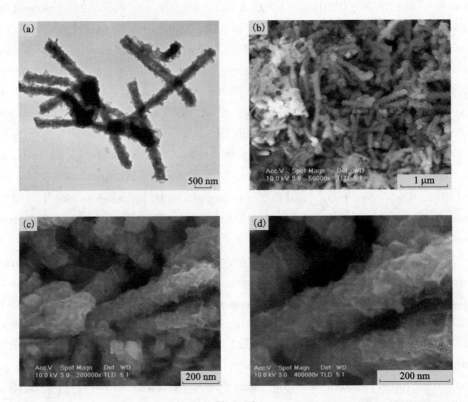

Figure 2

TEM image (a) of the CoTe nanotubes prepared by using H_2TeO_3 as Te source and with the surfactant SDS (1 mmol) processed at 140 ℃ for 24 h. Low-magnification and high-magnification SEM images (b) – (d) of the CoTe nanorods prepared by using Na_2TeO_3 as Te source with the absence of surfactants at 140 ℃ for 24 h.

500 nm. The morphology of as-prepared nanorods is relatively straight and the surfaces of the nanorods became smooth compared with the rough nanorods synthesized in the absence of surfactants. Figure 4(c) is an SEM image of as-prepared $CoTe_2$ nanotubes using CTAB (1 mmol) as surfactant. It can be found that the samples are built up of numerous nanotubes with mean diameters of about 60 nm and length 1 – 2 μm. The typical SEM image of as-prepared $CoTe_2$ product prepared by using 10 mL 1.0 M NaOH solution shown in Figure 4 (d) indicates flower-like hierarchical structures with interconnected thin leaf-like building blocks.

The concentration of NaOH is another important factor for preparation of $CoTe_2$ nanotubes. We consider that the reaction rate can be adjusted through precipitate slow-release method. When the molar ratio of Co and Te source is 1 to 2 without NaOH, the

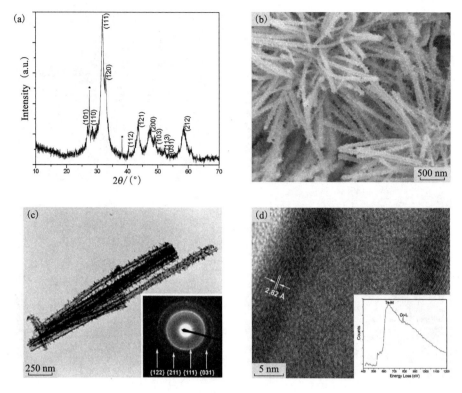

Figure 3

(a) XRD pattern of CoTe$_2$ nanotubes. (b) Low-magnification SEM image of as-prepared CoTe$_2$ nanotubes.
(c) TEM image and SAED patterns. (d) HRTEM image and EELS spectrum of CoTe$_2$ nanotubes.

product was the coexistence of Te and CoTe marked by asterisks and tilted cubes [Figure 5(a)], respectively. Figure 5(b) is the XRD pattern of as-prepared product obtained by using 10 mL 1.0 M NaOH, in which all characteristic diffraction peaks can be readily indexed to a pure orthorhombic phase CoTe$_2$. The concentration of NaOH solution can also control the structure of products through adjusting the reaction kinetics of the dissolution-deposition equation of Co(OH)$_2$ and further efficiently control the morphology of final products.

The magnetic properties of as-prepared products were measured with the Magnetic Property Measurement (MPMS XP-5, SQUID). The representative hysteresis loops of CoTe and CoTe$_2$ nanotubes and flower-like CoTe$_2$ hierarchical structures at 2 K are shown in Figure 6. The lower inset of Figure 6 is the enlarged hysteresis loops of as-prepared products. The saturation magnetization (Ms) of CoTe nanotubes is determined as about 13.4 emu/g with the magnetic coercivity (H$_c$) of about 283 Oe, indicating

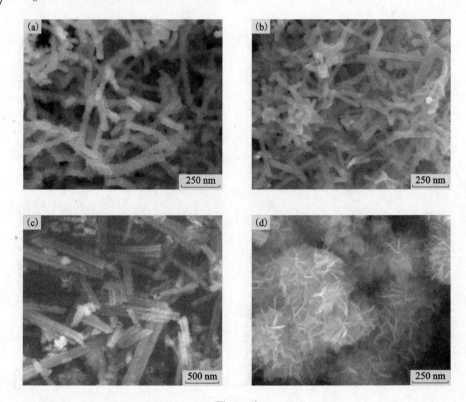

Figure 4

SEM images of CoTe$_2$ products prepared by using 10 mL 0.5 M NaOH solution with different surfactants at 140 ℃ for 48 h: (a) without surfactant; (b) sodium oleate; (c) CTAB. (d) SEM image of as-prepared CoTe$_2$ product obtained by using 10 mL 1.0 M NaOH solution with the surfactant SDS process at 140 ℃ for 48 h.

ferromagnetic behaviour for CoTe nanotubes at 2 K. The results might suggest that CoTe nanotubes are either a ferromagnetic substance or an alloy in which the magnetic moment of cobalt atoms is diluted by tellurium atoms. The M_s values of CoTe$_2$ nanotubes and flower-like CoTe$_2$ hierarchical structures are 4.4 emu/g and 6.7 emu/g, respectively. However, no hystereses are observed in the M – H curves for CoTe$_2$ nanostructures. The temperature dependences of magnetization of the flower-like CoTe$_2$ hierarchical structures in zero-field cooling (ZFC) and field cooling (FC) conditions under an applied field of 5000 Oe between 2 K and 300 K are shown in the upper inset of Figure 6. The magnetization of as-prepared product gradually increases with decreasing temperature in both ZFC and FC curves. Compared to the magnetic behavior of CoTe nanotubes, CoTe$_2$ nanostructures are no longer ferromagnetic while indicates paramagnetic behavior. We consider that it may be due to the increase of holes in the

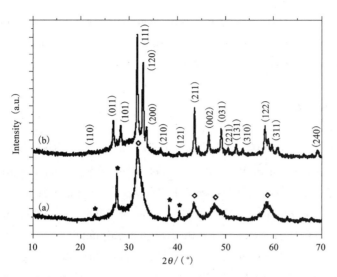

Figure 5 The evolution of the XRD patterns of the products obtained by using 0.001 mol Co(NO$_3$)$_2$ · 6H$_2$O and 0.002 mol H$_2$TeO$_3$ with the surfactant SDS (1 mmol) processed at 140 ℃ for 48 h
(a) without NaOH; (b) 10 mL 1.0 M NaOH

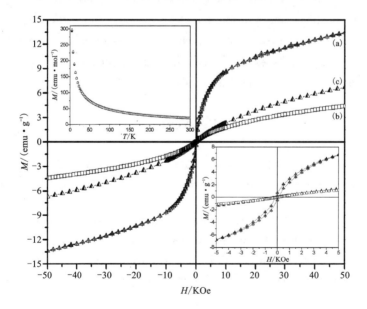

Figure 6
The hysteresis loops of as-prepared CoTe (a) and CoTe$_2$ (b) nanotubes and flower-like CoTe$_2$ hierarchical structures (c) at 2 K; the lower inset is the enlarged hysteresis loops of as-prepared products; the upper inset is the ZFC and FC magnetization curves of as-prepared flower-like CoTe$_2$ hierarchical structures under an applied field of 5000 Oe.

crystalline lattice of $CoTe_2$, which is in accordance with the previous report.[18] Furthermore, the magnetic properties of as-prepared products are also directly related to their size, shape, composition and structure.

4 Conclusion

In conclusion, uniform CoTe and $CoTe_2$ nanotubes have been successfully synthesized via a facile hydrothermal co-reduction route based on the precipitate slow-release controlled process. Results show H_2TeO_3 is a dominant factor for formation of CoTe nanotubes. Furthermore, the concentration of NaOH is another important factor for preparation of $CoTe_2$ nanotubes that cannot be obtained without NaOH. The concentration of NaOH affects the reaction kinetics through tuning the dissolution-deposition equation of $Co(OH)_2$. We consider that the reaction rate can be adjusted through precipitate slow-release method, which can regulate the kinetics of nucleation and growth of products and further efficiently control the morphology and structure of final products.[17] The synthetic strategy presented here may have good prospects in large-scale applications and provide an effective route to synthesize other telluride nanotubes. Owning to the excellent magnetic properties of as-prepared products, it is expected that the products exhibit some important applications in, e. g., magnetic semiconductors and other advanced materials.

References

[1] G. R. Patzke, F. Krumeich, R. Nesper. Oxidic nanotubes and nanorods-anisotropic modules for a future nanotechnology [J]. Angewandte Chemie International Edition, 2002, 41(14): 2446 – 2461.

[2] J. Wang, Y. Chen, W. J. Blau. Carbon nanotubes and nanotube composites for nonlinear optical devices [J]. Journal of Materials Chemistry, 2009, 19(40): 7425 – 7443.

[3] D. Yu, J. Q. Wu, Q. Gu, H. Park. Germanium telluride nanowires and nanohelices with memory-switching behavior [J]. Journal of the American Chemical Society, 2006, 128(25): 8148 – 8149.

[4] A. M. Qin, Y. P. Fang, P. F. Tao, J. Y. Zhang, C. Y. Su. Silver telluride nanotubes prepared by the hydrothermal method [J]. Inorganic chemistry, 2007, 46(18): 7403 – 7409.

[5] S. S. Garje, D. J. Eisler, J. S. Ritch, M. Afzaal, P. O'Brien, T. Chivers. A new route to antimony telluride nanoplates from a single-source precursor [J]. Journal of the American Chemical Society, 2006, 128(10): 3120 – 3121.

[6] S. H. Li, M. S. Toprak, H. M. A. Soliman, J. Zhou, M. Muhammed, D. Platzek, E. Müller. Fabrication of nanostructured thermoelectric bismuth telluride thick films by electrochemical deposition [J]. Chemistry of Materials, 2006, 18(16), 3627-3633.

[7] Z. F. Ding, S. K. Bux, D. J. King, F. L. Chang, T. -H. Chen, S. -C. Huang, R. B. Kaner. Lithium intercalation and exfoliation of layered bismuth selenide and bismuth telluride [J]. Journal of Materials Chemistry, 2009, 19(17): 2588-2592.

[8] K. Zweibel. The impact of tellurium supply on cadmium telluride photovoltaics [J]. Science, 2010, 328(5979): 699-701.

[9] J. Kim, C. Wang, T. Hughbanks. Synthesis and structures of new layered ternary manganese tellurides: $AMnTe_2$(A = K, Rb, Cs), $Na_3Mn_4Te_6$, and $NaMn_{1.56}Te_2$[J]. Inorganic Chemistry, 1999, 38 (2), 235-242

[10] B. Li, Y. Xie, J. X. Huang, Y. Liu, Y. T. Qian. Sonochemical synthesis of nanocrystalline copper tellurides Cu_7Te_4 and Cu_4Te_3 at room temperature [J]. Chemistry of Materials, 2000, 12(9): 2614-2616.

[11] X. H. Sun, B. Yu, G. Ng, M. Meyyappan. One-dimensional phase-change nanostructure: germanium telluride nanowire [J]. Journal of Physical Chemistry C, 2007, 111(6): 2421-2425.

[12] R. Blachnik, M. Lasocka, U. Walberecht. The system copper-tellurium [J]. Journal of Solid State Chemistry, 1983, 48(3): 431-438.

[13] Q. Peng, Y. J. Dong, Y. D. Li. Synthesis of uniform CoTe and NiTe semiconductor nanocluster wires through a novel coreduction method [J]. Inorganic Chemistry, 2003, 42 (7): 2174-2175.

[14] L. Zhang, Z. Ai, F. Jia, L. Liu, X. Hu, J. C. Yu. Controlled hydrothermal synthesis and growth mechanism of various nanostructured films of copper and silver tellurides [J]. Chemistry-A European Journal, 2006, 12(15): 4185-4190.

[15] A. Purkayastha, F. Lupo, S. Kim, T. Borca-Tasciuc, G. Ramanath. Low-temperature, template-free synthesis of single-crystal bismuth telluride nanorods [J]. Advanced Materials, 2006, 18(4): 496-500.

[16] H. Fan, Y. G. Zhang, M. F. Zhang, X. Y. Wang, Y. T. Qian. Glucose-assisted synthesis of CoTe nanotubes in situ templated by Te nanorods [J]. Crystal Growth & Design, 2008, 8 (8): 2838-2841.

[17] X. H. Liu, R. Yi, Y. T. Wang, G. Z. Qiu, N. Zhang and X. G. Li. Highly ordered snowflakelike metallic cobalt microcrystals [J]. The Journal of Physical Chemistry C, 2007, 111(1): 163-167.

[18] E. Uchida. Magnetic properties of cobalt telluride [J]. Journal of the Physical Society of Japan, 1955, 10(7): 517-522.

☆ R. R. Shi, X. H. Liu, Y. G. Shi, R. Z. Ma, B. P. Jia, H. T. Zhang, G. Z. Qiu. Selective synthesis and magnetic properties of uniform CoTe and $CoTe_2$ nanotubes [J]. Journal of Materials Chemistry, 2010, 20(36): 7634-7636.

Shape-Controlled Synthesis and Properties of Manganese Sulfide Microcrystals via a Biomolecule-Assisted Hydrothermal Process

Abstract: An effective biomolecule-assisted synthetic route has been successfully developed to prepare γ - manganese sulfide (MnS) microtubes under hydrothermal conditions. In the synthetic system, soluble hydrated manganese chloride was employed to supply Mn source and L - cysteine was used as precipitator and complexing reagent. Sea urchin-like γ - MnS and octahedron-like α - MnS microcrystals could also be selectively obtained by adjusting the process parameters such as hydrothermal temperature and reaction time. The phase structures, morphologies and properties of as-prepared products were investigated in detail by X-ray diffraction (XRD), scanning electron microscopy (SEM), energy dispersion spectroscopy (EDS), and photoluminescence spectra (PL). The photoluminescence studies exhibited the correlations between the morphology, size, and shape structure of MnS microcrystals and its optical properties. The formation mechanisms of manganese sulfide microcrystals were discussed based on the experimental results.

1 Introduction

Biomolecule-assisted synthesis has attracted more and more attention in recent years due to its special structures and fascinating self-assembling functions and important technological applications in the preparation of inorganic materials. In especially, biomolecules could be exploited to control nucleation and growth of inorganic materials and thus manipulate their structures and physical properties. Many kinds of biomolecules such as DNA,[1] protein,[2] glutathione[3] and virus[4] have been extensively utilized as templates for the fabrication of inorganic materials with complicated structures. Recently, a particularly significant breakthrough in highly ordered snowflakelike bismuth sulfide nanorods synthesis was made by Komarneni and co-workers via a simple biomolecule-assisted approach.[3] How to utilize the biomolecule templates to synthesize inorganic materials with desired shape or complicated structures is very important in biology, chemistry and material science. L-cysteine as a biomolecule not only has special structure and novel self-assembling functions, but also serves as sulfur source for the fabrication of metal sulfides.

Manganese sulfide (MnS) is an important magnetic semiconductor with band gap energy 3.7 eV that has been demonstrated to be of considerable applications in optoelectronic devices, magnetic semiconductors and luminescent fields.[5-6] It is well known that MnS has three different polymorphs: α, β and γ-MnS. Among these, green-colored α-MnS is the stable polymorph (alabandite) with rock-salt-type structure, while β and γ-MnS are pink metastable modifications with sphalerite (zinc blende, ZB) and wurtzite (W) structure, respectively.[7] Both tetrahedrally coordinated β and γ forms can only exist in a low-temperature range, and will transform into octahedrally coordinated α form at high temperature or pressure. In previous studies, considerable efforts in the field have been mostly placed on the control of the crystalline phases and morphologies of manganese sulfide. Recently, a variety of novel shapes such as MnS nanorod,[8] hollow spheres,[9] porous networks,[10] coral-shaped[11] and flowerlike hierarchical architectures[12] have been successfully synthesised by different methods such as hydrothermal, solvothermal, microwave irradiation, spray-produced and chemical bath deposition.[11,13] However, as far as we know, there has been no study on the preparation of tube-like γ-MnS microcrystals until now. In this paper, we have successfully developed a convenient biomolecule-assisted synthesis route to prepare γ-MnS microtubes under hydrothermal conditions. Sea urchin-like γ-MnS and octahedron-like α-MnS microcrystals could also be selectively obtained by adjusting hydrothermal temperature and reaction time. It is worthy to note that the current synthetic strategy can be used to synthesize other metal sulfides, and it will have a good prospect in the future large-scale application due to its high yields, simple reaction apparatus and low reaction temperature.

2 Experimental Section

All chemicals were of analytical grade, and were used without further purification.

2.1 Synthesis. In a typical procedure, $MnCl_2 \cdot 4H_2O$ (1 mmol) was put into Teflon-lined autoclave of 50 mL capacity and dissolved in 20 mL deionized water. And then L-cysteine (C_3H_7NS, 1 mmol) was added into the autoclave during magnetic stirring. Next, the autoclave was filled with deionized water up to 70% of the total volume, after 10 min stirring, sealed and maintained at 100–250 ℃ for a period of

2 – 24 h without shaking or stirring. The influence of surfactants sodium oleate on MnS microcrystals was also tentatively explored. The resulting products were filtered and washed with deionized water and anhydrous ethanol several times, and finally dried in vacuum at 60 ℃ for 4 h.

2.2　Characterization. The crystal structure of as-prepared products was characterized by X-ray powder diffraction (XRD) using a D/max2550 VB + X-ray diffractometer with Cu K_α radiation ($\lambda = 1.5418$ Å). The size and morphology of as-synthesized products were determined at 20 kV by a XL30 S-FEG scanning electron microscope (SEM). Energy-dispersive X-ray spectroscopy (EDS) was taken on the SEM. The room-temperature photoluminescence (PL) measurement was carried out on a Hitachi F – 4500 fluorescence spectrophotometer using the 245 nm excitation line of Xe light.

3　Results and Discussion

The phase structure and purity of as-prepared products were characterized by X-ray powder diffraction (XRD). Figure 1 shows the typical XRD pattern of as-prepared γ – MnS microtubes obtained with the absence of surfactants at 180 ℃ for 12 h. The main reflections in this pattern can be indexed as the hexagonal phase γ – MnS with lattice parameters $a = 3.9792$ Å and $c = 6.4469$ Å [space group: P6$_3$mc (186)], which are consistent with the values in the literature (JCPDS 40 – 1289). The peaks marked with asterisk could be attributed to the face-centered cubic phase [space group: Fm3m (225)] α – MnS with lattice parameters $a = 5.224$ (JCPDS 06 – 0518), which indicates trace amounts of α – MnS existed in γ – MnS under the current experimental conditions.

The size and morphology of as-prepared products were characterized by scanning electron microscope (SEM). Figure 2(a) shows the low-magnification SEM image of as-prepared γ – MnS microtubes obtained with the absence of surfactants at 180 ℃ for 12 h, indicating that a large amout of γ – MnS microtubes with diameter in the range of 0.5 – 2 μm and length up to 10 μm can be synthesized using this approach. Figure 2(b) indicates the representative SEM image with higher magnificaiton, in which many γ – MnS microtubes with hollow interior can be clearly observed. The outer diameter of the microtubes is about 500 nm. We think some rod-like microstructures may be microtubes due their thicker tube wall. With careful observation, the sphere-like and sea urchin-like microstructures can also be found. The hollow structures could

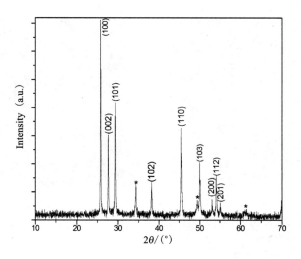

Figure 1

The XRD pattern of as-prepared γ - MnS microtubes obtained with the absence of surfactants at 180 ℃ for 12 h. (* : α - MnS)

be further confirmed via some broken microtubes. Figure 2 (c) shows a higher-magnification SEM image of as-prepared γ - MnS microtubes, which clearly shows a broken microtube, further confirming the γ - MnS microstructures with hollow interior. The chemical compositions of as-prepared γ - MnS microtubes have been investigated by means of energy dispersion spectroscopy (EDS). Results from EDS spectra [Figure 2(d)] show that the γ - MnS microtubes contain S and Mn. The atomic ratios are calculated to be 47.2 : 52.8 by the comparisons of relative areas under the peaks of Mn and S.

Furthermore, further studies suggested that the morphology and phase of final products strongly depended on reaction conditions such as reaction time, hydrothermal temperature and surfactant. For example, when the reaction time was decreased to 4 h, the product formed consisted of microtubes with mean diameter of 1 μm as well as some microspheres with a mean diameter of 2.5 μm [Figure 3(a)]. Prolonging the reaction time to 8 h, it is presented that the large quantity of microtubes marked by the arrow with length up to 10 μm and diameter in the range of 2 - 4 μm can be seen [Figure 3(b)]. Figure 3 (c) shows a typical SEM image of as-prepared γ - MnS microstructures obtained at 180 ℃ for 8 h. Herein, it is carefully observed that many similar subuliform structures radially grow and form similar sea urchin-like microcrystals. The formation of these subuliform structures may result from nucleation

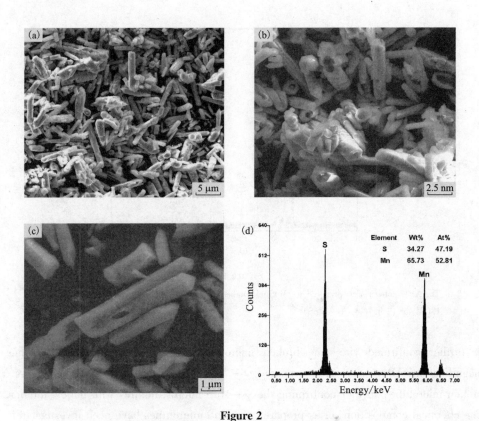

Figure 2

SEM images of as-prepared γ – MnS (a) – (c) microtubes with different magnifications. (d) EDS spectra of as-prepared γ – MnS microtubes; characteristic peaks for both S and Mn are observed. The molar ratios of S and Mn are shown inset.

and growth of MnS on the surface of microspheres under the function of stochastic diffusive force, and these new structures may grow up gradually and form mircotubes. When the reaction time is elongated to 24 h, the representative SEM image of as-prepared product is shown in Figure 3 (d). It is more obvious that plenty of mircotubes radially grow and form similar sea urchin-like microcrystals. The diameter and the length of mircotubes are about 1.2 μm and 8 μm, respectively. These mircotubes may form due to the further growth of subuliform structures by the function of the directive force.

Wurtzite structure γ – MnS can transform into rock salt (RS) structure α – MnS at high temperature. We explored the possibility for the synthesis of α – MnS microcrystals via elevated hydrothermal temperature. Figure 4(a) and 4(b) shows the evolution of

Figure 3 SEM images of as-prepared γ – MnS microstructures obtained at 180 ℃ for different reaction time

(a) 4 h; (b), (c) 8 h; (d) 24 h

the XRD patterns of the synthesized products obtained using L – cysteine as sulfide source and complexing reagent at 220 ℃ and 250 ℃ for 24 h, respectively. The main reflections in part (a) of Figure 4 can be indexed as the hexagonal phase [space group: P6$_3$mc (186)] of wurtzite structure γ – MnS with lattice constants $a = 3.9792$ Å and $c = 6.4469$ Å (JCPDS 40 – 1289). The peaks marked with asterisk could be attributed to the face-centered cubic phase [space group: Fm3m (225)] of rock salt structure α – MnS with lattice parameters $a = 5.224$ Å (JCPDS 06 – 0518). Compared to the product obtained at 180 ℃ (Figure 1), the peaks of α – MnS become stronger, accompanying the weaker γ – MnS peaks, which indicate the transition from γ – MnS to α – MnS with hydrothermal temperature elevation. When the reaction temperature is elevated to 250 ℃, all the reflections of XRD pattern can be finely indexed to a pure face-centered cubic phase [space group: Fm3m (225)] of rock salt structure α – MnS

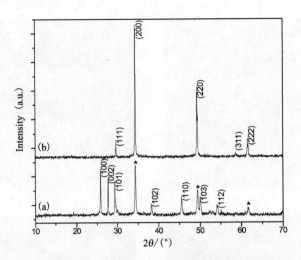

Figure 4 The evolution of the XRD patterns of as-prepared γ – MnS microstructures obtained at different hydrothermal temperature for 24 h

(a) 220℃; (b) 250℃. (* : α – MnS)

with lattice parameters $a = 5.224$ (JCPDS 06 – 0518). No impurity peaks were observed, which indicates that wurtzite structure γ – MnS has been converted completely into rock salt structure α – MnS under current experimental conditions.

Figure 5(a) shows a typical SEM image of γ – MnS microtubes obtained at 140℃ for 24 h. Herein, it is presented that the large quantity of γ – MnS microtubes with diameter in the range of 0.2 – 1 μm and length up to 8 μm can be synthesized. Compared to the product obtained at 180℃ for 12 h, both the average diameter and length of γ – MnS microtubes decreased for product obtained at 140℃ for 24 h. Figure 5(b) shows the typical SEM image of the γ – MnS product obtained at 220℃ for 24 h and indicates the sea urchin-like γ – MnS microcrystals were achieved under current conditions. It is clearly seen that plenty of microtubes radially grow and form similar sea urchin-like microcrystals. The diameter of microtubes is in the range of 0.5 – 1.5 μm and the length is up to 10 μm. When the reaction temperature was elevated to 250℃, the final products were mostly made up of octahedron-like microcrystals. Figure 5 (c) shows the low-magnification SEM image of as-prepared α – MnS microcrystals. The inset is the high-magnification SEM image. Herein, the corners and edges of octahedron-like α – MnS microcrystals with size in the range of 1 – 5 μm can be clearly observed. In our experiments, the influence of surfactant on the morphology of products was also explored. Figure 5(d) shows the typical SEM

Figure 5 SEM images of as-prepared γ – MnS microstructures obtained at different hydrothermal temperature for 24 h

(a) 140℃; (b) 220℃; (c) 250℃. Inset: higher-magnification image of γ – MnS octahedrons obtained from a selected area in (c). (d) SEM image of as-prepared γ – MnS microstructures obtained with the surfactant sodium oleate process at 180℃ for 24 h.

image of γ – MnS microcrystals synthesized with the surfactant sodium oleate at 180℃ for 24 h, in which the γ – MnS products are mixture of microspheres and short microtubes with diameter in the range of 1 – 4 μm and length up to 5 μm.

On the basis of the experimental results, the probable mechanism of the formation of the tube-, sea urchin- and octahedron-like MnS microcrystals is proposed. The influences of function groups such as —NH_2, —COOH and —SH of biomolecule L – cysteine are undoubtedly important to the morphology and size of MnS microcrystals. Burford and co-workers have proved the function groups in L – cysteine molecule such as —NH_2, —COOH and —SH have a strong tendency to interact with inorganic cations.[14] So when mixing the L – cysteine and $MnCl_2$ in solution, Mn^{2+} can coordinate with L – cysteine molecules to form precursor complexes. Zhang and co-workers have reported that the amino group can interact with the carboxyl group of the neighboring L – cysteine molecule to form dipeptide or polypeptide molecules in solution, which can serve as the template to form Bi_2S_3 microcrystals of special morphology.[15] Firstly, with increasing the temperature, the MnS nuclei might form by the high-temperature hydrolysis of the precursor complexes. Then, MnS nuclei grew quickly and formed spherical microcrystals due to Mn – L – cysteine complexes around it gradually reacting at its surface. Furthermore, we considered that long-chain polypeptide molecules can serve as biomolecule template to orient growth of manganese sulfide and form tubular microcrystals. With the prolongation of reaction time, continuously increased manganese sulfide may nucleate and form spots nanostructures on the surface of spherical microcrystals under the function of stochastic diffusive force and directive force, and polypeptide molecules may make the spot nanostructure crystallize in certain direction and form the sea urchin-like microcrystals with nanotube-based architecture. When the reaction temperature is elevated to 250℃, the phase of the products has completely transformed from γ forms to octahedrally coordinated α form and octahedron-like structure may be steady under current conditions.

The room-temperature photoluminescence (PL) spectra of as-prepared α – MnS products were measured on a fluorescence spectrophotometer using a Xe lamp with an excitation wavelength of 245 nm. Figure 6(a) to 6(c) presents the emission spectrum of as-prepared MnS microcrystals prepared at 180℃ for 12 h, 180℃ for 24 h and 250℃ for 24 h, respectively. It is obvious the as-prepared α – MnS microcrystals exhibited a maximum intensity emission at about 394 nm, which is similar to the bandgap of the bulk counterpart.[16] Compared to sea urchin-like γ – MnS microcrystals obtained at 180℃ for 24 h [Figure 6(b)], the intensity of the emission peaks of γ – MnS microtubes obtained at 180℃ for 12 h [Figure 6(a)] is improved, indicating the optical properties of tubular microcrystals are superior to sea urchin-like γ – MnS microcrystals. However, compared to octahedron-like α – MnS microcrystals prepared at 250℃ for 24 h, the intensity of the emission peaks are weaker for both tubular and sea urchin-like γ – MnS microcrystals, respectively. Although the photoluminescence emission mechanism of MnS is not fully understood, the emission peaks may be attributed to the recombination of charge carriers in deep traps of surface localized states and a photogenerated hole caused by surface defects.[17] Moreover, the optical properties are also directly influenced by the morphology, size and shape structure of final products.

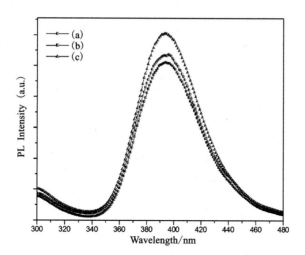

Figure 6 Photoluminescence (PL) of as-prepared products with an excitation wavelength of 245 nm at room temperature

(a) γ - MnS microtubes obtained at 180℃ for 12 h; (b) sea urchin-like γ - MnS microcrystals obtained at 180℃ for 24 h; (c) octahedron-like α - MnS microcrystals obtained at 250℃ for 24 h.

4 Conclusions

In summary, we have successfully developed a simple and convenient biomolecule-assisted hydrothermal route to prepare γ - MnS microtubes under mild conditions. Sea urchin-like γ - MnS and octahedron-like α - MnS microcrystals could also be selectively obtained by adjusting hydrothermal temperature and reaction time. The photoluminescence studies exhibited the optical properties of tubular microcrystals are superior to those of sea urchin-like γ - MnS microcrystals, and the optical properties are directly associated with the morphology, size, and structure of final products. It is worthy to note that the current synthetic strategy can be used to synthesize other metal sulfides, and it will have potential applications in the future large-scale synthesis due to its high yields, simple reaction apparatus and low reaction temperature.

References

[1] (a) A. Ongaro, F. Griffin, P. Beecher, L. Nagle, D. Iacopino, A. Quinn, G. Redmond, D.

Fitzmaurice. DNA-templated assembly of conducting gold nanowires between gold electrodes on a silicon oxide substrate [J]. Chemistry of Materials, 2005, 17(8): 1959 – 1964. (b) A. A. Zinchenko, K. Yoshikawa, D. Baigl. DNA-templated silver nanorings [J]. Advanced Materials, 2005, 17(23): 2820 – 2823. (c) L. Q. Dong, T. Hollis, B. A. Connolly, N. G. Wright, B. R. Horrocks, A. Houlton. DNA-templated semiconductor nanoparticle chains and wires [J]. Advanced Materials, 2007, 19(13): 1748 – 1751.

[2] (a) R. A. McMillan, J. Howard, N. J. Zaluzec, H. K. Kagawa, R. Mogul, Y. F. Li, C. D. Paavola, J. D. Trent. A self-assembling protein template for constrained synthesis and patterning of nanoparticle arrays [J]. Journal of the American Chemical Society, 2005, 127(9): 2800 – 2801. (b) T. Douglas, E. Strable, D. Willits, A. Aitouchen, M. Libera, M. Young. Protein engineering of a viral cage for constrained nanomaterials synthesis [J]. Advanced Materials, 2002, 14(6): 415 – 418.

[3] (a) Q. Y. Lu, F. Gao, S. Komarneni. Quantum dots for live cells, in vivo imaging, and diagnostics [J]. Science, 2005, 307(5709): 538 – 544. (b) Y. G. Zheng, Z. C. Yang, Y. Q. Li, J. Y. Ying. From glutathione capping to a crosslinked, phytochelatin-like coating of quantum dots [J]. Advanced Materials, 2008, 20(18): 3410 – 3415.

[4] (a) R. Tsukamoto, M. Muraoka, M. Seki, H. Tabata, I. Yamashita. Synthesis of CoPt and $FePt_3$ nanowires using the central channel of tobacco mosaic virus as a biotemplate [J]. Chemistry of Materials, 2007, 19(10): 2389 – 2391. (b) C. Radloff, R. A. Vaia, J. Brunton, G. T. Bouwer, V. K. Ward. Polyaniline nanofiber/gold nanoparticle nonvolatile memory [J]. Nano Letters, 2005, 5(6): 1077 – 1080.

[5] O. Goede, W. Heimbrodt. Optical properties of (Zn, Mn) and (Cd, Mn) chalcogenide mixed crystals and superlattices [J]. Physica Status Solidi B, 1988, 146(1): 11 – 62.

[6] Y. Zheng, Y. Cheng, Y. Wang, L. Zhou, F. Bao, C. Jia. Metastable γ – MnS hierarchical architectures: synthesis, characterization, and growth mechanism [J]. The Journal of Physical Chemistry B, 2006, 110(16): 8284 – 8288.

[7] C. Somuthawee, S. B. Bansall, F. A. Hummel. Phase equilibria in the systems ZnS – MnS, $ZnS – CuInS_2$, and $MnS – CuInS_2$ [J]. Journal of Solid State Chemistry, 1978, 25(4): 391 – 399.

[8] C. Zhang, F. Tao, G. Q. Liu, L. Z. Yao, W. L. Cai. Hydrothermal synthesis of oriented MnS nanorods on anodized aluminum oxide template [J]. Materials Letters, 2008, 62(2): 246 – 248.

[9] Y. Cheng, Y. S. Wang, C. Jia, F. Bao. MnS hierarchical hollow spheres with novel shell structure [J]. The Journal of Physical Chemistry B, 2006, 110(48): 24399 – 24402.

[10] F. Zuo, B. Zhang, X. Z. Tang, Y. Xie. Porous metastable γ – MnS networks: biomolecule-assisted synthesis and optical properties [J]. Nanotechnology, 2007, 18(21): 215608.

[11] L. Amirav, E. Lifshitz. Spray-produced coral-shaped assemblies of MnS nanocrystal clusters [J]. The Journal of Physical Chemistry B, 2006, 110(42): 20922 – 20926.

[12] J. G. Yu, H. Tang. Solvothermal synthesis of novel flower-like manganese sulfide particles [J]. Journal of Physics and Chemistry of Solids, 2008, 69(5): 1342 – 1345.

[13] N. Zhang, R. Yi, Z. Wang, R. R. Shi, H. D. Wang, G. Z. Qiu, X. H. Liu. Hydrothermal synthesis and electrochemical properties of alpha-manganese sulfide submicrocrystals as an attractive electrode material for lithium-ion batteries [J]. Materials Chemistry and Physics, 2008, 111(1): 13 – 16.

[14] N. Burford, M. D. Eelman, D. E. Mahony, M. Morash. Definitive identification of cysteine and glutathione complexes of bismuth by mass spectrometry: assessing the biochemical fate of bismuth pharmaceutical agents [J]. Chemical Communications, 2003, 1: 146 – 147.

[15] B. Zhang, X. C. Ye, W. Y. Hou, Y. Zhao, Y. Xie. Biomolecule-assisted synthesis and electrochemical hydrogen storage of Bi_2S_3 flowerlike patterns with well-aligned nanorods [J]. The Journal of Physical Chemistry B, 2006, 110(18): 8978 – 8985.

[16] S. H. Kan, I. Felner, U. Banin. Characterization, and magnetic properties of α – MnS nanocrystals [J]. Israel Journal of Chemistry, 2001, 41(1): 55 – 62.

[17] (a) X. H. Liu. A facile route to preparation of sea-urchinlike cadmium sulfide nanorod-based materials [J]. Materials Chemistry and Physics, 2005, 91(1): 212 – 216. (b) Y. Lin, J. Zhang, E. H. Sargent, E. Kumacheva. Photonic pseudo-gap-based modification of photoluminescence from CdS nanocrystal satellites around polymer microspheres in a photonic crystal [J]. Applied Physics Letters, 2002, 81(17): 3134 – 3136. (c) B. A. Simmons, S. C. Li, V. T. John, G. L. McPherson, A. Bose, W. L. Zhou, J. B. He. Morphology of CdS nanocrystals synthesized in a mixed surfactant system [J]. Nano Letters, 2002, 2(4): 263 – 268. (d) L. Spanhel, M. A. Anderson. Synthesis of porous quantum-size cadmium sulfide membranes: photoluminescence phase shift and demodulation measurements [J]. Journal of the American Chemical Society, 1990, 112(6): 2278 – 2284.

☆ J. H. Jiang, R. N. Yu, J. Y. Zhu, R. Yi, G. Z. Qiu, Y. H. He, X. H. Liu. Shape-controlled synthesis and properties of manganese sulfide microcrystals via a biomolecule-assisted hydrothermal process [J]. Materials Chemistry and Physics, 2009, 115(2 – 3): 502 – 506.

Conversion of Metal Oxide Nanosheets into Nanotubes

Abstract: In this part, structural relationship and conversion between two-dimensional (2D) nanosheets and one-dimensional (1D) nanotubes are reviewed. Nanotubes are spontaneously formed upon exfoliation of certain layered materials with a non-centrosymmetric or particular structure, such as $K_4Nb_6O_{17}$ and some perovskite-type Ruddlesden-Popper phase $k_2[A_{n-1}B_nO_{3n+1}]$ (A = Na, Ca, Sr, La, B = Ta, Ti). On the other hand, colloidal centrosymmetric nanosheets represented by titanium oxide, manganese oxide, and calcium niobate, can also be successfully converted into their corresponding nanotubes through a simple ion intercalation/deintercalation procedure at ambient temperature. The conversion validates the hypothesis, in which directly rolling a nanosheet yields a nanotube. The close relationship is of fundamental importance in revealing the formation mechanism of nanotubes and may be used to realize a customized synthesis of nanotubes from a wide range of layered materials.

1 Nanosheet and Nanotube

The coined term "nanosheet" is widely accepted as representing a type of nanomaterial in ultimate 2D anisotropy,[1-4] exhibiting molecular thickness (about 1 nm) but submicron-scale or micron-scale lateral dimensions, which is generally produced by swelling and exfoliation/delamination of a layered host into individual layers in a suitable medium (Figure 1). Nanosheet is typically obtained as a dispersed colloidal suspension. It is well known that some smectite clay minerals undergo spontaneous exfoliation in water. Similar behavior has been artificially achieved for several classes of layered materials such as certain layered oxides, dichalcogenides and metal phosphates by controlling layer-to-layer interactions via soft chemical procedures.[1-4] Graphene, created by exfoliating graphite, can be regarded as a new elementary nanosheet with high electrical and thermal conductivity and other peculiar electronic properties.[5]

Particularly, layered transition metal oxides (e. g., $Cs_{0.7}Ti_{1.825}O_4$, $K_{0.45}MnO_2$,

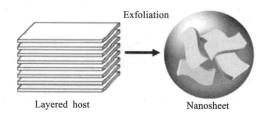

Figure 1 Schematic diagram showing the generation of colloidal nanosheets via exfoliation of layered hosts into single layers

$KCa_2Nb_3O_{10}$), prepared by a solid-state calcination method, can take up a high quantity of bulk organic ammonium cations between the layers after acid exchange. Under appropriate conditions, these layered hosts undergo osmotic swelling and exfoliation into nanosheets.[2-4] Figure 2 shows a representative structure of oxide nanosheets equivalent to one host layer. MO_6 (M = Ti, Mn, Nb) octahedra are linked via edge-sharing or corner-sharing to produce the host layer, and the thickness is about 0.75 nm, 0.48 nm, 1.44 nm, respectively. On the contrary, the lateral dimensions might be regarded as infinite in comparison with the thickness, e.g., up to several tens of microns. As a result, these oxide nanosheets have some unique characteristics: ① inorganic macroanions bearing negative charge; ② 2D single crystal, in which constituent elements are still regularly arranged; and ③ ultimate surface area.

Using oxide nanosheets with these unique characteristics as building blocks is attracting increasing attention. Research is being conducted to design nanocomposites or functional thin films utilizing these nanosheets.[6-11] In particular, as oxide nanosheet is negatively charged, nanosheets can be flocculated and restacked as wool-like deposits when an electrolyte is added into the colloidal suspension. Electrostatic interaction with various cationic materials, inorganic or organic, can be employed to prepare functional composite materials that have potential use in electrodes, fluorescent materials and photocatalysts by combining the properties of both the nanosheet and counter electrolyte.[6-9] Layer-by-layer assembly and the Langmuir-Blodgett method have also been used to build multilayer films based on nanosheets that, for example, demonstrate superior dielectric nature and magneto-optical properties.[10,11] In addition, exfoliation of layered double hydroxide (LDH), a natural counterpart to cationic clays, yielded positively charged nanosheets: inorganic macrocations.[12-14] Well-defined LDH nanosheets have been readily attained by synthesizing large LDH crystals in carbonate form via so-called homogeneous precipitation and subsequent exfoliation of the

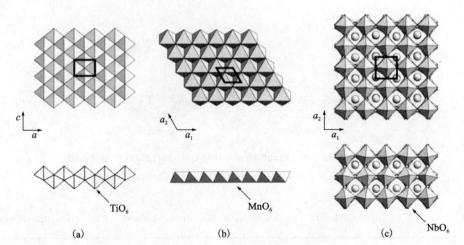

Figure 2 Examples of some oxide nanosheets
(a) Titanium oxide ($Ti_{0.91}O_2$); (b) Manganese oxide (MnO_2); (c) Calcium niobate ($Ca_2Nb_3O_{10}$). Bold lines indicate unit cell.

exchanged nitrate form in formamide.[12-14] A direct implication for the availability of LDH nanosheets is that direct electrostatic assembly of various anionic oxide nanosheets with cationic LDH nanosheets may be artificially realized at the molecular level.[15]

Interesting structural correlations between 2D nanosheets and 1D nanotubes have recently been revealed. As described in this book, nanotubes constructed from carbon,[16] boron nitride (BN),[17] metal chalcogenides (WS_2, MoS_2)[18, 19] and various oxides (TiO_2, VO_x),[20-23] have been intensively studied. Generally, nanotubes can be structurally classified into two distinct categories. One is referred to as the Matryoshka-doll/nesting type, represented by carbon, BN and chalcogenide nanotubes.[16-19] The other is referred to as the scroll type reported for TiO_2 and VO_x nanotubes.[20-23] As schematically shown in Figure 3, in the nesting-type nanotube, one concentric atomic plane is inserted into another. On the other hand, the scroll-type nanotube may be modeled as the rolling and folding up of an atomic plane onto itself. Under high-resolution transmission electron microscopy (HR-TEM), the number of layers on both walls of the hollow tube core is typically the same for a nesting nanotube. In contrast, a scrolled nanotube generally has one layer more or less on one of the sides. This difference in layer number can be effectively used to recognize the structure category of the nanotubes investigated. Also as shown in Figure 3, compared to the nesting-type BN nanotube, the TiO_2 nanotube is characteristic of the scroll type with one layer less on the right-hand side. Nevertheless, the category classification of a nanotube

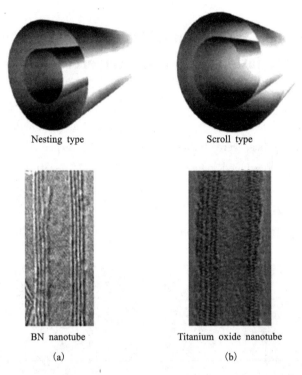

Figure 3 Structural models of nanotubes and typical examples
(a) Nesting-type / BN nanotube; (b) Scroll-type / titanium oxide nanotube.

is not always defined by the constructing material. For example, even though carbon nanotubes are traditionally well-known to be formed in a nesting structure, it was recently discovered that they could also be synthesized as a scroll type through a newly developed process.[24] This indicates that synthetic conditions may affect the structure category to which the resultant nanotubes belong.

Nanotubes are usually prepared under special conditions (chemical vapor deposition, hydrothermal synthesis, etc.). Nanotube formation is strongly dependent on synthetic parameters such as temperature, time, surfactants and templates used. Under certain particular synthetic conditions, both sheet-like crystals and nanotubes have been observed.[23] As these sheet-like objects are naturally curled or bent at the edge, it is plausible that rolling or folding up of a sheet-like object (atomic plane/ nanosheet) might generate a scroll-type nanotube, i. e., an artificial 2D to 1D structural transformation. However, the lack of definite data has prevented validation of the hypothetical process, in which directly rolling a nanosheet is intentionally realized and a nanotube is yielded.

2 Spontaneous Conversion of Nanosheets into Nanotubes During Exfoliation of Certain Layered Oxides

In their attempt to exfoliate $K_4Nb_6O_{17}$ when treating acid-exchanged $K_4Nb_6O_{17}$ with tetrabutylammonium ions (TBA^+), the Domen group[6] and Mallouk group[25] reported that abundant nanotubes were formed instead of the expected Nb_6O_{17} nanosheets [Figure 4(a)]. As-produced nanotubes are usually 15 – 30 nm in diameter and 0.1 – 1 μm in length. Furthermore, lateral curling or bending was observed for some nanosheets [Figure 4(b)], apparently occurring at the initial folding-up stage. The above observations strongly suggest that the nanotube formation might be induced by the

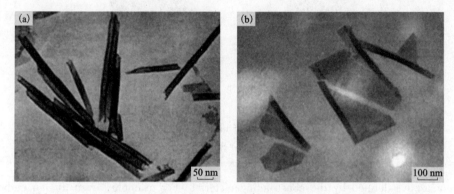

Figure 4　Nanotubes obtained via exfoliation of $K_4Nb_6O_{17}$

rolling up of exfoliated Nb_6O_{17} nanosheets. The driving force for such spontaneous rolling-up behavior appears to be the relief of built-in strain in an individual Nb_6O_{17} layer due to its non-centrosymmetric architecture. Host layers of $K_4Nb_6O_{17}$ (Nb_6O_{17} nanosheets) are not centrosymmetric in terms of arrangement and filling density of NbO_6 octahedra at the top and bottom sides (see Figure 5). As a result, the host material has two distinct interlayer environments. Being piled up in a bulk crystal, potassium ions (K^+) are intercalated between the sheets. It is considered that, upon exfoliation, the nanosheets would lose their flatness and roll up into a tube. Figure 6 is a schematic diagram showing the possible exfoliation and subsequent rolling-up scenario for $K_4Nb_6O_{17}$. As a first step, about 80% of the potassium ions between $K_4Nb_6O_{17}$ layers are substituted by protons via acid exchange. Solid acidity is generated and tetrabutylammonium ions come into the interlayers. High osmotic swelling is thus induced, promoting exfoliation. Taking into account the non-centrosymmetric nature of individual Nb_6O_{17} sheets, intercalation reactivity for tetrabutylammonium ions is supposed to differ in an adjacent double-layer pair. Sheet pairs with double-layer thickness were thus generated, reflecting the asymmetry at the top and bottom sides of $K_4Nb_6O_{17}$. These sheet pairs sandwich protons and potassium ions between the layers, which stabilizes the structure and maintains flatness. On the other hand, when exfoliation is further advanced, each in a pair is separated. Stress, derived from the asymmetry of a single-layer sheet, might cause curling/rolling up from the side with a higher filling density of NbO_6 octahedra, eventually forming a nanotube.

Figure 5   Crystal structure of $K_4Nb_6O_{17}$
There are two different interlayer environments.

It is noteworthy that the formation of Nb_6O_{17} nanotubes may be influenced by experimental parameters, such as the concentration and pH of Nb_6O_{17} nanosheet colloidal solution, ultrasonic treatment.[25] The balance between nanotube (rolling up) and nanosheet (unraveling) appears to be subtly decided by the history of conditions, inferring a comparatively small energy difference between nanosheets and nanotubes.

In addition, some protonated perovskite-type Ruddlesden-Popper phase $H_2[A_{n-1}B_nO_{3n+1}]$ (A = Na, Ca, Sr, La; B = Ta, Ti) were also found to yield nanotubes upon exfoliation.[26] Unlike $K_4Nb_6O_{17}$, these host compounds have a centrosymmetric lamellar structure. It is difficult to explain why spontaneous formation of nanotubes also occurs with these compounds, although possible clues and speculations have been pointed out. For example, calcium and sodium cations selectively occupy the A sites in $H_2CaNaTa_3O_{10}$, which might become a source of internal stress and induce the rolling-up behavior as illustrated for $K_4Nb_6O_{17}$. The influence of chemical exfoliation reagents and the cooperative distortion of MO_6

octahedra in $H_2SrTa_2O_7$ and $H_2La_2Ti_3O_{10}$ have also been speculated.

3 Conversion of nanosheets into Nanotubes via a Designed Soft Chemical Procedure

The previous section described nanotube formation induced by structural features of the exfoliated nanosheets, i. e., a spontaneous process. However, the spontaneous formation of nanotubes was found to be extremely limited. Conversely, a large variety of lamellar solids has been successfully exfoliated/delaminated into single-layer nanosheets via a soft chemical procedure, as shown in Figure 6. These nanosheets exhibit top and bottom symmetry, namely centrosymmetry. It has become clear that these nanosheets can also be converted into nanotubes when a suitable soft chemical process is applied.[27]

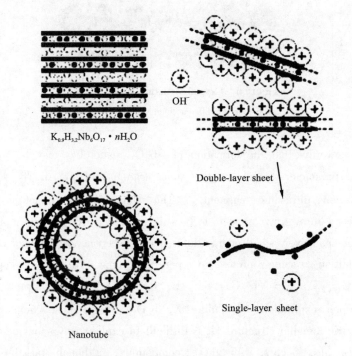

Figure 6 Exfoliation of $K_4Nb_6O_{17}$ and nanotube formation

The conversion of centrosymmetric oxide nanosheets into nanotubes may be achieved by the following procedure. The colloidal nanosheets undergo flocculation/restacking by changing ionic strength. Typically, when the colloidal suspension is poured into a concentrated NaOH aqueous solution (e. g., 1 mol/dm^3), wool-like precipitate is instantly yielded. The resulting material is filtered and rinsed with copious distilled water, and then shaken in pure water. A well-dispersed suspension is generally achieved after extended (e. g., 36 h) continuous shaking. The final product, which is collected from centrifugation and air-dried at room temperature, contains some nanotubes.

Figure 7(a) depicts a typical TEM image of the starting $Ti_{0.91}O_2$ nanosheets. The nanosheet exhibits a uniform and faint contrast, reflecting unilamellar thickness. The flocculation of colloidal nanosheets in NaOH solution leads to a turbostratic aggregate [Figure 7(b)]. X-ray diffraction characterizations of the flocculated products indicates the absence of three-dimensional order. This suggests that the flocculation results in turbostratic restacking of single-layer nanosheets, which accommodate Na ions and water molecules between the neighboring sheets. Based on microscopic characterization, turbostratic stacks of 5 – 20 nanosheets appear dominant in the flocculated product. After filtering and continuous shaking in water, restacked nanosheets are partially delaminated again and become well dispersed. Most of the dispersed nanosheets exhibit lateral curling. Some needle-shaped objects are observed by scanning electron microscopy (SEM) [Figure 7(c)]. TEM observation further demonstrates that the needle-shaped crystallites are nanotubes with an apparent hollow core [Figure 7(d), Figure 7(e)]. The diameter is found to remain constant for different tilting angles (0 – 30°) along the tube axis, identifying a true tubular nature rather than wrinkling of the nanosheet. HR-TEM images also indicate that the tube walls are usually very thin, comprising only 3 – 6 layers [Figure 7(e)].

Similarly, as-converted manganese oxide and calciumniobium oxide nanotubes are displayed in Figure 8. All three types of nanotubes have an outer diameter of 15 – 60 nm and a wall thickness of 3 – 6 layers. The dimensions imply that the nanotube may be constructed by wrapping a single 2D nanosheet which is 140 – 1130 nm in lateral size, consistent with the lateral size of the starting nanosheets (0.1 – 1 μm). This fact indicates that there is no fracture or breakage of the starting nanosheets during the rolling up. Based on microscopy statistics, approximately 20% of titanium oxide and manganese oxide nanosheets are transformed into nanotubes. As niobium oxide nanosheets are thicker (1.44 nm) than titanium oxide (0.75 nm) and manganese oxide (0.48 nm) nanosheets and are supposed to be more rigid, their conversion into nanotubes brings a lower yield (about 10%).

The formation of 1D nanotubes from 2D nanosheets under this designed protocol

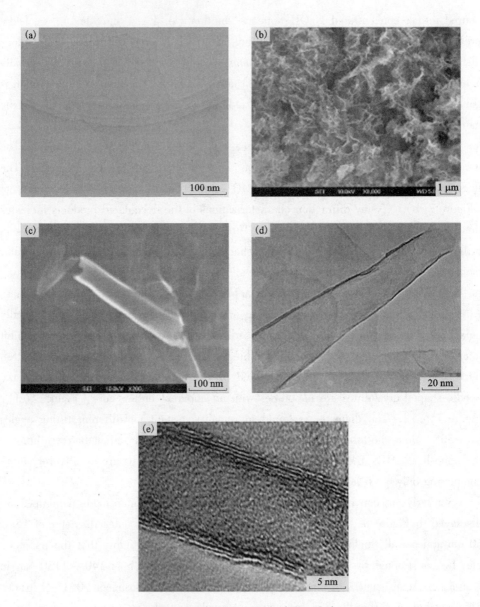

Figure 7 Converted titanium oxide ($Ti_{0.91}O_2$) nanotube

(a) 2D nanosheet; (b) flocculated aggregate; (c) SEM image of a needle-like object; (d) hollow nanotube (TEM image); (e) HR-TEM image of a nanotube.

may be explained as schematically illustrated in Figure 9. The starting unilamellar nanosheets are stabilized in a suspension with tetrabutylammonium ions. During the flocculation using NaOH solution, Na ions are substituted for tetrabutylammonium ions.

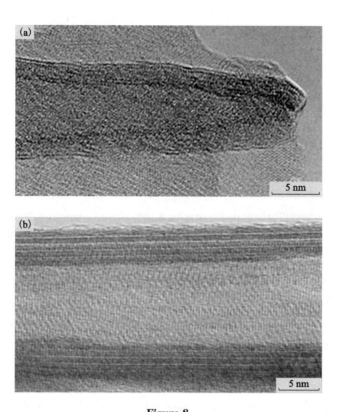

Figure 8

(a) Manganese oxide (MnO_2) nanotube; (b) Calcium niobate oxide ($Ca_2Nb_3O_{10}$) nanotube.

Colloidal 2D nanosheets become unstable and are restacked incorporating Na ions. The Na ions are intercalated into the intersheet gallery and play a role in pinning up adjacent sheets. This results in an aggregate of fine crystallites of 5 – 20 turbostratic layered nanosheets. When shaken in water, the intersheet Na ions are gradually deintercalated/extracted. This deintercalation process is realized through the exchange with protons and water molecules (H_3O^+), causing a variance in bonding characteristics between adjacent sheets. During shaking, the intersheet spacing might also be expanded due to the increase in water content. The electrostatic interaction between neighboring sheets is thus significantly reduced. As a result, the turbostratic restacked nanosheets become very loosely bonded and are granted a tendency to delaminate into individual nanosheets again. In pure water, the solution conditions are not perfectly suitable for attaining a well-dispersed colloid and individual nanosheets tend to flocculate and restack. However, a dilute nanosheet concentration makes it difficult to find counterparts to

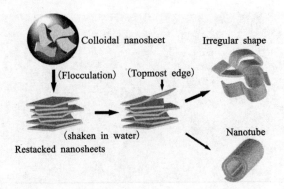

Figure 9 Schematic diagram illustrating a possible mechanism for the conversion of oxide nanosheets ($Ti_{0.91}O_2$, MnO_2, $Ca_2Nb_3O_{10}$) into nanotubes

restack, and the nanosheet flocculates within itself to gain stabilization energy. In other words, it needs to be folded. If some particular optimum geometrical requirement is fulfilled, nanotube formation takes place. In a major case, the nanosheets form irregularly shaped objects. It might be reasonable to suggest that the deintercalation of Na ions would initially start at the edge sites in the turbostratic layered nanosheets. The reduction of electrostatic interaction therefore starts at the lateral edges of the nanosheets. Due to the reduced interaction with underlying layered sheets, the lateral edges of the topmost sheet might gradually curl up, similar to a thin film free from the underlying substrate.[28] As the edge curling-up process advances, a topmost sheet might be completely peel off and roll up into a multilayer nanotube.

The above conversion process is not related to the type of structure a nanosheet might have. In a sense, it resembles a more general natural phenomenon, such as that seen when wood is planed and the shavings roll up at the edge once they are planed away from the wood. Utilizing this simple but important technique, it is possible to convert any type of single-layer nanosheets into nanotubes from various layered host materials.

4 Conclusion

A close structural correlation was found between two types of nanomaterials which differ considerably in morphology, i. e., 2D nanosheet and 1D nanotube. The relationship is of fundamental importance in revealing the formation mechanism of nanotubes. On the other hand, in spite of a yield concern, the rolling-up process itself might be regarded as a new apporach to generating nanotubes, and will be useful for widening the scope of application of nanosheets. For example, $H^+/K_4Nb_6O_{17}$ nanotubes are reported to exhibit high catalyst activity in the photolysis of water. [25] $C_3F_7 - Azo^+/K_4Nb_6O_{17}$ hybrid nanotubes show interesting photo-response property. [29] Titanium oxide nanotubes are being studied for ion-exchange activity and functional thin-film fabrication. [30, 31] Further striking developments are expected.

References

[1] A. J. Jacobson. Colloidal dispersions of compounds with layer and chain structures [C]. Materials Science Forum. 1994, 152: 1 – 12.

[2] T. Sasaki, M. Watanabe, H. Hashizume, H. Yamada, H. Nakazawa. Macromolecule – like aspects for a colloidal suspension of an exfoliated titanate. Pairwise association of nanosheets and dynamic reassembling process initiated from it [J]. Journal of the American Chemical Society, 1996, 118(35), 8329 – 8335.

[3] Y. Omomo, T. Sasaki, L. Wang. Redoxable nanosheet crystallites of MnO_2 derived via delamination of a layered manganese oxide [J]. Journal of the American Chemical Society, 2003, 125(12): 3568 – 3575.

[4] Y. Ebina, T. Sasaki, M. Harada. Restacked perovskite nanosheets and their Pt-loaded materials as photocatalysts [J]. Chemistry of Materials, 2002, 14(10): 4390 – 4395.

[5] K. S. Novoselov, A. K. Geim, S. V. Morozov. Electric field effect in atomically thin carbon films [J]. Science, 2004, 306(5696): 666 – 669.

[6] R. Abe, K. Shinohara, A. Tanaka. Preparation of porous niobium oxide by the exfoliation of $K_4Nb_6O_{17}$ and its photocatalytic activity [J]. Journal of Materials Research, 1998, 13(4): 861 – 865.

[7] T. Sasaki, Y. Ebina, T. Tanaka. Layer-by-layer assembly of titania nanosheet /polycation composite films [J]. Chemistry of Materials, 2001, 13(12): 4661 – 4667.

[8] H. Xin, R. Ma, L. Wang. Photoluminescence properties of lamellar aggregates of titania nanosheets accommodating rare earth ions [J]. Applied Physics Letters, 2004, 85(18): 4187 – 4189.

[9] K. Akatsuka, Y. Ebina, M. Muramatsu. Photoelectrochemical properties of alternating multilayer films composed of titania nanosheets and Zn porphyrin [J]. Langmuir, 2007, 23(12): 6730 – 6736.

[10] M. Osada, Y. Ebina, K. Takada, T. Sasaki. Gigantic Magneto-Optical Effects in Multilayer Assemblies of Two-Dimensional Titania Nanosheets [J]. Advanced Materials, 2006, 18(3): 295 – 299.

[11] M. Osada, Y. Ebina, H. Funakubo, S. Yokoyama, T. Kiguchi, K. Takada. High-κ Dielectric Nanofilms Fabricated from Titania Nanosheets [J]. Advanced Materials, 2006, 18 (8): 1023-1027.

[12] L. Li, R. Ma, Y. Ebina, N. Iyi, T. Sasaki. Positively charged nanosheets derived via total delamination of layered double hydroxides [J]. Chemistry of Materials, 2005, 17(17): 4386-4391.

[13] Z. Liu, R. Ma, M. Osada, N. Iyi, Y. Ebina, K. Takada, T. Sasaki. Synthesis, anion exchange, and delamination of Co – Al layered double hydroxide: assembly of the exfoliated nanosheet/polyanion composite films and magneto-optical studies [J]. Journal of the American Chemical Society, 2006, 128(14): 4872-4880.

[14] R. Ma, Z. Liu, L. Li, N. Iyi, T. Sasaki. Exfoliating layered double hydroxides in formamide: a method to obtain positively charged nanosheets [J]. Journal of Materials Chemistry, 2006, 16 (39): 3809-3813.

[15] L. Li, R. Ma, Y. Ebina, K. Fukuda, K. Takada, T. Sasaki. Synthesis, anion exchange, and delamination of Co – Al layered double hydroxide: assembly of the exfoliated nanosheet/polyanion composite films and magneto-optical studies [J]. Journal of the American Chemical Society, 2006, 128(14): 4872-4880..

[16] S. Iijima. Helical microtubules of graphitic carbon [J]. Nature, 1991, 354(6348): 56-58.

[17] N. G. Chopra, R. J. Luyken, K. Cherrey, V. H. Crespi, M. L. Cohen, S. G. Louie, A. Zettl. Boron nitride nanotubes [J]. Science, 1995, 269(5226): 966-967.

[18] R. Tenne, L. Margulis, M. Genut, G. Hodes. Polyhedral and cylindrical structures of tungsten disulphide [J]. Nature, 1992, 360(6403): 444-446..

[19] Y. Feldman, E. Wasserman, D. J. Srolovitz, R. Tenne. High-rate, gas-phase growth of MoS2 nested inorganic fullerenes and nanotubes [J]. Science, 1995, 267(5195): 222-225.

[20] M. E. Spahr, P. Bitterli, R. Nesper, M. Muller, F. Krumeich, H. U. Nissen. Redox-Active Nanotubes of Vanadium Oxide [J]. Angewandte Chemie International Edition, 1998, 37(9): 1263-1265.

[21] T. Kasuga, M. Hiramatsu, A. Hosono, T. Sekino, K. Niihara. Titania nanotubes prepared by chemical processing [J]. Advanced Materials, 1999, 11(15): 1307-1311.

[22] G. R. Patzke, F. Krumeich, R. Nesper. Oxidic nanotubes and nanorods-anisotropic modules for a future nanotechnology [J]. Angewandte Chemie International Edition, 2002, 41(14): 2446-2461.

[23] X. Chen, X. M. Sun, Y. D. Li. Self-assembling vanadium oxide nanotubes by organic molecular templates [J]. Inorganic Chemistry, 2002, 41(17): 4524-4530.

[24] L. M. Viculis, J. J. Mack, R. B. Kaner. A chemical route to carbon nanoscrolls [J]. Science, 2003, 299(5611): 1361-1361.

[25] G. B. Saupe, C. C. Waraksa, H. Kim, Y. J. Han, D. M. Kaschak, D. M. Skinner, T. E. Mallouk. Nanoscale tubules formed by exfoliation of potassium hexaniobate [J]. Chemistry of

Materials, 2000, 12(6): 1556 - 1562.

[26] R. E. Schaak, T. E. Mallouk. Self-assembly of tiled perovskite monolayer and multilayer thin films [J]. Chemistry of Materials, 2000, 12(9): 2513 - 2516.

[27] R. Ma, Y. Bando, T. Sasaki. Directly rolling nanosheets into nanotubes [J]. The Journal of Physical Chemistry B, 2004, 108(7): 2115 - 2119.

[28] O. G. Schmidt, K. Eberl. Nanotechnology: Thin solid films roll up into nanotubes [J]. Nature, 2001, 410(6825): 168 - 168.

[29] Z. W. Tong, S. Takagi, T. Shimada, H. Tachibana, H. Inoue. Photoresponsive multilayer spiral nanotubes: intercalation of polyfluorinated cationic azobenzene surfactant into potassium niobate [J]. Journal of the American Chemical Society, 2006, 128(3): 684 - 685.

[30] R. Ma, T. Sasaki, Y. Bando. Alkali metal cation intercalation properties of titanate nanotubes [J]. Chemical Communications, 2005, 7: 948 - 950.

[31] R. Ma, T. Sasaki, Y. Bando. Layer-by-layer assembled multilayer films of titanate nanotubes, Ag-or Au – loaded nanotubes, and nanotubes/nanosheets with polycations [J]. Journal of the American Chemical Society, 2004, 126(33): 10382 - 10388.

☆ R. Z. Ma, T. Sasaki. Conversion of matel oxide nanosheets into nanotubes [J]. Inorganic and Metallic Nanotubular Materials, T. Kijima (Ed.), Topics in Applied Physics, 2010, 117: 135 - 146.

Chapter III

Cobalt and Nickel Oxide Nanorings

Cobalt Hydroxide Nanosheets and Their Thermal Decomposition to Cobalt Oxide Nanorings

Abstract: We demonstrate herein that single-crystalline β-cobalt hydroxide $[\beta\text{-Co(OH)}_2]$ nanosheets can be successfully synthesized in large quantities by a facile hydrothermal synthetic method with aqueous cobalt nitrate as the cobalt source and triethylamine as both an alkaline and a complexing reagent. This synthetic method have good prospects for the future large-scale production of single-crystalline β-Co(OH)$_2$ nanosheets owing to its high yield, low-cost, and simple reaction apparatus. Single-crystalline porous nanosheets and nanorings of cobalt oxide (Co_3O_4) were obtained by a thermal-decomposition method with single-crystalline β-Co(OH)$_2$ nanosheets as the precursor. A probable mechanism of formation of β-Co(OH)$_2$ nanosheets, porous Co_3O_4 nanosheets, and Co_3O_4 nanorings was proposed on the basis of the experimental results.

1 Introduction

Cobalt hydroxide has attracted increasing attention in recent years because of its novel electric and catalytic properties and important technological applications.[1-3] In particular, cobalt hydroxide can be used to enhance the electrochemical performance when added to nickel oxyhydroxide electrodes (NOEs) by enhancing the electrode conductivity and chargeability.[4] It is well known that cobalt hydroxide has two polymorphs: α-Co(OH)$_2$ and β-Co(OH)$_2$. These two phases are all-layered and have the same hexagonal structures, except that the β form is isostructural with brucite-like compounds and consists of a hexagonal packing of hydroxyl ions with Co^{II} occupying alternate rows of octahedral sites.[5] α-Co(OH)$_2$, however, is isostructural with hydrotalcite-like compounds that consist of stacked Co(OH)$_{2-x}$ layers intercalated with various anions (e.g., nitrate, carbonate, etc.) in the interlayer space to restore charge neutrality. α-Co(OH)$_2$ thus has a larger interlayer spacing (> 7.0 Å, depending on intercalated anions) than the brucite-like β-Co(OH)$_2$ (4.6 Å); because of that, the α form has higher electrochemical activity. However, the hydrotalcite-like phase [α - Co(OH)$_2$] is metastable and easily undergoes a phase transformation into the more stable brucite-like compounds in strongly alkaline media. β - Co(OH)$_2$ is often selected as additives of alkaline secondary batteries owing to its stability in alkaline electrolytes and enhanced conductivity when charged to β - CoOOH.[6, 7]

Several chemical and electrochemical methods have been employed to prepare cobalt hydroxide, for example, force precipitation of Co(NO$_3$)$_2$, direct precipitation with liquid ammonia[8] and potassium hydroxide,[9] urea hydrolysis,[10] and electrochemical synthesis.[11] Sampanthar and Zeng[12] reported the synthesis of butterfly-like β - Co(OH)$_2$ nanocrystals by the ethylenediamine-mediated approach. Li et al.[13] prepared β - Co(OH)$_2$ nanostructures consisting of a mixture of nanoflakes and nanorods by the CoC$_2$O$_4$ · 2H$_2$O conversion method. Recently, Liu et al.[14] synthesized single-crystalline nanosheets of α - Co(OH)$_2$ and β - Co(OH)$_2$ by using hexamethylenetetramine as a hydrolysis agent. Hou et al.[15] also synthesized single-crystalline β - Co(OH)$_2$ nanosheets by homogeneous precipitation with sodium hydroxide as alkaline reagent in the presence of poly(vinylpyrrolidone). Although many attempts have been made on the synthesis of cobalt hydroxides, the control of their morphology, size, and crystallinity still remains a highly sophisticated challenge to materials scientists and chemists.

Spinel cobalt oxide (Co$_3$O$_4$) is an important magnetic p-type semiconductor that has been demonstrated to have considerable application as, for example, solid-state sensors, ceramic pigments, heterogeneous catalysts, rotatable magnets, electrochromic devices, and in energy storage.[16-19] Several methods have been successfully applied in the synthesis of Co$_3$O$_4$ nanoparticles, such as spray pyrolysis, chemical vapor deposition, sol-gel techniques, forced hydrolysis, and so on.[20-23] Recently a variety of novel shapes such as Co$_3$O$_4$ nanoboxes,[24] nanocubes,[25] nanofibers,[26] nanorods,[27] and nanotubes[28] have been reported. However, as far as we know, there is no study on the preparation of Co$_3$O$_4$ nanorings until now.

Herein we demonstrated that single-crystalline β - Co(OH)$_2$ nanosheets can be successfully synthesized in large quantities by a facile hydrothermal synthetic method with triethylamine as both an alkaline and complexing reagent under mild conditions. The influences of hydrothermal temperature, reaction time, amount of triethylamine, and concentration of cobalt nitrate on the size and shape of nanosheets was carefully investigated. Single-crystalline porous nanosheets and nanorings of spinel cobalt oxide

(Co_3O_4) were successfully obtained by a thermal-decomposition method with single-crystalline β - Co(OH)$_2$ nanosheets as the precursor. The mechanism of formation of porous nanosheets and nanorings of Co_3O_4 is also discussed on the basis of the experimental results. Notably, the current synthetic strategy can be used to synthesize other metal hydroxides nanosheets and prepare corresponding metal oxides nanorings by calcination at appropriate temperatures, and it has good prospect for future large-scale applications owing to its high yields, simple reaction apparatus, and low reaction temperature.

2 Experimental Section

All chemicals used in this work, such as aqueouscobalt nitrate [$Co(NO_3)_2 \cdot 6H_2O$] and triethylamine, were analytical-grade regents from the Beijing Chemical Factory, China. They were used without further purification.

2.1 Synthesis. In a typical procedure, $Co(NO_3)_2 \cdot 6H_2O$ (0.291g, 1 mmol) was placed in a 50 mL teflon-lined autoclave and dissolved in deionized water (20mL) to form a pink solution at room temperatur. Triethylamine (0.5 - 1.5 mL) was then added dropwise with magnetic stirring, and the solution immediately turned black. Next, the autoclave was filled with deionized water up to 80% of the total volume, after 10 min of stirring, sealed and maintained at 100 - 180℃ for 2 - 24 h without shaking or stirring. The resulting products were filtered and washed with deionized water and anhydrous ethanol several times, and finally dried under vacuum at 60℃ for 4 h. As-prepared cobalt hydroxide was calcined to produce porous nanosheets and nanorings of Co_3O_4 in air at 400 - 600℃ for 2 h.

2.2 Characterization. The samples obtained were characterized on a Brucker D8-advance powder X-ray diffractometer with Cu_{K_α} radiation (λ = 1.5418 Å). The operation voltage and current were kept at 40 kV and 40 mA, respectively. The size and morphology of as-synthesized products were determined at 20 kV by an XL30 S - FEG scanning electron microscope and at 160 kV by a JEM - 200CX transmission electron microscope and a JEOL JEM - 2010F high-resolution transmission electron microscope. SAED was further performed to determine the crystallinity. DSC and TGA were carried out with a NETZSCH STA -449C simultaneous TG-DTA/DSC apparatus at a heating rate of 10 K/min in flowing air. A nitrogen adsorption system (Coulter SA 3100 plus) was employed to record the adsorption-desorption isotherm at liquid-nitrogen temperature of 196℃. The electrochemical properties of as-prepared Co_3O_4 as cathode were evaluated by using two-electrode cells with lithium metal as anode. The cathode was prepared by compressing a mixture of Co_3O_4/acetylene black/polyvinylidene fluoride (PVDF) with weight ratio 75∶15∶15. The cathode were dried for 24 h at 80 ℃ in a vacuum oven and cut into a disk (1.0 cm^2). The electrolyte solution was 1 M $LiPF_6$ dissolved in a 1∶1 mixture of ethylene carbonate/diethyl carbonate. The cell was assembled in an Ar-filled glovebox with porous polypropylene (Celgard 2500) as a separator. The electrode capacity was measured by a galvanostatic charge/discharge experiment with a current density of 100 mA/g at a potential between 0 V and 3.0 V.

3 Results and Discussion

X-ray diffraction (XRD) was carried out to determine the chemical composition and crystallinity of as-prepared products. Figure 1 shows the typical XRD patterns of β-Co(OH)$_2$ nanosheets prepared by using a 1.5 mL solution of triethylamine at 180 ℃ for 24 h. All diffraction peaks in this pattern can be indexed as the hexagonal cell of brucite-like, β-Co(OH)$_2$ with lattice constants a = 3.182 Å and c = 4.658 Å [space group: $P\bar{3}m1$ (No. 164)], which are consistent with the values in the literature (JCPDS 30-0443). The (001) peak is taller and far narrower than other peaks in the reflections, which implies the highly preferentially oriented growth of the β-Co(OH)$_2$

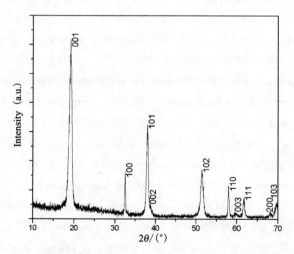

Figure 1 XRD pattern of as-prepared β-Co(OH)$_2$ nanosheets obtained by using 1.5 mL triethylamine at 180 ℃ for 24 h.

nanosheets. Diffraction peaks of α - Co(OH)$_2$ or impurities were not observed, which indicates the high purity of the final products successfully synthesized under current experimental conditions.

The size and morphology of as-prepared product were examined by scanning electron microscopy (SEM) and transmission electron microscopy (TEM). The SEM images [Figure 2(a)] indicate that a large quantity of hexagonal β - Co(OH)$_2$ nanosheets with good uniformity were achieved by using this approach. These nanosheets had a mean width of about 120 nm with little deviation. The inset shows a high-magnification SEM image. Here, the corners and edges of β - Co(OH)$_2$ nanosheets can be clearly observed. The average thickness and edge size of these hexagonal nanosheets are about 15 nm and 60 nm, respectively. SEM observations also indicated that almost 100% of as-prepared products are uniformity hexagonal β - Co(OH)$_2$ nanosheets. Figure 2(b) is the representative TEM image, in which the β - Co(OH)$_2$ nanosheets are hexagonal and quasi-hexagonal with the angles of adjacent edges of 120° and widths in the range of 100 - 170 nm. The inset of Figure 2(b) shows the selected area electron diffraction (SAED) pattern of β - Co(OH)$_2$ nanosheets, which reveals that the β - Co(OH)$_2$ nanosheets are single-crystalline hexagonal structures lying on their {001} crystal planes, consistent with the XRD result presented above.

High-resolution transmission electron microscopy (HRTEM) provided further insight into the nanostructure of as-prepared hexagonal β - Co(OH)$_2$ nanosheets. Figure 2(c) shows a typical image of an individual hexagonal β - Co(OH)$_2$ nanosheet with a mean width of about 120 nm and edge length of about 65 nm. The inset of Figure 2(c) displays a magnified part of image. The interlayer spacing was calculated to be about 0.27 nm, which corresponds to the interlayer distance of the (100) crystal plane in β - Co(OH)$_2$. Figure 2(d) shows a side-view HRTEM image of an individual β - Co(OH)$_2$ nanosheet with a thickness of about 15 nm. The well-resolved lattice fringes in the nanosheet can be clearly observed, which confirms that the nanosheets were formed with a single-crystalline structure. Further magnification clearly shows that the interlayer spacing is about 0.46 nm [Figure 2(d), inset], which corresponds to the interlayer distance of the (001) crystal plane of brucite-like β - Co(OH)$_2$ and shows that the growth direction of β - Co(OH)$_2$ is along the (001) crystal plane.

To confirm the morphology of as-prepared product further, a tilted-angle investigation was carried out to obtain the TEM images from different viewing angles. Some of the nanorod-like β - Co(OH)$_2$ were virtually hexagonal nanosheets in shape,

Figure 2

(a) Low-magnification and high-magnification (inset) SEM images of as-synthesized β-Co(OH)$_2$ nanosheets. (b) TEM image of the hexagonal β-Co(OH)$_2$ nanosheets with angles of adjacent edges of 120°. Inset: SAED pattern of the β-Co(OH)$_2$ nanosheets taken from an individual hexagonal nanosheet. (c) HRTEM image of the individual β-Co(OH)$_2$ nanosheets. (d) Side view of the hexagonal β-Co(OH)$_2$ nanosheets. Inset in (c) and (d): Further-magnified images showing the crystal lattice.

as shown by the areas marked with a circle of Figure 3. Figure 3(b) is a perpendicular view of the β-Co(OH)$_2$ nanocrystals with a perfect rodlike shape. When the copper grid was tilted through $-10° - 20°$, parts of the rodlike pattern disappear, and the hexagonal sheetlike pattern appears gradually. When it was tilted to 20°, nearly hexagonal nanosheets resulted [Figure 3(d)]. Furthermore, the hexagonal sheetlike pattern can theoretically transform into the rodlike pattern because the two patterns are just a result of the different viewing angles of the same product. Such a transformation can also be clearly seen in Figure 3. This behavior means that the as-prepared products are indeed hexagonal β-Co(OH)$_2$ nanosheets, which agrees well with the SEM

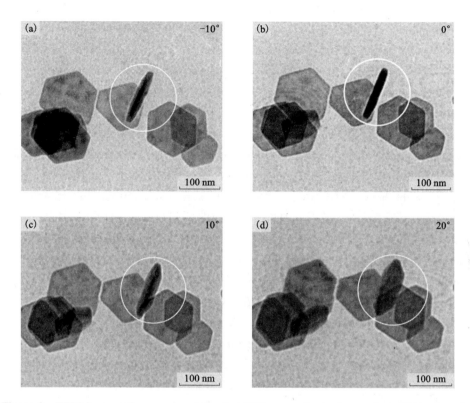

Figure 3 TEM images of as-prepared $\beta - Co(OH)_2$ nanosheets observed at different angles (a) $-10°$, (b) $0°$, (c) $10°$, (d) $20°$.

results.

Furthermore, further studies suggested that hydrothermal temperature, reaction time, amount of triethylamine, and concentration of cobalt nitrate all influence the size and morphology of $\beta - Co(OH)_2$ nanosheets. For example, the size of $Co(OH)_2$ nanosheets was small and the crystallinity was poor at low temperature, whereas the average size and crystallinity of nanosheets gradually increased with the elevation of hydrothermal temperature. Figure 4 (a) shows a typical TEM image of the product prepared at 100℃ over 24 h. Here, the hexagonal and quasi-hexagonal nanosheets can be clearly observed in addition to some nanoparticles. When the reaction temperature was elevated to 140℃, the final products were mostly made up of nanosheets with the widths in the range of 60 – 140 nm [Figure 4 (b)]. When the reaction time was decreased to 2 h, the product formed consisted of nanosheets with widths in the range of 70 – 150 nm as well as some nanoparticles with a mean size of 15 nm [Figure 4(c)].

Figure 4(d) shows the TEM image of products obtained by using 0.5 mL solution of triethylamine at 180℃ for 24 h, which clearly shows the nanosheets as fine hexagons with widths in range of 100 – 260 nm. Generally, the size and morphology of the β – Co(OH)$_2$ nanosheets could also be manipulated by changing the concentration of cobalt nitrate concentration. Figure 4(e) shows the TEM image of products prepared by using 0.5 mmol Co(NO$_3$)$_2 \cdot$ 6H$_2$O, in which hexagonal Co(OH)$_2$ nanosheets with a mean length of about 100 nm and a thickness of about 10 nm can be clearly observed. Furthermore, a few large nanosheets were also found. When the concentration of cobalt nitrate doubled, that is, when 2 mmol Co(NO$_3$)$_2 \cdot$ 6H$_2$O was used, the length and thickness of the hexagonal Co(OH)$_2$ nanosheets were about 150 nm and 15 – 20 nm, respectively [Figure 4(f)]. However, further elevation of the cobalt nitrate concentration to five or ten times the original resulted in morphologies of the hexagonal nanosheets that tended to be irregular, which implied that the nucleation and growth behavior were out of the kinetic control.[29]

The thermal behavior of hexagonal β – Co(OH)$_2$ nanosheets oxidized to spinel Co$_3$O$_4$ was investigated with thermogravimetric analysis (TGA) and differential scanning calorimetry(DSC) in the temperature range of 25 – 800℃ (in Figure 5). The gradual mass loss in the range 25 – 150℃ can be attributed to evaporation of the adsorbed triethylamine and water species on nanosheet surfaces. The major weight loss profile exhibits a well-defined decrease between 150℃ and 400℃ with an inflection point at about 170℃. The weight loss was measured to be about 13.7% over and above that of adsorbed triethylamine and water species, in good agreement with the theoretical value (13.6%). The DSC curve showed an endothermic peak at 172℃, which corresponds to the dominant mass loss.

Cobalt oxide can be obtained by a thermal-decomposition method with cobalt hydroxide as the precursor, on the basis of the TGA and DSC results. Single-crystalline porous nanosheets and nanorings of cobalt oxide (Co$_3$O$_4$) were selectively obtained by calcination of as-prepared hexagonal β – Co(OH)$_2$ nanosheets obtained by using 1.5 mL solution of triethylamine at 180℃ for 24 h in air at 400℃ and 600℃, respectively for 2 h. Figure 6 shows the XRD patterns of porous Co$_3$O$_4$ nanosheets and Co$_3$O$_4$ nanorings. All the reflections of the XRD pattern can be indexed to the pure face-centered-cubic phase [space group: Fd3m (No. 227)] of spinel cobalt oxide with lattice constants a = 8.065 Å (JCPDS 74 – 1657). No impurity peaks were observed, which indicates that brucite-like Co(OH)$_2$ was completely converted into spinel structure Co$_3$O$_4$ by calcinations in air at 400℃ and 600℃ for 2 h. When the calcination

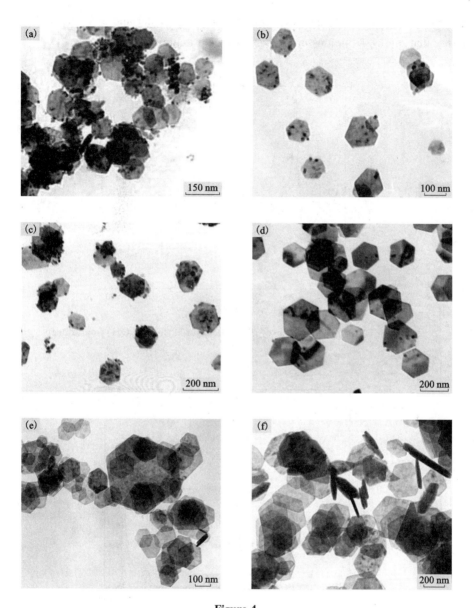

Figure 4

(a) - (c) TEM images of hexagonal β - Co(OH)$_2$ nanosheets synthesized by using 1.5 mL triethylamine at different reaction temperatures and time: (a) 100℃, 24 h; (b) 140℃, 24 h; (c) 180℃, 2 h. (d) TEM images of hexagonal β - Co(OH)$_2$ nanosheets synthesized by using 0.5 mL triethylamine at 180℃ for 24 h. (e) - (f) TEM images of hexagonal β - Co(OH)$_2$ nanosheets synthesized by using different cobalt nitrate concentrations at 180℃ for 24 h: (e) 0.5 mmol; (f) 2 mmol.

temperature was increased, the diffraction peaks became taller and much narrower, thus showing that the crystallinities of the samples were improved.

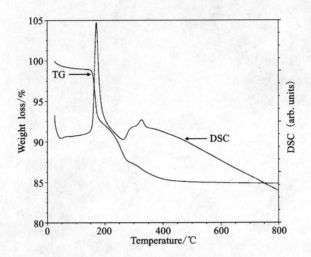

Figure 5　DSC and TGA curves of the hexagonal β – Co(OH)$_2$ nanosheets

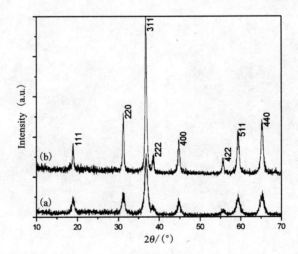

Figure 6

XRD patterns of porous Co$_3$O$_4$ nanosheets (a) and Co$_3$O$_4$ nanorings (b) obtained by calcination of as-synthesized β – Co(OH)$_2$ nanosheets in air at 400℃ and 600℃, respectively, for 2 h.

Figure 7(a) shows a typical TEM image of the porous Co_3O_4 nanosheets obtained by calcination of as-prepared $\beta-Co(OH)_2$ nanosheets in air at 400 ℃ for 2 h. As with the $\beta-Co(OH)_2$ nanosheets, the porous Co_3O_4 nanosheets also consist of hexagonal and quasi-hexagonal with the size ranging from 80 nm to 150 nm. However, under careful observation, the Co_3O_4 nanosheets were found to be composed of many pores with a mean diameter of 3 nm. The inset of Figure 7(a) is an SAED pattern taken from an individual Co_3O_4 nanosheet; the pattern consists of many spots, which shows that each porous Co_3O_4 nanosheet is a single crystal. All spots are identified as diffraction from spinel Co_3O_4. It is interesting that the as-prepared $\beta-Co(OH)_2$ nanosheets could be converted into Co_3O_4 nanorings by calcination in air at 600 ℃ for 2 h. A typical TEM image of Co_3O_4 nanorings is shown in Figure 7(b). It is clear that some of the Co_3O_4 nanocrystals exhibit ringlike structures with an average size of about 100 nm. The inset shows an SAED pattern taken from a mass of the Co_3O_4 nanorings; the pattern reveals the satisfactory crystallinity of the sample, which can be indexed to the face-centered-cubic phase of spinel Co_3O_4. A typical HRTEM image of an individual Co_3O_4 nanoring is shown in Figure 7(c). The magnified image (inset) shows that the nanoring is structurally uniform with an interlayer spacing about 0.46 nm, which corresponds to the value of the (111) lattice plane of the spinel Co_3O_4. The Brunauer-Emmett-Teller (BET) method with nitrogen adsorption was carried out to investigate the surface-area data of as-prepared Co_3O_4 samples, which is critical for their technological application. The N_2 - adsorption-desorption isotherm of as-prepared Co_3O_4 can be categorized as type Ⅳ with a distinct hysteresis loop. The BET surface-area data was calculated to be about 18.153 m^2/g for the porous Co_3O_4 nanosheets [Figure 7(d)] and 13.734 m^2/g for the Co_3O_4 nanorings [Figure 7(e)].

Figure 7(f) shows the initial discharge curve and the cycle performance of the Li – Co_3O_4 cell made by the as-prepared Co_3O_4 at a current density of 100 mA/g at room temperature. The porous Co_3O_4 nanosheets had the most initial insertion capacity at 1301 mAh/g, and their capacity came to about 1011 mAh/g after the second cycle. Subsequently, the discharge capacity of the Li – Co_3O_4 cell appeared to increase slightly, which usually proceeds with the activation process for electrochemical reaction of lithium. The as-prepared porous Co_3O_4 nanosheets exhibited excellent cyclability. After 21 cycles, the as-prepared porous Co_3O_4 nanosheets electrode maintained a capacity of 1001 mAh/g, which corresponds to about 99% of the second discharge capacity. The initial capacity of as-prepared Co_3O_4 nanorings reached 1298.7 mAh/g, whereas the capacity dropped rapidly to 676 mAh/g after 21 cycle. The capacity retention of the

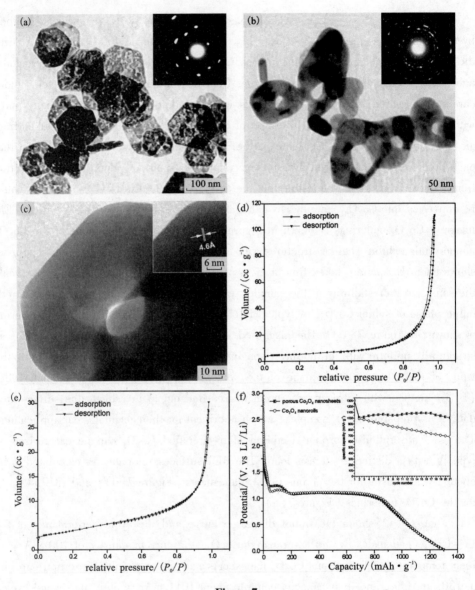

Figure 7

(a) TEM image of porous Co_3O_4 nanosheets. Inset: SAED pattern of the porous Co_3O_4 nanosheets taken from an individual nanosheet, which indicates that each porous nanosheet is a single crystal. (b) TEM image and SAED pattern (inset) of Co_3O_4 nanorings. (c) HRTEM image of an individual Co_3O_4 nanoring. Inset: higher-magnification of Co_3O_4 nanoring obtained from a selected area in (c). N_2-adsorption-desorption isotherm of porous Co_3O_4 nanosheets (d) and Co_3O_4 nanorings (e). Squares = absorption, inverted triangles = desorption. (f) The first discharge curve of the Li – Co_3O_4 cells made by the as-prepared porous Co_3O_4 nanosheets (filled squares) and Co_3O_4 nanorings (empty circles) at a current density of 100 mA/g. Inset: cycle performance of the Li – Co_3O_4 cells at a current density of 100 mA/g at room temperature.

porous Co_3O_4 nanosheets was much better than that of the Co_3O_4 nanorings, which can be ascribed to the surface area and crystallinity of samples. The porous Co_3O_4 nanosheets have a larger surface area and higher vacancy owing to poor crystallinity, which probably makes the lithium ions easy to extract and insert into the porous Co_3O_4 electrode and results in the increase in recharge ability.[30]

On the basis of the experimental results, a probable mechanism of formation of the hexagonal β - $Co(OH)_2$ nanosheets, porous Co_3O_4 nanosheets, and Co_3O_4 nanorings is proposed. The formation of hexagonal β - $Co(OH)_2$ nanosheets may mainly comprises two processes: ① formation of β - $Co(OH)_2$ crystal nuclei and ② subsequent crystal growth from these nuclei to form hexagonal nanosheets. Triethylamine may play a key role in the crystal growth process. Triethylamine was used both as alkaline reagent to provide an alkaline environment and as a complexing reagent to influence the morphology of the final products. Our experimental results suggest that triethylamine may provide strong kinetic control over the growth rates of various faces of β - $Co(OH)_2$ by being selectively adsorbed on the crystal planes, which results in the formation of hexagonal β - $Co(OH)_2$ nanosheets.[31] However, it is still not clear how triethylamine influences the growth of different crystal planes of β - $Co(OH)_2$. We also used diethanolamine and triethanolamine as alkaline and complexing reagents to prepare $Co(OH)_2$, but the morphology of the final products was irregular. The spinel Co_3O_4 nanorings may begin with the nanopores at the core of the hexagonal β - $Co(OH)_2$ nanosheets. The crystal structure of the hexagonal β - $Co(OH)_2$ nanosheets includes Co - OH layers and counter anions between the Co - OH layers. With elevated calcination temperature, the Co - OH layers are converted into cobalt oxide through pyrolysis and dehydration process. Thus, the spaces of the OH layers and counter anions are converted into nanopores. With elongation of reaction time, the hexagonal β - $Co(OH)_2$ nanosheets gradually dehydrated and shrank and are finally converted into porous Co_3O_4 nanosheets and nanorings at different calcination temperature.[32] The calcination temperature was especially important in the formation of Co_3O_4 nanorings. It is difficult to obtain the Co_3O_4 nanorings at low calcination temperature, whereas the products of Co_3O_4 obtained at 400℃ for 2 h were mostly made up of nanosheets with porous structures. Hou et al.[15] also prepared porous Co_3O_4 nanosheets by thermal decomposition of hexagonal β - $Co(OH)_2$ nanosheets in air at 450℃ for 5 h. Our synthetic strategy could be employed to fabricate porous nanosheets and nanorings of other metal oxide by calcination of the corresponding metal hydroxide nanosheets under appropriate conditions.

4 Conclusion

In summary, we have developed a simple method for the synthesis of single-crystalline hexagonal β-Co(OH)$_2$ nanosheets in large quantities by choosing triethylamine as both an alkaline and a complexing reagent under mild conditions. This novel synthetic method can be carried out to synthesize other high-quality hydroxide nanosheets, and it should have potential applications in the future large-scale synthesis owing to its high yield, simple reaction apparatus and low reaction temperature. Notably, single-crystalline porous Co$_3$O$_4$ nanosheets and Co$_3$O$_4$ nanorings were selectively obtained by thermal decomposition of the single-crystalline β-Co(OH)$_2$ nanosheets in air at 400℃ and 600℃ respectively, for 2 h. This strategy may become a general method for the fabrication of porous nanosheets and nanorings of other metal oxide by calcinations of the corresponding metal hydroxide nanosheets under appropriate conditions.

References

[1] (a) Z. P. Xu, H. C. Zeng. Interconversion of brucite-like and hydrotalcite-like phases in cobalt hydroxide compounds [J]. Chemistry of Materials, 1999, 11(1): 67-74. (b) P. V. Kamath, G. H. A. Therese, J. Gopalakrishnan. On the existence of hydrotalcite-like phases in the absence of trivalent cations [J]. Journal of Solid State Chemistry, 1997, 128(1): 38-41.

[2] (a) L. Cao, F. Xu, Y-Y. Liang, H-L. Li. Preparation of the novel nanocomposite Co(OH)$_2$/ultra-stable Y zeolite and its application as a supercapacitor with high energy density [J]. Advanced Materials, 2004, 16(20): 1853-1857. (b) V. Pralong, A. Delahaye-Vidal, B. Beaudoin, J-B. Leriche, J-M. Tarascon. Electrochemical behavior of cobalt hydroxide used as additive in the nickel hydroxide electrode [J]. Journal of the Electrochemical Society, 2000, 147(4): 1306-1313.

[3] L. Zhang, A. K. Dutta, G. Jarero, P. Stroeve. Nucleation and growth of cobalt hydroxide crystallites in organized polymeric multilayers [J]. Langmuir, 2000, 16(17): 7095-7100.

[4] (a) M. Butel, L. Gautier, C. Delmas. Cobalt oxyhydroxides obtained by "chimie douce" reactions: structure and electronic conductivity properties [J]. Solid State Ionics, 1999, 122(1): 271-284. (b) M. Oshitani, H. Yufu, K. Takashima, S. Tsuji, Y. Matsumaru. Development of a pasted nickel electrode with high active material utilization [J]. Journal of the Electrochemical Society, 1989, 136(6): 1590-1593.

[5] (a) C. Mockenhaupt, T. Zeiske, H. D. Lutz. Crystal structure of brucite-type cobalt hydroxide β-Co{O(H, D)}$_2$ - neutron diffraction, IR and Raman spectroscopy [J]. Journal of Molecular

Structure, 1998, 443(1): 191-196. (b) P. Benson, G. W. D. Briggs, W. F. K. Wynne-Jones. The cobalt hydroxide electrode-I. structure and phase transitions of the hydroxides [J]. Electrochimica Acta, 1964, 9(3): 275-280.

[6] V. Pralong, A. Delahaye-Vidal, B. Beaudoin, B. Gérand, J-M. Tarascon. Oxidation mechanism of cobalt hydroxide to cobalt oxyhydroxide [J]. Journal of Materials Chemistry, 1999, 9(4): 955-960.

[7] (a) F. Lichtenberg, K. Kleinsorgen. Stability enhancement of the CoOOH conductive network of nickel hydroxide electrodes [J]. Journal of Power Sources, 1996, 62(2): 207-211. (b) K. Watanabe, T. Kikuoka, N. Kumagai. Physical and electrochemical characteristics of nickel hydroxide as a positive material for rechargeable alkaline batteries [J]. Journal of Applied Electrochemistry, 1995, 25(3): 219-226.

[8] T. N. Ramesh, M. Rajamathi, P. V. Kamath, Ammonia induced precipitation of cobalt hydroxide: observation of turbostratic disorder [J]. Solid State Sciences, 2003, 5(5): 751-756.

[9] P. Elumalai, H. N. Vasan, N. Munichandraiah. Electrochemical studies of cobalt hydroxide-an additive for nickel electrodes [J]. Journal of Power Sources, 2001, 93(1): 201-208.

[10] M. Dixit, G. N. Subbanna, P. V. Kamath. Homogeneous precipitation from solution by urea hydrolysis: a novel chemical route to the α - hydroxides of nickel and cobalt [J]. Journal of Materials Chemistry, 1996, 6(8): 1429-1432.

[11] R. S. Jayashree, P. V. Kamath. Electrochemical synthesis of α - cobalt hydroxide [J]. Journal of Materials Chemistry, 1999, 9(4): 961-963.

[12] J. T. Sampanthar, H. C. Zeng. Arresting butterfly-like intermediate nanocrystals of β - Co(OH)$_2$ via ethylenediamine-mediated synthesis [J]. Journal of the American Chemical Society, 2002, 124(23): 6668-6675.

[13] X. L. Li, J. F. Liu, Y. D. Li. Low-temperature conversion synthesis of M(OH)$_2$(M = Ni, Co, Fe) nanoflakes and nanorods [J]. Materials Chemistry and Physics, 2003, 80(1): 222-227.

[14] Z. Liu, R. Ma, M. Osada, K. Takada, T. Sasaki. Selective and controlled synthesis of α - cobalt and β - cobalt hydroxides in highly developed hexagonal platelets [J]. Journal of the American Chemical Society, 2005, 127(40): 13869-13874.

[15] Y. Hou, H. Kondoh, M. Shimojo, T. Kogure, T. Ohta. High-yield preparation of uniform cobalt hydroxide and oxide nanoplatelets and their characterization [J]. The Journal of Physical Chemistry B, 2005, 109(41): 19094-19098.

[16] M. Andok, T. Kobayashi, S. Iijima, M. Haruta. Optical recognition of CO and H$_2$ by use of gas-sensitive Au - Co$_3$O$_4$ composite films [J]. Journal of Materials Chemistry, 1997, 7(9): 1779-1783.

[17] P. Nkeng, J. -F. Koenig, J. L. Gautier, P. Chartier, G. Poillerat. Enhancement of surface areas of Co$_3$O$_4$ and NiCo$_2$O$_4$ electrocatalysts prepared by spray pyrolysis [J]. Journal of

Electroanalytical Chemistry, 1996, 402(1): 81-89.

[18] (a) M. T. Verelst, O. Ely, C. Amiens, E. Snoeck, P. Lecante, A. Mosset, M. Respaud, J. M. Broto, B. Chaudret. Synthesis and characterization of CoO, Co_3O_4, and mixed Co/CoO nanoparticules [J]. Chemistry of Materials, 1999, 11(10): 2702-2708. (b) S. Takada, M. Fujii, S. Kohiki. S. Kohiki, Intraparticle magnetic properties of Co_3O_4 nanocrystals [J]. Nano Letters, 2001, 1(7): 379-382. (c) S. A. Makhlouf, Magnetic properties of Co_3O_4 nanoparticles [J]. Journal of Magnetism and Magnetic Materials, 2002, 246(1): 184-190.

[19] D. Barreca, C. Massignan, S. Daolio, M. Fabrizio, C. Piccirillo, L. Armelao, E. Tondello. Composition and microstructure of cobalt oxide thin films obtained from a novel cobalt (II) precursor by chemical vapor deposition [J]. Chemistry of Materials, 2001, 13(2): 588-593.

[20] M. E. Baydi, G. Poillerat, J. L. Rehspringer, J. L. Gautier, J. -F. Koenig, P. Chartier. A sol-gel route for the preparation of Co_3O_4 catalyst for oxygen electrocatalysis in alkaline medium [J]. Journal of Solid State Chemistry, 1994, 109(2): 281-288.

[21] E. Fujii, H. Torii, A. Tomozawa, R. Takayama, T. Hirao. Preparation of cobalt oxide films by plasma-enhanced metalorganic chemical vapour deposition [J]. Journal of Materials Science, 1995, 30(23): 6013-6018.

[22] J. L. Gautier, E. Rios, M. Gracia, J. F. Marco, J. R. Gancedo. Characterisation by X-ray photoelectron spectroscopy of thin $Mn_xCo_{3-x}O_4$ ($0 \leqslant x \leqslant 1$) spinel films prepared by low-temperature spray pyrolysis [J]. Thin Solid Films, 1997, 311(1): 51-57.

[23] E. Matijevic. Preparation and properties of uniform size colloids [J]. Chemistry of Materials, 1993, 5(4): 412-426.

[24] T. He, D. Chen, X. Jiao, Y. Wang. Co_3O_4 Nanoboxes: Surfactant-templated fabrication and microstructure characterization [J]. Advanced Materials, 2006, 18(8): 1078-1082.

[25] (a) X. Liu, G. Qiu, X. Li. Shape-controlled synthesis and properties of uniform spinel cobalt oxide nanocubes [J]. Nanotechnology, 2005, 16(12): 3035. (b) R. Xu, H. C. Zeng. Self-generation of tiered surfactant superstructures for one-pot synthesis of Co_3O_4 nanocubes and their close-and non-close-packed organizations [J]. Langmuir, 2004, 20(22): 9780-9790. (c) R. Xu, H. C. Zeng. Mechanistic investigation on salt-mediated formation of free-standing Co_3O_4 nanocubes at 95 ℃ [J]. The Journal of Physical Chemistry B, 2003, 107(4): 926-930. (d) J. Feng, H. C. Zeng, Size-controlled growth of Co_3O_4 nanocubes [J]. Chemistry of Materials, 2003, 15(14): 2829-2835.

[26] H. Guan, C. Shao, S. Wen, B. Chen, J. Gong, X. Yang. A novel method for preparing Co_3O_4 nanofibers by using electrospun PVA/cobalt acetate composite fibers as precursor [J]. Materials Chemistry and Physics, 2003, 82(3): 1002-1006.

[27] Y. Liu, G. Wang, C. Xu, W. Wang. Fabrication of Co_3O_4 nanorods by calcination of precursor powders prepared in a novel inverse microemulsion [J]. Chemical Communications, 2002, 14: 1486-1487.

[28] X. Shi, S. Han, R. J. Sanedrin, C. Galvez, D. -G. Ho, B. Hernandez, F. Zhou, M.

Selke. Formation of cobalt oxide nanotubes: Effect of intermolecular hydrogen bonding between Co (III) complex precursors incorporated onto colloidal templates [J]. Nano Letters, 2002, 2 (4): 289 – 293.

[29] X. Sun, Y. Li. Size-controllable luminescent single crystal CaF_2 nanocubes [J]. Chemical Communications, 2003, 14: 1768 – 1769.

[30] X. Wang, X. Chen, L. Gao, H. Zheng, Z. Zhang, Y. Qian. One-dimensional arrays of Co_3O_4 nanoparticles: synthesis, characterization, and optical and electrochemical properties [J]. The Journal of Physical Chemistry B, 2004, 108(42): 16401 – 16404.

[31] X. Sun, S. Dong, E. Wang. High-yield synthesis of large single-crystalline gold nanoplates through a polyamine process [J]. Langmuir, 2005, 21(10): 4710 – 4712.

[32] (a) X. H. Liu, R. Yi, N. Zhang, R. R. Shi, X. G. Li, G. Z. Qiu. Rationally synthetic strategy: from nickel hydroxide nanosheets to nickel oxide nanorolls [J]. Nanotechnology, 2005, 16(8): 1400 – 1405; (b) E. Hosono, S. Fujihara, I. Honma, H. Zhou. Fabrication of morphology and crystal structure controlled nanorod and nanosheet cobalt hydroxide based on the difference of oxygen-solubility between water and methanol, and conversion into Co_3O_4 [J]. Journal of Materials Chemistry, 2005, 15(19): 1938 – 1945.

☆ X. H. Liu, R. Yi, N. Zhang, R. R. Shi, X. G. Li, G. Z. Qiu. Cobalt hydroxide nanosheets and their thermal decomposition to cobalt oxide nanorings, Chemistry-An Asian Journal, 2008, 3(4): 732 – 738.

Rationally Synthetic Strategy: From Nickel Hydroxide Nanosheets to Nickel Oxide Nanorolls

Abstract: We demonstrate herein that single-crystal β-Ni(OH)$_2$ nanosheets can be selectively synthesized in large quantities through a facile hydrothermal synthetic method using aqueous nickel nitrate as the nickel source and sodium hydroxide as alkaline reagent. This attempt is based on the treatment of aqueous nickel nitrate with high concentrations of sodium hydroxide solution under hydrothermal condition. Single-crystalline NiO nanorolls can be obtained via a thermal decomposition method using single-crystal β-Ni(OH)$_2$ nanosheets as the precursor. The influences of alkaline concentration and reaction temperature are carefully investigated in this paper. The morphologies and phase structures of as-prepared products are investigated in detail by X-ray diffraction (XRD), transmission electron microscopy (TEM), and selected area electron diffraction (SAED). The probable formation mechanism of the NiO nanorolls is proposed on the basis of the experimental results.

1 Introduction

Over the past decades, nickel hydroxide has attracted great interest especially due to its critical electrochemical function and great potential applications in rechargeable Ni-based alkaline batteries (e.g., Ni/Cd, Ni/H$_2$, Ni/MH, Ni/Fe, and Ni/Zn), that are widely used in various fields ranging from portable electronics to electric vehicles and satellites.[1,2] Nickel hydroxide has a hexagonal layered structure with two most common polymorphs, α-Ni(OH)$_2$ and β-Ni(OH)$_2$.[3] The α-Ni(OH)$_2$ is categorized as being isostructural with hydrotalcite-like compounds and consists of stacked Ni(OH)$_{2-x}$ layers intercalated with various anions or water molecules.[4] The β-Ni(OH)$_2$ is categorized as being isostructural with brucite-like [Mg(OH)$_2$] structure and consists of an ordered stacking of well oriented 'NiO$_2$' slabs.[5] However, the β-Ni(OH)$_2$ exhibits superior stability compared to the α-Ni(OH)$_2$ and the β-Ni(OH)$_2$ can be easily derived from α-Ni(OH)$_2$ in alkaline medium.[6] Therefore, the β-Ni(OH)$_2$ is most widely used as the active material of positive electrodes on the basis of its chemical and thermal stability.

It is well known that the electrochemical properties, such as utilization and capacity, of nickel hydroxide electrodes sensitively depend on their morphology and size. Nickel hydroxide nanostructures may provide some immediate advantages over their solid counterparts for their applications.[7, 8] However, there have been only a few reports for the synthesis of nickel hydroxide nanostructures. Gedanken et al.[9] have stated they prepared nanosized α - $Ni(OH)_2$ with the aid of ultrasound radiation. Li et al.[10] have reported the synthesis of β - $Ni(OH)_2$ nanomaterials with the mixture structures of nanosheets and nanorods by the $NiC_2O_4 \cdot 2H_2O$ conversion method. Recently, Liu et al.[11] have synthesized nanosized β - $Ni(OH)_2$ by solid-state reaction using $NiC_2O_4 \cdot 2H_2O$ as precursor.

Nickel oxides have been demonstrated to be of considerable applications in various fields. For example, they can be used as gas sensors,[12] catalysis,[13] magnetic materials,[14] fuel cells,[15] and p-type semiconductors[16, 17] due to their wide bandgap energy from 3.5 eV to 4.0 eV. Various methods such as evaporation,[18] sputtering,[19, 20] electrodeposition,[21] and sol-gel techniques[22] have been attempted to synthesize nickel oxide. Recently a variety of novel shapes such as rods,[23] ribbons,[24] and sheets[25] have been reported through chemical reactions of precursors at room or slightly elevated temperatures. However, it remains a challenge to control the morphology and size of final products to fine-tune their properties.

In this paper, we demonstrated that single-crystal β - $Ni(OH)_2$ nanosheets could be selectively synthesized through a facile hydrothermal synthetic method using sodium hydroxide as alkaline reagent under mild conditions. By adjusting the alkaline concentration and reaction temperature, the morphologies, phases, and sizes of final products can be easily controlled. Furthermore, single-crystalline NiO nanorolls can be successfully obtained via a thermal decomposition method using single-crystal β - $Ni(OH)_2$ nanosheets as the precursor. The synthetic strategy surely can be applied in fabricating other metal hydroxides nanosheets and be carried out to prepare corresponding metal oxides nanorolls by calcination at appropriate temperature, and it will have a good prospect in the future large-scale application due to its high yield, simple reaction apparatus and low reaction temperature.

2 Experimental Section

All chemicals in this work, such as aqueous nickel nitrate [$Ni(NO_3)_2 \cdot 6H_2O$]

and sodium hydroxide (NaOH), were AR regents from the Beijing Chemical Factory, China. They were used without further purification.

2.1 Methods. In a typical procedure, 2.91 g $Ni(NO_3)_2 \cdot 6H_2O$ was put into a Teflon-lined autoclave of 50 mL capacity and dissolved in 40 mL 2 – 10 M NaOH solution. After stirring for 10 min, the autoclave was sealed and heated at 140 – 180 ℃ for 24 h. The autoclave was allowed to cool down to room temperature naturally after heat treatment. The resulting products were filtered, washed with distilled water and absolute ethanol, and finally dried in vacuum at 60 ℃ for 4 h.

As-prepared nickel hydroxide was calcined to produce nickel oxide nanorolls in air at 500 ℃ for 2 h.

2.2 Characterization. The obtained samples were characterized on a Brucker D8-advance X-ray powder diffractometer (XRD) with Cu K_α radiation ($\lambda = 1.5418$ Å). The operation voltage and current were kept at 40 kV and 40 mA, respectively. TEM patterns were recorded on a Hitachi Model H – 800 transmission electron microscope at an accelerating voltage of 200 kV. The samples were dispersed in absolute ethanol in an ultrasonic bath. Then the suspensions were dropped onto Cu grids coated with amorphous carbon films. Selected area electron diffraction (SAED) was further performed to identify the crystallinity.

3 Results and Discussion

Figure 1 shows the evolution of the X-ray diffractions (XRD) patterns (2θ scan) from products 1 to 5. From this Figure, it is clear that all the reflections of the XRD pattern can be finely indexed to a hexagonal phase [space group: P$\bar{3}$m1 (164)] of β – Ni(OH)$_2$ (JCPDS 74 – 2075). Diffraction peaks of α – Ni(OH)$_2$ or impurities cannot be observed, which indicates that the products are pure phase compounds. With the elevation of alkaline concentrations (2 M, 4 M, 6 M, 8 M and 10 M), the XRD patterns of samples in the whole process show that the crystallinity of the products is continuously improved.

In our experiments, the influences of reaction temperature on the shapes of product were also explored. Product 6 obtained after 24 h at 140 ℃ using 2 M NaOH solution as alkaline agent is the coexistence of hydrotalcite-like α – Ni(OH)$_2$ and brucite-like β – Ni(OH)$_2$ phases (Figure 2). The Bragg reflections are broad, indicative of smaller sizes and poor crystallinity. When the reaction temperature is lifted to 180 ℃, the XRD pattern of product 7 obtained using 10 M NaOH solution as alkaline agent, is shown in

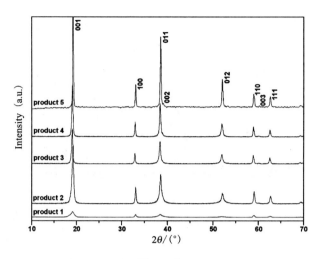

Figure 1

The evolution of XRD patterns of products 1 (2 M), 2 (4 M), 3 (6 M), 4 (8 M) and 5 (10 M) obtained through a facile hydrothermal synthetic method at 160 ℃ for 24 h.

Figure 2. All of the peaks of the XRD pattern can be readily identified as the pure hexagonal β - Ni(OH)$_2$ phase, similar to Figure 1. However, the diffraction peaks were higher and far narrower than the peaks that were maintained in lower temperature and concentrations of sodium hydroxide solution, implying the high orientation and well-crystal nanostructure.

The size and morphology of the β - Ni(OH)$_2$ nanocrystals were further examined by transmission electron microscopy (TEM). Figure 3(a) shows a typical TEM image of product 1 prepared at 160 ℃ for 24 h, using 2 M NaOH solution as alkaline agent. Figure 3(b) is a higher magnification TEM image that was obtained from a selected area of the Figure 3(a). Herein, it is presented that the large quantity of nanosheets with sizes in the range of 15 - 60 nm can be seen besides the rod-like nanocrystals with diameters of 5 - 15 nm and lengths up to about 70 nm, implying that β - Ni(OH)$_2$ nanosheets could be achieved by using this effective approach. The TEM image of product 3 is shown in Figure 3(c), which shows β - Ni(OH)$_2$ nanosheets have irregular morphologies with sizes ranging from 25 nm to 70 nm. When the alkaline concentration is lifted to 10 M, product 5 is mostly made up of sheet-like irregular hexagonal with size in the range of 35 - 140 nm and thickness up to about 15 nm [in Figure 3(d)]. The inset in Figure 3(d) shows the select area electron diffraction (SAED) pattern collected from the hexagonal nanosheet marked by the circle. The

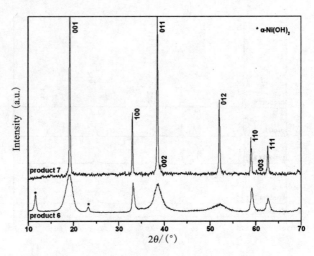

Figure 2 The XRD patterns of products 6 (2 M) and 7 (10 M) obtained at 140℃ and 180℃ for 24 h, respectively [* : α – Ni(OH)$_2$]

SAED pattern was obtained by directing the electron beam perpendicular to the nanosheet surface. We suggest that the surface of hexagonal β – Ni(OH)$_2$ nanosheets is the {0001} planes. It is also observed that the SAED pattern consists of many spots, which shows that the nanosheet is single crystal. All spots are identified as the diffraction from hexagonal β – Ni(OH)$_2$.

Further TEM investigations indicated that reaction temperature was also found to have influences on the morphologies of final products. Figure 4(a) shows TEM image of product 6 obtained under the same condition as product 1 except that the hydrothermal treatment temperature was 140℃. Compared to product 1 obtained at 160℃, the product 6 is mostly comprised of the large quantity of small nanorods with average diameters of 10 – 20nm and lengths up to about 100 nm. The TEM image of product 7 obtained at 180℃ is shown in Figure 4(b). It is obvious that the exclusive nanosheet with sizes ranging from 80 nm to 200 nm contains the coexistence of hexagonal and quasi-circular. Compared to product 5 obtained at 160℃, both the average size and crystallinity of nanosheets increased for product 7 obtained at 180℃, which agreed well with the XRD results. It is worthy to note that the average thickness of the nanosheets of product 7 obtained at 180℃ for 24 h is thinner than those of product 5 obtained at 160℃ for 24 h, nevertheless repeated TEM study confirmed that the exclusive nanosheets were prevalent.

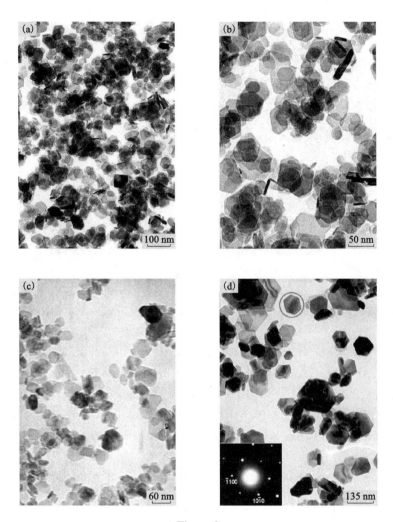

Figure 3

From (a) to (d): TEM images of the hexagonal β – Ni(OH)$_2$ nanosheets prepared by the hydrothermal method with different alkaline concentrations at 160℃ for 24 h, (a), (b) product 1; (c) product 3, (d) product 5. (a) Low-magnification and (b) high-magnification TEM images of product 1. Inset of (d) shows a SAED pattern of β – Ni(OH)$_2$ nanosheets of product 5 taken on an individual hexagonal nanosheet labeled by the circle.

It is well known that nickel oxide could be obtained via a thermal decomposition method using nickel hydroxide as the precursor. We demonstrated here that single-crystal nickel oxide nanorolls could be selectively synthesized by calcination of as-prepared nickel hydroxide nanosheets in air at 500℃ for 2 h. Figure 5 shows the

Figure 4

(a) The TEM images of product 6 (2 M) obtained at 140℃ for 24 h. (b) The TEM images of products 7, that is the same as products 5 except that the reaction temperature is raised to 180℃.

evolution of the X-ray diffractions (XRD) patterns (2θ scan) from product 1' to product 7'. All of the peaks in Figure 5 can be indexed to a face-centered cubic phase [space group: Fm3m (225)] NiO with lattice parameters a = 4.17 (JCPDS 78 – 0643). Diffraction peaks of Ni(OH)$_2$ or impurities cannot be observed, which indicates that Ni(OH)$_2$ has been converted completely to NiO by calcination at 500℃ for 2 h.

The size and morphology of as-prepared products of NiO were examined with transmission electron microscopy. Figure 6(a) shows the overviews of TEM image of product 1' of NiO nanocrystals obtained by calcination of product 1 in air at 500℃ for 2 h. The inset of Figure 6(a) is a higher magnification TEM image. Compared to product 1 of β – Ni(OH)$_2$ nanocrystals, the product 1' of NiO nanocrystals is also a coexistence of nanosheets and nanorods. However, with careful observation, the NiO nanosheets can be found to be composed of many pores with diameters of 2 – 5 nm. The TEM image of product 3' is shown in Figure 6(b). The inset is a higher magnification TEM image obtained from a selected area of the Figure 6(b). Herein, the NiO nanosheets with porous structures can be clearly observed. It is interesting that the β – Ni(OH)$_2$ nanosheets prepared at 160℃ for 24 h, using 10 M NaOH solution as alkaline agent, could be converted into NiO nanorolls by calcination at 500℃ for 2 h. A typical TEM image of product 5' is shown in Figure 6(c). It is clear that some of the

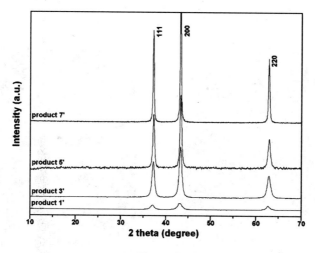

Figure 5

The evolution of XRD patterns of products 1′, 3′, 5′, 7′ obtained by calcination of as-prepared products 1, 3, 5, 7 in air at 500 ℃ for 2 h, respectively.

NiO nanocrystals ehibit roll-like structures with sizes in the range of 30 – 100 nm. The magnified image of an individual NiO nanoroll is shown in inset of Figure 6(c), and the morphology of nanoroll is clearly observed. Besides the NiO nanorolls, about 30% of nanocrystals appear porous structures. Compared to product 5′, the product 7′ obtained by calcination of product 7 at 500 ℃ for 2 h is mostly comprised of the NiO nanorolls with average sizes of 70nm, as shown in Figure 6(d). The inset of Figure 6(d) is the select area electron diffractions (SAED) pattern recorded on an individual NiO nanoroll, and the SAED pattern can be identified as the face-centered cubic NiO, which indicates that each NiO nanoroll is a single crystal.

On the basis of the experiment results, we proposed a probable formation mechanism of the NiO nanorolls helping to control the morphology of final products. We considered the face-centered cubic NiO nanorolls began with the pore at the core of the hexagonal β – Ni(OH)$_2$ nanosheets. With the calcination temperature increased, the hexagonal β – Ni(OH)$_2$ nanosheets gradually dehydrated and shrinked from the pore and finally the hexagonal β – Ni(OH)$_2$ nanosheets were converted into face-centered cubic NiO nanorolls. In our synthetic system, it was very important to prepare the thinner β – Ni(OH)$_2$ nanosheets with higher crystallization. The thicker β – Ni(OH)$_2$ nanosheet with poor crystallization were disadvantageous for us to get the NiO nanorolls. Furthermore, the temperature of calcination was also important in the formation of NiO

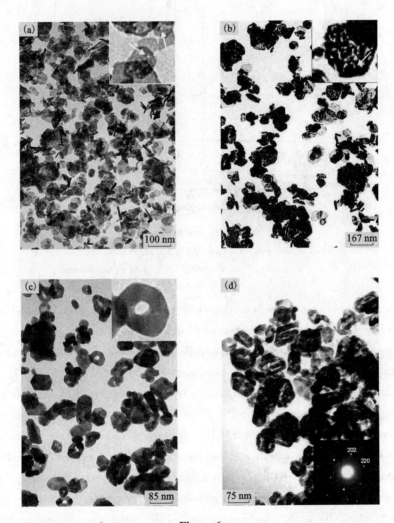

Figure 6

From (a) to (d): TEM images of NiO nanocrystallines (a) product 1′; (b) product 3′; (c) product 5′; (d) product 7′). The inset of (a) is a high magnification image of product 1′, that shows that the NiO nanosheets of product 1′ are composed of many small pores. The inset of (b) is the high magnification image taken from the section in (b). The inset of (c) is a typical TEM image of an individual NiO nanoroll obtained from (a) selected area of (c). Inset of (d) shows a SAED pattern of NiO nanorolls of product 7′ taken on an individual NiO nanoroll, which indicates that these NiO nanorolls are single-crystallized with face-centered cubic structure.

nanorolls. We could not obtain the NiO nanorolls at low temperature conditions, though the pure NiO could be prepared by calcinations of the β-Ni(OH)$_2$ nanosheets in air at 300℃, whereas the final products of NiO obtained at 300℃ for 2 h were mostly made up of nanosheets with porous structures. The NiO nanosheets were also observed by Zhu et al.[25] via thermal decomposition of β-Ni(OH)$_2$ nanosheets at 400℃ for 2 h.

The formation mechanism could surely be employed to synthesize other kinds of face-centered cubic metal oxides nanorolls by calcinations of the corresponding hexagonal metal hydroxides nanosheets, such as Co(OH)$_2$, Ca(OH)$_2$, Mg(OH)$_2$, and Cd(OH)$_2$. At present, we are synthesizing other roll-like metal oxide and sheet-like metal hydroxide nanoscale materials, and we are also paying considerable attention to their electrochemical properties and their potential applications in catalysts and electrode materials.

4 Conclusions

In summary, we have successfully synthesized single-crystal β-Ni(OH)$_2$ nanosheets in large quantities through a facile hydrothermal synthetic method, using sodium hydroxide as alkaline reagent under mild conditions. The experimental results demonstrated that it was possible to control the morphologies, phases and sizes of Ni(OH)$_2$ nanosheets by adjusting the reaction temperature and alkaline concentration. It is worthy of note that the single-crystal NiO nanorolls could be obtained by thermal decomposition of the single-crystal β-Ni(OH)$_2$ nanosheets in air at 500℃ for 2 h. This strategy is convenient and low cost for large scale synthesis of β-Ni(OH)$_2$ nanosheets and NiO nanorolls. Owing to the excellent chemical and physical properties of the nickel hydroxide (oxide), it is expected that the β-Ni(OH)$_2$ nanosheets and NiO nanorolls will exhibit some important applications in, e.g., alkaline batteries, sensors, semiconductors, catalytic, etc.

References

[1] S. R. Ovshinsky, M. A. Fetcenko, J. Ross. A nickel metal hydride battery for electric vehicles [J]. Science, 1993, 260(5105): 176-181.

[2] H. M. French, M. J. Henderson, A. R. Hillman, E. Vieil. Ion and solvent transfer discrimination at a nickel hydroxide film exposed to LiOH by combined electrochemical quartz crystal microbalance (EQCM) and probe beam deflection (PBD) techniques [J]. Journal of

Electroanalytical Chemistry, 2001, 500(1): 192 - 207.

[3] P. Oliva, J. Leonardi, J. F. Laurent, C. Delmas, J. J. Braconnier, M. Figlarz, F. Fievet, A. de Guibert. Review of the structure and the electrochemistry of nickel hydroxides and oxy-hydroxides [J]. Journal of Power sources, 1982, 8(2): 229 - 255.

[4] F. Bardé, M. R. Palacin, Y. Chabre, O. Isnard, J. -M. Tarascon. In situ neutron powder diffraction of a nickel hydroxide electrode [J]. Chemistry of Materials, 2004, 16(20): 3936 - 3948.

[5] G. N. Subbanna, P. áVishnu Kamath, On the existence of a nickel hydroxide phase which is neither α nor β [J]. Journal of Materials Chemistry, 1997, 7(11): 2293 - 2296.

[6] H. Bode, K. Dehmelt, J. Witte. Zur kenntnis der nickelhydroxidelektrode - I. Über das nickel (II)-hydroxidhydrat [J]. Electrochimica Acta, 1966, 11(8): 1079 - 1087.

[7] X. Han, X. Xie, C. Xu, D. Zhou, Y. Ma. Morphology and electrochemical performance of nano-scale nickel hydroxide prepared by supersonic coordination-precipitation method [J]. Optical Materials, 2003, 23(1): 465 - 470.

[8] X. Liu, L. Yu. Influence of nanosized $Ni(OH)_2$ addition on the electrochemical performance of nickel hydroxide electrode [J]. Journal of Power Sources, 2004, 128(2): 326 - 330.

[9] P. Jeevanandam, Y. Koltypin, A. Gedanken. Synthesis of nanosized α - nickel hydroxide by a sonochemical method [J]. Nano Letters, 2001, 1(5): 263 - 266.

[10] X. L. Li, J. F. Liu, Y. D. Li. Low-temperature conversion synthesis of $M(OH)_2$ (M = Ni, Co, Fe) nanoflakes and nanorods [J]. Materials Chemistry and Physics, 2003, 80(1): 222 - 227.

[11] X. Liu, L. Yu. Synthesis of nanosized nickel hydroxide by solid-state reaction at room temperature [J]. Materials Letters, 2004, 58(7): 1327 - 1330.

[12] C. B. Alcock, B. Li, J. W. Fergus, L. Wang. New electrochemical sensors for oxygen determination [J]. Solid State Ionics, 1992, 53: 39 - 43.

[13] D. Levin, J. Y. Ying. Oxidative dehydrogenation of propane by non-stoichiometric nickel molybdates [J]. Studies in Surface Science and Catalysis, 1997, 110: 367 - 373.

[14] A. C. Felic, F. Lama, M. Piacentini. Photoacoustic spectroscopy of diluted magnetic semiconductors [J]. Journal of Applied physics, 1996, 80: 6925 - 6930.

[15] P. Tomczyk, G. Mordarski, J. Obł. Kinetics of the oxygen electrode reaction in molten Li + Na carbonate eutectic: Part 5. Linear voltammetric and chronoamperometric data for the reduction processes at NiO monocrystalline electrodes [J]. Journal of Electroanalytical Chemistry, 1993, 353(1): 177 - 193.

[16] G. Boschloo, A. Hagfeldt. Spectroelectrochemistry of nanostructured NiO [J]. The Journal of Physical Chemistry B, 2001, 105(15): 3039 - 3044.

[17] D. Adler, J. Feinleib. Electrical and optical properties of narrow-band materials [J]. Physical Review B, 1970, 2(8): 3112.

[18] A. Agrawal, H. R. Habibi, R. K. Agrawal, J. P. Cronin, D. M. Roberts. Effect of

deposition pressure on the microstructure and electrochromic properties of electron-beam-evaporated nickel oxide films [J]. Thin Solid Films, 1992, 221(1): 239 -253.

[19] W. N. Wang, Y. Itoh, I. W. Lenggoro, K. Okuyama. Nickel and nickel oxide nanoparticles prepared from nickel nitrate hexahydrate by a low pressure spray pyrolysis [J]. Materials Science and Engineering: B, 2004, 111(1): 69 -76.

[20] K. Yoshimura. T. Miki, S. Tanemura, Nickel oxide electrochromic thin films prepared by reactive DC magnetron sputtering [J]. Japanese Journal of Applied Physics, 1995, 34(5A): 2440 -2416.

[21] C. Natarajan, H. Matsumoto, G. Nogami. Improvement in electrochromic stability of electrodeposited nickel hydroxide thin film [J]. Journal of the Electrochemical Society, 1997, 144(1): 121 -126.

[22] A. Šurca, B. Orel, B. Pihlar, P. Bukovec. Optical, spectroelectrochemical and structural properties of sol-gel derived Ni-oxide electrochromic film [J]. Journal of Electroanalytical Chemistry, 1996, 408(1): 83 -100.

[23] W. Wang, Y. Liu, C. Xu, C. Zheng, G. Wang. Synthesis of NiO nanorods by a novel simple precursor thermal decomposition approach [J]. Chemical Physics Letters, 2002, 362 (1): 119 -122.

[24] K. Matsui, B. K. Pradhan, T. Kyotani, A. Tomita. Formation of nickel oxide nanoribbons in the cavity of carbon nanotubes [J]. The Journal of Physical Chemistry B, 2001, 105(24): 5682 -5688.

[25] Z. H. Liang, Y. J. Zhu, X. L. Hu. β - nickel hydroxide nanosheets and their thermal decomposition to nickel oxide nanosheets [J]. The Journal of Physical Chemistry B, 2004, 108 (11): 3488 -3491.

☆ X. H. Liu, G. Z. Qiu, Z. Wang, X. G. Li. Rationally synthetic strategy: from nickel hydroxide nanosheets to nickel oxide nanorolls, Nanotechnology, 2005, 16(8): 1400 -1405.

Chapter IV

Rare-Earth Oxide and Oxysulfate Hollow Spheres

Controllable Fabrication and Optical Properties of Uniform Gadolinium Oxysulfate Hollow Spheres

Abstract: Uniform gadolinium oxysulfate ($Gd_2O_2SO_4$) hollow spheres were successfully fabricated by calcination of corresponding Gd – organic precursor obtained via a facile hydrothermal process. The $Gd_2O_2SO_4$ hollow spheres have a mean diameter of approximately 550 nm and shell thickness in the range of 30 – 70 nm. The sizes and morphologies of as-prepared $Gd_2O_2SO_4$ hollow spheres could be deliberately controlled by adjusting the experimental parameters. Eu – doped $Gd_2O_2SO_4$ hollow spheres have also been prepared for the property modification and practical applications. The structure, morphology, and properties of as-prepared products were characterized by XRD, TEM, HRTEM, SEM and fluorescence spectrophotometer. Excited with ultraviolet (UV) pump laser, successful downconversion (DC) could be achieved for Eu – doped $Gd_2O_2SO_4$ hollow spheres.

1 Introduction

Hollow spheres have been attracting great attention due to their superior properties such as high specific surface area, low density, high permeability and therefore show promising potential applications in various fields such as lithium batteries, catalysis and sensing, drug controlled release and delivery, and photonic building blocks, etc.[1-6] Plenty of chemical and physicochemical strategies such as Ostwald ripening,[7] Kirkendall diffusion,[8] chemically induced self-transformation,[9] template-assisted synthesis,[10] and spray drying followed by annealing[11, 12] have been applied for the design and controlled fabrication of various micro/nanospheres with hollow interiors. In particular, template-assisted synthesis has been demonstrated to be the most effective and versatile synthesis method. The templates can be generally divided into hard templates[13-15] and soft templates,[16-18] which have been widely used to fabricate hollow spheres. Among them, biomolecules, as attractive templates for the synthesis of metal and inorganic compound nanostructures, have been exploited for the precise control of the size and shape of various micro/nanomaterials, owing to the well-defined chemical and structural heterogeneity. [19-22] In spite of these pioneering work, it is still challenging and imperative to exploit an efficient but simple way for the synthesis of hollow spheres.

Rare-earth oxysulfate ($RE_2O_2SO_4$) have aroused great interest in recent years due to the unique magnetic[23] and luminescent properties[24,25] as well as significant applications in large volume oxygen storage. [26,27] $RE_2O_2SO_4$ is also an important matrix compound for luminescent rare-earth ions to fabricate downconversion (DC) or upconversion (UC) phosphors due to the incompletely filled 4f electron shell of rare-earth ions. [28-30] $RE_2O_2SO_4$ could be synthesized by the thermal decomposition of the corresponding hydrous sulfates ($RE_2(SO_4)_3 \cdot nH_2O$), layered rare-earth hydroxides intercalated with dodecyl sulfate (DS) ions, and layered rare-earth hydroxylsulfate ($RE_2(OH)_4SO_4 \cdot nH_2O$). [31-34] Nevertheless, the size and morphology of $RE_2O_2SO_4$ products prepared by the above methods are not well controlled and no particular shape or uniform size can be achieved. Recently, we reported a unique synthetic process to prepare $Y_2O_2SO_4$ hollow structure, which was mainly intended for the use of photoluminescence host materials. [35]

For $Gd_2O_2SO_4$, due to its unique half-filled outer electron shell in rare-earth elements, it is promising in combining magnetic and luminescent properties. A peculiar hollow structure further endows $Gd_2O_2SO_4$ to be a multifunctional nanomaterial for biomedical applications, such as magnetic resonance imaging, drug delivery host carriers and diagnostic analysis. This brings far-reaching impact than the availability of $Y_2O_2SO_4$ hollow structure. Herein, we present a facile biomolecule-assisted route to prepare uniform $Gd_2O_2SO_4$ hollow spheres via the calcination of corresponding spherical Gd – organic precursor obtained by using L – cysteine (Cys) as a biomolecule template. The size and morphology of as-prepared $Gd_2O_2SO_4$ hollow spheres can be deliberately controlled by adding different surfactants with varied amount. The formation process of the hollow spheres is elucidated by monitoring the species change and crystal structure evolution with elevated annealing temperature. Eu – doped $Gd_2O_2SO_4$ hollow spheres have also been successfully synthesized and the luminescence properties of as-prepared products were studied in detail.

2 Experimental Section

All the reagents are of analytical grade and used as starting materials without further purification.

2.1 Preparation of gadolinium oxysulfate hollow spheres. In a typical synthetic procedure of $Gd_2O_2SO_4$ hollow spheres, 1 mmol of hydrated gadolinium nitrate ($Gd(NO_3)_3 \cdot 6H_2O$), 2.0 mmol of L-Cys (L-cysteine) and 0.3 g of PVP (polyvinylpyrrolidone) were dissolved in 20 ml deionized water under vigorous magnetic stirring. Then the resulting solution was transferred into Teflon-lined stainless steel autoclave of 50 ml capacity and maintained at 140℃ for 24 h. After cooling to room temperature naturally, the resulting precipitates were washed with distilled water and anhydrous alcohol for several times, and dried at 50℃ for 4 h. Finally, the precursors can be transformed into $Gd_2O_2SO_4$ hollow spheres by calcination the Gd-organic precursor at 600℃ for 2 h. Furthermore, the 5% Eu-doped $Gd_2O_2SO_4$ hollow spheres were also obtained by similar process.

2.2 Characterization. X-ray diffraction patterns were recorded by a D/max2550 VB+ diffractometer with $Cu_{K\alpha}$ radiation ($\lambda = 0.15405$ nm) in the 2θ range of 10°–70°. The morphology of as-prepared products was examined by a field emission scanning electron microscopy (FE-SEM, Sirion 200) with an accelerating voltage of 15 kV. The energy dispersive spectrometer (EDS) was taken on the SEM. Transmission electron microscopy (TEM) images, selected area electron diffraction (SAED), high-resolution TEM (HRTEM) and the elemental mapping were recorded on a Tecnai G2 F20 transmission electron microscope with an accelerating voltage of 200 kV. Thermogravimetric and differential scanning calorimetry (TG-DSC) were carried out using a simultaneous thermal analysis (STA, NETZSCH STA 449C) in a temperature range of 25–650℃ at a heating rate of 10℃/min under an air flow. Fourier transform infrared (FT-IR) spectroscopy were obtained on a Nicolet Nexus 6700 instrument. Baird PS-6 Inductively Coupled Plasma Atomic Emission Spectrometer (ICP-AES) were used to evaluate the element content. The photoluminescence (PL) excitation and emission spectras were obtained on a fluorescence spectrophotometer (Hitachi F-4500) at room temperature.

3 Results and Discussion

X-ray diffraction (XRD) was carried out to illuminate the change and evolution of chemical composition and crystal structures. Figure 1(a) shows the XRD patterns of the Gd-organic precursor and corresponding $Gd_2O_2SO_4$ obtained by calcination at 600℃ for 2 h. No diffraction peaks were verified, indicating the initial precursor with broad featureless peaks was amorphous or non-crystalline. After annealing at 600℃ for 2 h, the precursor was converted into a single phase of $Gd_2O_2SO_4$, and no other impurity phases can be observed. All the reflections can be indexed to the literature values (JCPDS 29 - 0613). The crystal structure of $Gd_2O_2SO_4$ can commonly be depicted as an alternative stacking of $Gd_2O_2^{2+}$ and anion groups of sulfate (SO_4^{2-}) layers along the a-axis, as shown in the inset of Figure 1(a). The $Gd_2O_2^{2+}$ layer consists of [GdO_4] tetrahedra linked together by shared of edges. Every [SO_4] tetrahedra unit is coordinated with two Gd atoms.[36] The thermal decomposition behaviors of Gd - organic precursor was investigated in the temperature range of 25 -650℃ at a heating rate of 10℃/min in air. As shown in Figure 1(b), the weight loss in the temperature range from 25 to 200℃ was about 5.9% by mass, which can be associated with evaporation of physically absorbed water and organic residues on the Gd - organic precursor surfaces. The subsequent weight loss took place rapidly at a much higher temperature range. The continuous stages of weight loss in the range of 200 to 600℃ were 18.2% and 14.1% by mass. The tremendous decrease of weight can be attributed to the oxidation or combustion of the initial precursor and crystallization into $Gd_2O_2SO_4$. Corresponding to the two remarkable mass loss, the DSC curve of the sample displayed three major exothermal peaks in the gravimetric gain region centered at 274℃, 516℃ and 535℃ respectively. As shown in the TG curve, little weight change can be observed at temperatures higher than 600℃, suggesting that the relatively stable compound was obtained. Therefore, the hydrothermal products were annealed at 600℃ for the crystallization of $Gd_2O_2SO_4$ hollow spheres.

Fourier transform infrared (FT-IR) spectroscopy was employed to investigate the structural and functional group information of the Gd - organic precursors and powders calcined at different temperatures. As shown in the Figure 2(a), the FT-IR spectra reveal the existence of absorbed water, crystal water, hydroxyl groups (~3410 cm^{-1} and 1640 cm^{-1}), carbonates anions (~1580 cm^{-1} and 1415 cm^{-1}) and sulfates anions (~680 cm^{-1}) in the Gd - organic precursors.[37] The weak peaks at 2965 cm^{-1} and 2927 cm^{-1} are assigned to the —C—H vibration mode of —CH_2.[38] As the temperature of calcination increasing to 200℃ and 400℃, the broaden band at 3410 cm^{-1} becomes weaker and weaker while the small peak at 1640 cm^{-1} disappears at 400℃, which can be attributed to the removal of absorbed water and crystal water from the Gd - organic precursors. A similar behavior of carbonates absorption bands at 1580 cm^{-1} and 1415 cm^{-1} can be observed, suggesting that the carbonate anions in the precursors decomposed or vaporized with increasing the temperature. These results are

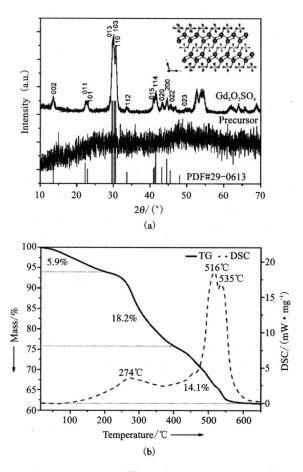

Figure 1

(a) XRD patterns of as-prepared Gd – organic precursor and corresponding $Gd_2O_2SO_4$. The inset depicts the corresponding crystal structure of $Gd_2O_2SO_4$. The Gd, O, and S species are represented by violet, red, and yellow balls, respectively. (b) TG and DSC curves of as-prepared Gd – organic precursor annealing from 25 to 650℃ at a heating rate of 10℃/min in air.

in good agreement with the results of TG-DSC analysis. Both the broaden band at 3410 cm^{-1} and carbonates absorption bands are significantly reduced at a higher calcination temperature of 600℃; while a broaden sulfates absorption band at 1130 cm^{-1} appears at 400℃ and splits into three narrow and sharp peaks at 1198 cm^{-1}, 1121 cm^{-1} and 1063 cm^{-1} at 600℃. The broaden sulfates absorption band at 680 cm^{-1} in the precursors becomes weaker and splits into three narrow and sharp peaks at 663 cm^{-1}, 621 cm^{-1} and 603 cm^{-1} in the final products. These two group of narrow and sharp sulfates absorption bands are assigned to the deformation vibrations and the asymmetric stretching of SO_4^{2-} anions, respectively.[39] These results are in accordance with those obtained from TG-DSC, XRD and ICP analysis, illustrating the composition and structural evolution of the $Gd_2O_2SO_4$ products.

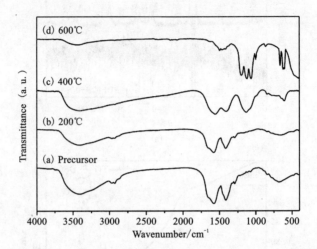

Figure 2 FT-IR spectras of the Gd – organic precursors (a) and the powders after calcinating at 200℃ (b), 400℃ (c) and 600℃ (d) for 2 h

Scanning electron microscopy (SEM) and transmission electron microscopy (TEM) were employed to characterize the sizes and morphologies of as-prepared products. Figure 3(a) and 3(b) show the spherical Gd – organic precursors with a smooth surface and an average size of approximately 650 nm. After calcinating the Gd – organic precursors at 600℃ for 2 h, as shown in Figure 3(c) and 3(d), $Gd_2O_2SO_4$ hollow spheres with relatively rough surfaces were obtained. The average diameter of the hollow spheres was estimated to be approximately 550 nm with a slightly decreasing in comparison with that of the precursor, implying the tendency to shrink after calcination. The strongly contrast between the dark periphery and greyish center of $Gd_2O_2SO_4$ spheres reveals that these spheres were of hollow structures, and the shell thickness was about 60 nm. The inset in Figure 3(d) represents a typical selected area electron diffraction (SAED) pattern, which can be indexed to the monoclinic structure of $Gd_2O_2SO_4$, consistent with the XRD result presented above. Figure 3(e) displays the corresponding high-resolution TEM (HRTEM) image, in which the lattice fringes were measured to be about 0.27 and 0.30 nm, corresponding to the interplanar spacings between (112) and (013) crystallographic planes, respectively. The current synthetic route could be adopted as a general strategy for the preparation of a series of rare-earth oxysulfate hollow spheres.

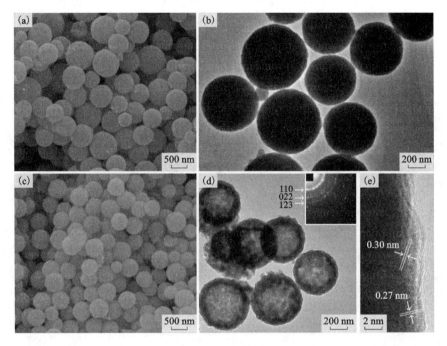

Figure 3

(a) SEM and (b) TEM images of spherical Gd – organic precursors. (c) SEM and (d) TEM images of $Gd_2O_2SO_4$ hollow spheres. The inset in (d) is corresponding SAED pattern; (e) HRTEM image of $Gd_2O_2SO_4$ hollow sphere

L – Cys, as a biomolecule template, possesses abundant functional groups, such as —SH, —NH_2, and —COOH, which can coordinate to Gd^{3+} and form homogeneous Gd – organic coordination compound on the basis of metal-ligand interaction in the solution,[3,4,35] leading to the formation of spherical precursors through aggregation and coagulation. Calcination temperature-depended formation mechanism of hollow spheres was investigated in detail. After calcinating the solid spherical Gd – organic precursors at 200 ℃ for 2 h, dark periphery and slightly greyish center of the spheres could be observed in the product. As the temperature of calcination increasing to 400 ℃, the area of the greyish center of the spheres increased. Finally, the spheres with apparent hollow structure were obtained at the calcination temperature of 600 ℃. We consider that the formation mechanism of the hollow spheres may involve two steps: First, a dense rigid shell formed in the surface of the solid spheres as the existence of the a large temperature gradient (ΔT) along the radial direction at initial stage of calcination.[40] Then in the subsequent calcination, as the adhesion force (Fa) surpasses the contraction force (Fc), the inner part shrinks outward, a hollow cavity in the center of the spheres were obtained.[41] The organic substances were all burnt out at 600 ℃ and the Gd – organic precursors were gradually crystallized into $Gd_2O_2SO_4$ at the peripheries, meanwhile, the hollow structure was formed.

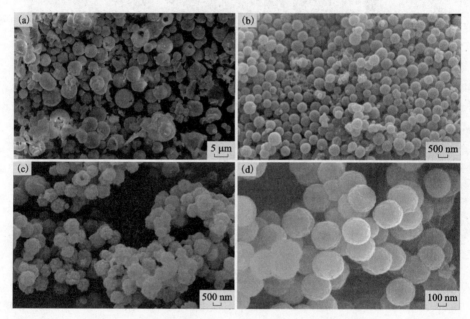

Figure 4　SEM images of as-prepared $Gd_2O_2SO_4$ hollow spheres obtained by using different surfactants

(a) without any surfactants; (b) 0.15 g PVP; (c) 0.6 g PVP; (d) 1 mmol CTAB

It was generally believed that surfactants played an important role in the control of morphologies and sizes of nanomaterials. Xia et al. studied the metal crystal growth kinetic process by using the different surfactants, such as cetyltrimethyl ammonium bromide (CTAB), polyvinylpyrrolidone (PVP), polyethylene glycol (PEG) and so on, to maneuver the surface energies and growth rates for different facets.[42, 43] The ratio between growth rates of different facets determined the growth habit of a nanocrystal, leading to the formation of different sizes and morphologies of nanomaterial. PVP had been widely introduced into the shape controlled synthesis of nanomaterials, such as and nanowires, nanosheets, nanospheres and so forth.[44, 45] In this paper, we have studied the effect of surfactants on the synthesis of $Gd_2O_2SO_4$

hollow spheres. Figure 4(a) shows the SEM image of as-prepared $Gd_2O_2SO_4$ without using any surfactants. Although $Gd_2O_2SO_4$ hollow spheres with broken shell could be observed in the absence of surfactant, the products had a tendency to agglomerate into block, and the size also reached the micrometer range. As shown in Figure 4(b), when 0.15 g PVP was introduced into the synthesis of $Gd_2O_2SO_4$ hollow spheres. The resulting product was mainly uniform spherical particles with smooth surfaces. However, with increasing the amount of PVP to 0.6 g [Figure 4(c)], the surface of hollow spheres became relatively rough. Thus, 0.3 g PVP was chosen as an optimal amount in the typical synthetic procedure of $Gd_2O_2SO_4$ hollow spheres. The exact mechanism of the function of PVP on the morphology and size of $Gd_2O_2SO_4$ hollow spheres is yet to be fully understood, it is believed that the strong interaction between the surfaces of Gd – organic precursors and PVP through coordination bonding with the O and N atoms of the pyrrolidone ring played a major role in determining the product morphology and size.[45] We also found that the CTAB as surfactant has similar functions in the synthesis of $Gd_2O_2SO_4$ hollow spheres, as shown in Figure 4(d). CTAB was used instead of PVP while other synthetic parameters were kept unchanged. The resulting product was mainly uniform $Gd_2O_2SO_4$ spheres with rough surface and the average size decreased to approximately 350 nm. These results further proved the indispensable role of surfactants in the formation of $Gd_2O_2SO_4$ hollow spheres.

The introduction of other rare-earth ions such as Eu^{3+} ions into $Gd_2O_2SO_4$ host lattice caused little change both on morphology and crystal phase. As shown in Figure 5, when 5% Eu^{3+} was added into the $Gd_2O_2SO_4$ host lattice, the morphology of final products, as well as the organic precursor, remained unchanged compared with the pure $Gd_2O_2SO_4$. The crystalline nature of $Gd_2O_2SO_4$: Eu hollow spheres was confirmed by HRTEM. Figure 5(c) clearly shows the lattice fringes were measured to be about 0.18 nm, corresponding to the interplanar spacing of (024) crystallographic plane, which fairly well agree with the standard interplanar spacing. The result of X – ray diffraction analyses further proved that the introduction of 5% Eu^{3+} ions into the $Gd_2O_2SO_4$ host lattice has no significant change on the crystal structure, owing to the same trivalent state and similar ionic radius of Gd^{3+} ions ($r_{(Gd^{3+})}$ = 0.0938 nm) and Eu^{3+} ions ($r_{(Eu^{3+})}$ =0.095 nm). The elemental maps of the 5% Eu – doped $Gd_2O_2SO_4$ hollow spheres obtained on TEM were displayed in Figure 5(d), which clearly demonstrates a homogeneous distribution of Gd, Eu, S and O elements. The energy dispersive spectrometer (EDS) spectrum reveals that the as-obtained product mainly contains Gd, Eu, S and O elements (Au signals were come from the spray-gold treatment to enhance

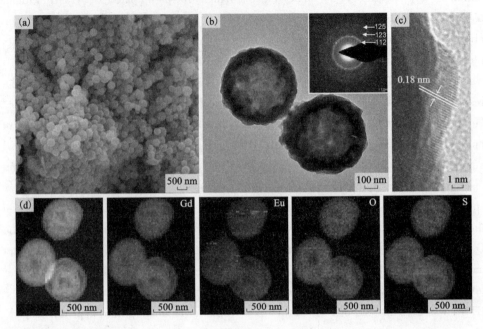

Figure 5

(a) SEM and (b) TEM images of as-prepared 5% Eu – doped $Gd_2O_2SO_4$ hollow spheres. Inset is the corresponding SAED pattern. (c) HRTEM image of 5% Eu – doped $Gd_2O_2SO_4$ hollow spheres; (d) STEM HAADF and elemental maps of Gd, Eu, O and S of 5% Eu – doped $Gd_2O_2SO_4$ hollow spheres

the electrical conductivity of the material). The molar ratio of Eu: Gd was about 3.23: 96.77, which was consistent with the ratio of used reagents in synthetic process. The above results confirm that successful doping could be achieved through current synthetic strategy.

The excitation spectra of the 5% Eu – doped $Gd_2O_2SO_4$ phosphors was recorded in the wavelength range of 200 – 500 nm at room temperature, as shown in Figure 6(a), one can see that a broad absorption band with a maximum at around 270 nm exists, which is resulted from the typical $^8S_{7/2} \rightarrow {}^6I_{7/2}$ transition of the Gd^{3+} ions.[46] Furthermore, other two comparatively weak peaks centered at 394 nm and 465 nm can be respectively assigned to the typical f-f transition of Eu^{3+} ions, corresponding to the $^7F_0 \rightarrow {}^5L_6$ and $^7F_0 \rightarrow {}^5D_2$ transitions.[37] Excitation spectra of the 5% Eu – doped $Gd_2O_2SO_4$ phosphors was taken by monitoring the wavelength of 617 nm.

The emission spectrums of 5% Eu – doped $Gd_2O_2SO_4$ under 270 nm light excitation shown in Figure 6(b) demonstrate the characteristic $^5D_0 \rightarrow {}^7F_J$ (J = 1, 2, 3, 4) and

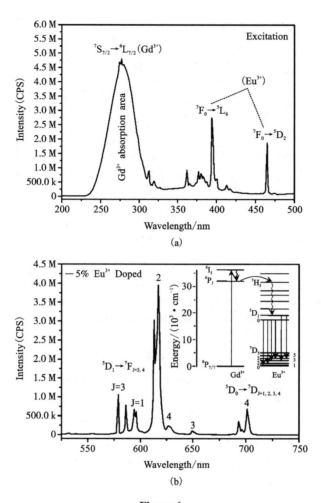

Figure 6

(a) Excitation spectrum of 5% Eu-doped $Gd_2O_2SO_4$ hollow spheres. (b) Emission spectrum of 5% Eu-doped $Gd_2O_2SO_4$ hollow spheres. Inset is corresponding scheme of the energy level and energy transition of 5% Eu-doped $Gd_2O_2SO_4$ hollow spheres

$^5D_1 \rightarrow ^7F_J$ (J = 3, 4) transitions of Eu^{3+} ions, indicating the effective cooperative luminescence between Gd^{3+} and Eu^{3+}. The strongest emission which splits into two peaks centered at 613 nm and 617 nm can be attributed to the forced electric dipole $^5D_0 \rightarrow ^7F_2$ transition of Eu^{3+} ions. All the other emission peaks are easily assigned to the $^5D_1 \rightarrow ^7F_3$ (579, 586 nm), $^5D_0 \rightarrow ^7F_1$ (594, 596 nm), $^5D_1 \rightarrow ^7F_4$ (627 nm), $^5D_0 \rightarrow ^7F_3$ (649 nm), $^5D_0 \rightarrow ^7F_4$ (693, 701 nm) transition of Eu^{3+} ions, respectively.[47-50] In

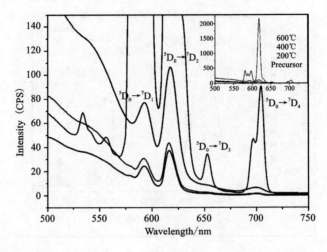

Figure 7

Partially magnified emission spectrums of the Gd – organic precursors and the powders after calcinating at 200 ℃, 400 ℃ and 600 ℃ for 2 h under a laser with wavelength of 270 nm on a Hitachi F – 2500 at room temperature. The inset depicts the corresponding full emission spectra.

this process, trivalent Gd^{3+} ions, as sensitizer, absorb ultraviolet excitation light and subsequently transfer energy to the neighboring Eu^{3+} ions act as activator, resulting in the overall red emission of Eu^{3+}. The detailed energy level and transfer scheme was shown in inset of Figure 6(b). Upon excitation by 270 nm, Gd^{3+} ions will be excited into $^6I_{7/2}$ state from ground state in the first step and then fast relax from this high excitation state to the ^6P_J state. Secondly, the Gd^{3+} ions in the ^6P_J state can easily transfer the excitation energy to the Eu^{3+} ions (^5H_J) because of the energy level match between ^6P_J state and ^5H_J state.[50] Fast non-radiative relaxation from ^5H_J state to the $^5D_1$ or $^5D_0$ state occurs. The electron on high excitation $^5D_1$ and $^5D_0$ states further relaxes radiatively to the ground-state to generate different wavelength visible emissions. Furthermore, as shown in Figure 7, the emission intensity of the Gd – organic precursors with poor crystallinity can be negligible comparing to the final products with high crystallinity.

4 Conclusion

In summary, uniform gadolinium oxysulfate hollow spheres have been successfully achieved by a facile hydrothermal process combining with a calcination of Gd – organic precursors. Based on the experimental results, we found both the amount and the type of surfactants play an important role for the formation of $Gd_2O_2SO_4$ hollow spheres. Eu – doped $Gd_2O_2SO_4$ hollow spheres have also been successfully synthesized with little change both on size and crystal phase. Optical properties reveal that the Eu – doped $Gd_2O_2SO_4$ hollow spheres can be used to down-convert UV light to visible light under the UV excitation. It is expected that the uniform $Gd_2O_2SO_4$ hollow spheres have potential applications in various research field, such as large volume oxygen storage, drug delivery host carriers, optical/display devices and luminescence probes.

References

[1] X. W. Lou, L. A. Archer, Z. Yang. Hollow micro-/nanostructures: synthesis and applications [J]. Advanced Materials, 2008, 20(21): 3987 – 4019.

[2] G. Tian, Z. J. Gu, X. X. Liu, L. J. Zhou, W. Y. Yin, L. Yan, S. Jin, W. L. Ren, G. M. Xing, S. J. Li. Facile fabrication of rare-earth-doped Gd_2O_3 hollow spheres with upconversion luminescence, magnetic resonance, and drug delivery properties [J]. The Journal of Physical Chemistry C, 2011, 115(48): 23790 – 23796.

[3] X. Liu, D. Zhang, J. Jiang, N. Zhang, R. Ma, H. Zeng, B. Jia, S. Zhang, G. Qiu. General synthetic strategy for high-yield and uniform rare-earth oxysulfate ($RE_2O_2SO_4$, RE = La, Pr, Nd, Sm, Eu, Gd, Tb, Dy, Y, Ho, and Yb) hollow spheres [J]. RSC Advance, 2012, 2(25): 9362 – 9365.

[4] G. Chen, W. Ma, X. Liu, S. Liang, G. Qiu, R. Ma. Controlled fabrication andoptical properties of uniform CeO_2 hollow spheres [J]. RSC Advances, 2013, 3(11): 3544 – 3547.

[5] B. Wang, W. Meng, M. Bi, Y. Ni, Q. Cai, J. Wang. Uniform magnesium silicate hollow spheres as high drug-loading nanocarriers for cancer therapy with low systemic toxicity [J]. Dalton Transactions, 2013, 42(24): 8918 – 8925.

[6] H. Zhang, L. Zhou, O. Noonan, D. J. Martin, A. K. Whittaker, C. Yu. Tailoring the void size of iron oxide@ carbon yolk-shell structure for optimized lithium storage [J]. Advanced Functional Materials, 2014, 24(27): 4337 – 4342.

[7] H. C. Zeng. Synthetic architecture of interior space for inorganic nanostructures [J]. Journal of Materials Chemistry, 2006, 16(7): 649 – 662.

[8] F. Zhang, Y. Shi, X. Sun, D. Zhao, G. D. Stucky. Formation of hollow upconversion rare-earth fluoride nanospheres: nanoscale kirkendall effect during ion exchange [J]. Chemistry of Materials, 2009, 21(21): 5237 – 5243.

[9] J. G. Yu, H. Guo, S. A. Davis, S. Mann. Fabrication of hollow inorganic microspheres by chemically induced self-transformation [J]. Advanced Functional Materials, 2006, 16(15): 2035 – 2041.

[10] J. Zhang, Y. Li, F. F. An, X. Zhang, X. Chen, C. S. Lee. Preparation and size control of sub - 100 nm pure nanodrugs [J]. Nano Letters, 2014, 15(1): 313 - 318.

[11] L. Zhou, H. Xu, H. Zhang, J. Yang, S. B. Hartono, K. Qian, J. Zou, C. Yu. Cheap and scalable synthesis of α-Fe_2O_3 multi-shelled hollow spheres as high-performance anode materials for lithium ion batteries [J]. Chemical Communications, 2013, 49: 8695 - 8697.

[12] Z. Padashbarmchi, A. H. Hamidian, H. Zhang, L. Zhou, N. Khorasani, M. Kazemzad, C. Yu. A systematic study on the synthesis of α-Fe_2O_3 multi-shelled hollow spheres [J]. RSC Advances, 2015, 5(14): 10304 - 10309.

[13] M. M. Titirici, M. Antonietti, A. Thomas. A generalized synthesis of metal oxide hollow spheres using a hydrothermal approach [J]. Chemistry of Materials, 2006, 18(16): 3808 - 3812.

[14] V. Salgueiriño-Maceira, M. Spasova, M. Farle. Water-stable, magnetic silica-cobalt/cobalt oxide-silica multishell submicrometer spheres [J]. Advanced Functional Materials, 2005, 15(6): 1036 - 1040.

[15] Y. Xia, R. Mokaya. Hollow spheres of crystalline porous metal oxides: A generalized synthesis route via nanocasting with mesoporous carbon hollow shells [J]. Journal of Materials Chemistry, 2005, 15(30): 3126 - 3131.

[16] C. E. Fowler, D. Khushalani, S. Mann. Interfacial synthesis of hollow microspheres of mesostructured silica [J]. Chemical Communications, 2001, 19: 2028 - 2029.

[17] H. T. Schmidt, A. E. Ostafin. Liposome directed growth of calcium phosphate nanoshells [J]. Advanced Materials, 2002, 14(7): 532 - 535.

[18] H. P. Hentze, S. R. Raghavan, C. A. McKelvey, E. W. Kaler. Silica hollow spheres by templating of catanionic vesicles [J]. Langmuir, 2003, 19(4): 1069 - 1074.

[19] M. Knez, A. M. Bittner, F. Boes, C. Wege, H. Jeske, E. Maiβ, K. Kern. Biotemplate synthesis of 3 - nm nickel and cobalt nanowires [J]. Nano Letters, 2003, 3(8): 1079 - 1082.

[20] Q. Lu, F. Gao, S. Komarneni. Biomolecule-assisted synthesis of highly ordered snowflakelike structures of bismuth sulfide nanorods [J]. Journal of the American Chemical Society, 2004, 126(1): 54 - 55.

[21] P. Zhao, T. Huang, K. Huang. Fabrication of indium sulfide hollow spheres and their conversion to indium oxide hollow spheres consisting ofmultipore nanoflakes [J]. The Journal of Physical Chemistry C, 2007, 111(35): 12890 - 12897.

[22] B. Li, Y. Xie, Y. Xue. Controllable synthesis of CuS nanostructures from self-assembled precursors with biomolecule assistance [J]. The Journal of Physical Chemistry C, 2007, 111(33): 12181 - 12187.

[23] W. Paul. Magnetism and magnetic phase diagram of $Gd_2O_2SO_4$ I. Experiments [J]. Journal of Magnetism and Magnetic Materials, 1990, 87(1): 23 - 28.

[24] L. Song, P. Du, Q. Jiang, H. Cao, J. Xiong. Synthesis and luminescence of high-brightness $Gd_2O_2SO_4$: Tb^{3+} nanopieces and the enhanced luminescence by alkali metal ions co-doping

[J]. Journal of Luminescence, 2014, 150: 50 - 54.

[25] X. Wei, W. Wang, K. Chen. Preparation and characterization of ZnS: Tb, Gd and ZnS: Er, Yb, Gd nanoparticles for bimodal magnetic-fluorescent imaging [J]. Dalton Transactions, 2013, 42(5): 1752 - 1759.

[26] M. Machida, K. Kawamura, K. Ito. Novel oxygen storage mechanism based onredox of sulfur in lanthanum oxysulfate/oxysulfide [J]. Chemical Communications, 2004, 6: 662 - 663.

[27] M. Machida, T. Kawano, M. Eto, D. Zhang, K. Ikeue. Ln dependence of the large-capacity oxygen storage/release property of Ln oxysulfate/oxysulfide systems [J]. Chemistry of Materials, 2007, 19(4): 954 - 960.

[28] F. Auzel. Upconversion and anti-stokes processes with f and d ions in solids [J]. Chemical Reviews, 2004, 104(1): 139 - 174.

[29] F. Wang, X. Liu. Recent advances in the chemistry of lanthanide-doped upconversion nanocrystals [J]. Chemical Society Reviews, 2009, 38(4): 976 - 989.

[30] J. Lian, X. Sun, J. G. Li, B. Xiao, K. Duan. Characterization and optical properties of (Gd_{1-x}, Pr_x)$_2O_2S$ nano-phosphors synthesized using a novel co-precipitation method [J]. Materials Chemistry and Physics, 2010, 122(2): 354 - 361.

[31] T. Kijima, T. Shinbori, M. Sekita, M. Uota, G. Sakai. Abnormally enhanced Eu^{3+} emission in $Y_2O_2SO_4$: Eu^{3+} inherited from their precursory dodecylsulfate-templated concentric-layered nanostructure [J]. Journal of Luminescence, 2008, 128(3): 311 - 316.

[32] T. Kijima, T. Isayama, M. Sekita, M. Uota, G. Sakai. Emission properties of Tb^{3+} in $Y_2O_2SO_4$ derived from their precursory dodecylsulfate-templated concentric-and straight-layered nanostructures [J]. Journal of Alloys and Compounds, 2009, 485(1): 730 - 733.

[33] M. Machida, K. Kawamura, K. Ito, K. Ikeue. Large-capacity oxygen storage by lanthanide oxysulfate/oxysulfide systems [J]. Chemistry of Materials, 2005, 17(6): 1487 - 1492.

[34] J. Liang, R. Ma, F. Geng, Y. Ebina, T. Sasaki. $Ln_2(OH)_4SO_4 \cdot nH_2O$ (Ln = Pr to Tb; n ~ 2): A new family of layered rare-earth hydroxides rigidly pillared by sulfate ions [J]. Chemistry of Materials, 2010, 22(21): 6001 - 6007.

[35] G. Chen, F. Chen, X. Liu, W. Ma, H. Luo, J. Li, R. Ma, G. Qiu. Hollow spherical rare-earth-doped yttrium oxysulfate: A novel structure for upconversion [J]. Nano Research, 2014, 7(8): 1093 - 1102.

[36] S. Zhukov, A. Yatsenko, V. Chernyshev, V. Trunov, E. Tserkovnaya, O. Antson, J. Hölsä, P. Baulés. Structural study of lanthanum oxysulfate ($LaO)_2SO_4$ [J]. Materials Research Bulletin, 1997, 32(1): 43 - 50.

[37] J. Lian, X. Sun, Z. Liu, J. Yu, X. Li. Synthesis and optical properties of (Gd_{1-x}, Eu_x)$_2O_2SO_4$ nano-phosphors by a novel co-precipitation method [J]. Materials Research Bulletin, 2009, 44(9): 1822 - 1827.

[38] Y. Ru, Q. Jie, L. Min, G. Liu. Synthesis of yttrium aluminum garnet (YAG) powder by homogeneous precipitation combined with supercritical carbon dioxide or ethanol fluid drying

[J]. Journal of the European Ceramic Society, 2008, 28(15): 2903-2914.

[39] R. S. Jayasree, V. P. Mahadevan Pillai, V. U. Nayar, I. Odnevall, G. Keresztury. Raman and infrared spectral analysis of corrosion products on zinc $NaZn_4Cl(OH)_6SO_4 \cdot 6H_2O$ and $Zn_4Cl_2(OH)_4SO_4 \cdot 5H_2O$ [J]. Materials Chemistry and Physics, 2006, 99(2): 474-478.

[40] J. Guan, F. Mou, Z. Sun, W. Shi. Preparation of hollow spheres with controllable interior structures by heterogeneous contraction [J]. Chemical Communications, 2010, 46: 6605-6607.

[41] L. Z, D. Zhao, X. W. Lou. Double-shelled $CoMn_2O_4$ hollow microcubes as high-capacity anodes for lithium-ion batteries [J]. Advanced Materials, 2012, 24(6): 745-748.

[42] Y. Xia, Y. Xiong, B. Lim, S. E. Skrabalak. Shape-controlled synthesis of metal nanocrystals: simple chemistry meets complex physics? [J]. Angewandte Chemie International Edition, 2009, 48(1): 60-103.

[43] M. Jin, G. He, H. Zhang, J. Zeng, Z. Xie, Y. Xia. Shape-controlled synthesis of copper nanocrystals in an aqueous solution with glucose as a reducing agent and hexadecylamine as a capping agent [J]. Angewandte Chemie International Edition, 2011, 50(45): 10560-10564.

[44] J. Wang, X. Wang, Q. Peng, Y. Li. Synthesis and characterization of bismuth single-crystalline nanowires and nanospheres [J]. Inorganic Chemistry, 2004, 43(23): 7552-7556.

[45] F. Zhou, X. Zhao, H. Xu, C. Yuan. CeO_2 spherical crystallites: synthesis, formation mechanism, size control, and electrochemical property study [J]. The Journal of Physical Chemistry C, 2007, 111(4): 1651-1657.

[46] Y. Liu, D. Tu, H. Zhu, R. Li, W. Luo, X. Chen. A strategy to achieve efficient dual-mode luminescence of Eu^{3+} in lanthanides doped multifunctional $NaGdF_4$ nanocrystals [J]. Advanced Materials, 2010, 22(30): 3266-3271.

[47] J. B. Lian, X. D. Sun, X. D. Li. Synthesis, characterization and photoluminescence properties of $(Gd_{1-x}, Eu_x)_2O_2SO_4$ sub-microphosphors by homogeneous precipitation method [J]. Materials Chemistry and Physics, 2011, 125(3): 479-484.

[48] Y. Song, H. You, Y. Huang, M. Yang, Y. Zheng, L. Zhang, N. Guo. Highly uniform and monodisperse Gd_2O_2S: Ln^{3+} (Ln = Eu, Tb) submicrospheres: solvothermal synthesis and luminescence properties [J]. Inorganic Chemistry, 2010, 49(24): 11499-11504.

[49] Q. Chen, Y. Shi, L. An, S. Wang, J. Chen, J. Shi. A novel co-precipitation synthesis of a new phosphor Lu_2O_3: Eu^{3+} [J]. Journal of the European Ceramic Society, 2007, 27(1): 191-197.

[50] R. T. Wegh, H. Donker, K. D. Oskam, A. Meijerink. Visible quantum cutting in $LiGdF_4$: Eu^{3+} through downconversion [J]. Science, 1999, 283(5402): 663-666.

☆ F. S. Chen, G. Chen, T. Liu, N. Zhang, X. H. Liu, H. M. Luo, J. H. Li, L. M. Chen, R. Z. Ma, G. Z. Qiu. Controllable Fabrication and Optical Properties of Uniform Gadolinium Oxysulfate Hollow Spheres [J]. Scientific Reports, 2015, DOI: 10.1038/srep17934.

Hollow Spherical Rare-Earth-Doped Yttrium Oxysulfate: A Novel Structure for Upconversion

Abstract: A facile biomolecule-assisted hydrothermal route followed by calcinations has been employed for the preparation of monoclinic yttrium with other rare earth ions (Yb^{3+}, Eu^{3+} or Er^{3+}). The formation of hollow spheres may involve Ostwald ripening. The resulting hybrid materials were used for upconversion application. The host crystal structure allows the easy co-doping of two different rare-earth metal ions without significantly changing the host lattice. The luminescent properties were affected by the ratio and concentration of dopant rare earth metal ions due to energy transfer and the symmetry of crystal field. The type of luminescent center and the crystallinity of samples were also shown to have a significant influences on the optical properties of as-prepared products.

1 Introduction

Rare-earth ions exhibit unique luminescent properties, including the ability of converting the near infrared long-wavelength excitation radiation into shorter visible wavelengths by photon upconversion (UC) due to the incompletely filled 4f electron shell.[1-4] Compared to organic fluorophores and semiconducting nanocrystals, rare-earth-based nanomaterials display higher photochemical stability, sharp emission bandwidths, large anti-Stokes shifts and Stokes shifts (up to 500 nm) that separate discrete emission peaks from the ultraviolet/infrared excitation. Generally, doping is a widely applied strategy in materials design that involves incorporating atoms or ions into host lattices to yield hybrid materials with desirable properties and functions.[5] In recent years, rare-earth-doped UC materials have been increasingly employed as luminescent optical labels in biological assays and medical imaging and have become promising alternatives to organic fluorophores and quantum dots, etc.[6-9] Nonetheless, the reported rare-earth-doped systems have mainly been based on the crystalline $NaREF_4$,[10-12] $LiYF_4$,[13] REF_3[14] (RE is short for rare-earth metal). Liu and co-workers have reported their excellent research on the simultaneous phase and size control of the NaREF$_4$ crystal structure and the UC performance of nanocrystals doped with other rare-earth ions.[5, 10] A similar NaGdF$_4$: Yb, Tm@ NaGdF$_4$: Eu core/shell structure was also achieved for both UC and downconversion (DC) applications.[15] However, there have been few reports on rare-earth-based other structures as alternative materials for the UC application.

Here we describe the use of rare-earth oxysulfate (RE$_2$O$_2$SO$_4$) materials for UC application. Both sensitizer and emitter ions were doped into the host crystal structure in order to achieve excellent optical emission properties of the resulting products. RE$_2$O$_2$SO$_4$ is an important matrix compound for luminescence materials, and it has attracted increasing attention in recent years due to its unique luminescent and specific magnetic properties as well as significant applications in large volume oxygen storage.[16-22] RE$_2$O$_2$SO$_4$ can been obtained by a high-temperature treatment of the corresponding hydrous sulfates [RE$_2$(SO$_4$)$_3 \cdot n$H$_2$O] or layered hydroxides intercalated with dodecyl sulfate (DS) ions.[20-22] Layered rare-earth hydroxysulfate [RE$_2$(OH)$_4$SO$_4 \cdot n$H$_2$O] can also be used for the preparation of RE$_2$O$_2$SO$_4$ by a thermal-induced dehydroxylation process.[23] Nevertheless, most of RE$_2$O$_2$SO$_4$ products prepared by the above methods are bulk materials with irregular morphology. In the present work, uniform rare-earth-doped yttrium oxysulfate (Y$_2$O$_2$SO$_4$) hollow spheres were successfully synthesized and were introduced for the first time for UC. Due to their hollow spherical structure with its interior space, Y$_2$O$_2$SO$_4$ hollow spheres are expected to provide additional advantages by virtue of their low density, high specific surface area, and high surface-to-volume ratio facilitating technological applications in various fields.[24]

It has been reported that replacing the relatively small Y^{3+} ion in NaREF$_4$ with the larger Gd^{3+} ion cause a phase change. Compared with NaREF$_4$-based crystal structure, RE$_2$O$_2$SO$_4$ shows a better structural compatibility when replacing Y^{3+} ion with other rare-earth metal ions. Two different rare-earth metal ions can be easily co-doped into the crystal structure of RE$_2$O$_2$SO$_4$ without changing the host structure or significantly changing the lattice constant. The basic requirements for UC process were studied by doping different luminescent centers such as Eu^{3+} and Er^{3+} ions and also by comparison of luminescent features between well-crystalline products and amorphous precursors. The relationship between luminescent properties and the ratio and concentration of dopant RE ions was studied in detail. The RE$_2$O$_2$SO$_4$ material system offers promising opportunities to afford hybrid materials with desirable optical properties by the multi-doping process.

2 Experimental Section

Y(NO$_3$)$_3$ · 6H$_2$O of analytical grade from Sinopharm Chemical Reagent Co., Ltd and Yb(NO$_3$)$_3$ · 5H$_2$O, Er(NO$_3$)$_3$ · 5H$_2$O, Eu(NO$_3$)$_3$ · 5H$_2$O of analytical grade from Aladdin Chemical Reagent Co., Ltd were used as starting materials without further purification.

2.1 Preparation of rare-earth-doped Y$_2$O$_2$SO$_4$ hollow spheres. In a typical procedure, 1 mmol (based on the metal) of RE(NO$_3$)$_3$ · nH$_2$O (Y:Yb:Eu = 90:9:1) was added into a 50 mL beaker with 25 mL de-ionized water, and then 2 mmol L-cysteine and 0.3 g polyvinylpyrrolidone (PVP) were added into the solution under continuous stirring for 5 min. Then the solution was transferred into a 50 mL Teflon-lined stainless steel autoclave and held at a temperature of 140–180℃ for 24 h. Subsequently, the system was allowed to cool down to room temperature naturally. The resulting precipitate was collected by filtration and washed several times with absolute ethanol and distilled water in sequence. The final product was dried in a vacuum box at 50℃ for 12 h. As prepared rare-earth-based precursor were calcinated to produce Y$_2$O$_2$SO$_4$: Yb, Eu/Er hollow spheres in air at 650℃ for 2 h. The corresponding Y$_2$O$_2$SO$_4$: Yb, Er materials were prepared in the same way.

2.2 Characterization. The crystal structures of obtained specimens were characterized on a X-ray diffractometer (XRD, D/max2550 VB+) with Cu K_α radiation (λ = 1.5418 Å). The particle sizes and morphologies of the products were characterized by a field-emission scanning electron microscope (FE-SEM, Sirion 200) and transmission electron microscope (TEM, Tecnai G2 F20). High resolution transmission electron microscope (HRTEM, JEOL 2010) was employed to check the lattice features. Energy dispersive spectrometty (EDS) and element mapping were obtained using the JEOL 2010 instrument. The thermal behavior of precursor compound to Y$_2$O$_2$SO$_4$: Yb, Eu/Er crystals was evaluated by thermogravimetric analysis (TG), using a simultaneous thermal analysis (STA, NETZSCH STA 449C). The photoluminescence (PL) properties were measured on a fluorescence spectrometer (FluoroMax® -4 fluorescence spectrometer) and a Hitachi F-4500 fluorescence spectrophotometer at room temperature.

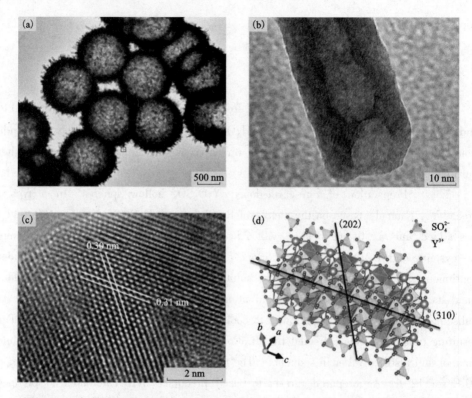

Figure 1 (a) TEM image of $Y_2O_2SO_4$: Yb, Eu hollow spheres and (b) magnified TEM image of a single nanotube on the surface of hollow spheres; (c) HRTEM image; (d) crystal structure of $Y_2O_2SO_4$

3 Results and Discussion

Typical TEM images of the $Y_2O_2SO_4$: Yb, Eu hollow spheres are shown in Figure 1. Figure 1 (a) shows the $Y_2O_2SO_4$: Yb, Eu hollow spheres with average diameter of 1 μm. The sharp contrast between the dark periphery and greyish center of these spheres indicates apparent hollow feature. Figure 1(b) is a higher magnification image of the area marked with a red square in Figure 1(a). Careful observation shows that the shells of the hollow spheres are composed of small tubular nanostructures. The HRTEM image shown in Figure 1(c) confirms the crystalline nature of $Y_2O_2SO_4$: Yb, Eu hollow spheres. The measured interplanar spacing are about 0.30 nm and 0.31 nm, corresponding to the (202) and (310) plane, respectively. However, the standard interplanar spacing of the (202) and (310) planes of undoped $Y_2O_2SO_4$ are 2.93 Å and 2.97 Å, respectively. The introduction of other atoms such as Yb, Eu or Er in a large amount (10 mol% in total) cause a change of about 3% in unit-cell parameters in host structure. The phase of uncalcined rare-earth-based precursor and calcined products of $Y_2O_2SO_4$: Yb, Eu were identified by powder XRD, which clearly indicates that precursor is amorphous, and no peaks can be observed in the XRD pattern. Thus, a calcination temperature of 650℃ was selected for the crystallization of hollow spheres based on the thermal behaviour analysis of rare-earth-based precursor. After calcination, all the reflections of products can be indexed to the monoclinic structure of $Y_2O_2SO_4$ (JCPDS 53 – 0168) showing that a hybrid rare-earth-doped $Y_2O_2SO_4$ can be achieved by our synthetic strategy. The monoclinic $RE_2O_2SO_4$ materials adopt the same space group C2/c (No. 15) with similar unit-cell parameters: RE and O atoms occupy the 8f sites and S atoms occupy the 4e sites. The crystal structures of $Y_2O_2SO_4$ is commonly described in terms of alternative stacking of a $Y_2O_2^{2+}$ layer and a layer of sulfate (SO_4^{2-}) anion along the a-axis, as shown in Figure 1(d). The $Y_2O_2^{2+}$ layer consists of YO_4 tetrahedra linked together by shared of edges. Every SO_4^{2-} anion in $Y_2O_2SO_4$ is coordinated to two Y^{3+} ions.[16, 25] The lattice parameters a, b, and c will be inevitably affected by doping with a second ion with different radius. As is well known, the choice of the host lattice determines the distance between the dopant ions, their relative spatial position, their coordination numbers, and the type of anions surrounding the dopant. The properties of the host lattice and its interaction with the dopant ions therefore have a strong influence on the UC process. The layered crystal structure offers relatively large lattice constants which can facilitate the further promising application in PL by means of doping.

The functional groups, such as —SH, —NH_2, —COOH of the L – cysteine present in our synthesis mixture have a strong tendency to coordinate with rare-earth ions. On the basis of such metal-ligand interactions, the coordinated precursor aggregates into spherical products driven by the minimization of interfacial energy under hydrothermal conditions. The dependence of the formation of hollow spheres was investigated. The hydrothermal product obtained at 140℃ for 8 h was composed of solid spheres. After reaction for 12 h, core-shell or yolk-shell structures could be observed in products. Finally, the spheres turned into hollow structure with further increase of reaction duration. We consider that the formation mechanism of the hollow spheres may involve the Ostwald ripening process, which is consistent with that has been discussed in detail in our previous work.[26-28] Our synthetic route can be adopted as general strategy for the preparation of rare-earth oxysulfate hollow spheres.[26] More recently, our research extended its use to the synthesis of transition metal sulfide (NiS_2) hollow spheres based on the Ostwald ripening process.[28] $RE_2O_2SO_4$ hollow spheres could be obtained by calcination of corresponding amorphous precursors. It is worth noting that not only can the hollow structure be obtained through calcination of the precursor, but also in addition the superficial microstructure of the hollow spheres after calcination can be varied by changing hydrothermal conditions such as temperature. $Y_2O_2SO_4$: Yb, Eu hollow spheres obtained by hydrothermal synthesis at 140℃ consisted of nanoparticles, as shown in Figure 2(a) and 2(b). The hollow feature can be identified by TEM. On increasing the hydrothermal temperature to 160℃, the surface of hollow sphere became relatively rough [Figure 2(c) and 2(d)]. Meanwhile, the average size of as-prepared hollow spheres also increased. Finally, $Y_2O_2SO_4$: Yb, Eu hollow spheres composed of nanotubes with larger average size were formed at 180℃ [Figure 2(e) and Figure 2(f)].

The elemental maps of the $Y_2O_2SO_4$: Yb, Eu obtained using the TEM are displayed in Figure 3, which clearly demonstrates a homogeneous distribution of Y, Yb, Eu, O and S elements. This confirms that successful doping can be achieved through our synthetic strategy. The line-scan analysis in Figure 3(a) reveals the profile of three main elements of Y, O and S across the radial direction, again indicating the apparent hollow feature. These results further confirm that host structure of the rare-earth-oxysulfate is composed of $Y_2O_2SO_4$ and the Yb^{3+} ions and Eu^{3+} ions were indeed incorporated into the host structure of the hollow spheres without segregation. Detailed EDS analysis taken on the TEM was also employed to demonstrate the metal composition of as-prepared composites. The molar ratio of Y:Yb:Eu is about 88.11:11.06:0.83, which agrees with the initial reagent concentration ratio of 90:9:1 in the precursor mixture.

As is well known, Yb^{3+} is often chosen as a sensitizer for UC due to its large absorption cross-section at about 980 nm. Nonetheless, the UC process relies on the energy transfer between both sensitizer and activator ions. Thus, the energy level matching is of critical importance. The PL properties of $Y_2O_2SO_4$: Yb, Eu indicate a DC emission process instead of UC emission. PL spectra of the $Y_2O_2SO_4$: Yb, Eu phosphors were recorded at room temperature to illustrate this relationship as shown in Figure 4. We can see a broad absorption band with a peak maximum centered at around 260 nm, which is results from the charge-transfer (CT) transitions between O^{2-} and Eu^{3+} ions. Specifically, it arises from the transition of 2p electrons of O^{2-} to the empty 4f orbitals of Eu^{3+} ions.[29-31] Additionally, the two weak peaks located at 395 nm and 460 nm can be assigned to the typical $^7F_0 \rightarrow {}^5L_6$ and $^7F_0 \rightarrow {}^5D_2$ transitions of Eu^{3+} ions,

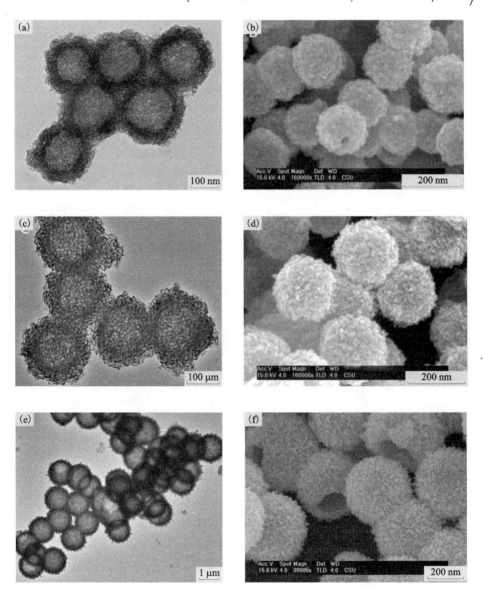

Figure 2

TEM images of $Y_2O_2SO_4$: Yb, Eu hollow spheres after calcination of precursors prepared at different hydrothermal temperature: (a) 140℃, (c) 160℃ and (e) 180℃, respectively; (b), (d) and (f) are the corresponding SEM images.

respectively. Excitation spectra of the $Y_2O_2SO_4$: Yb, Eu phosphors were taken by monitoring the wavelength of 616 nm. In addition, the intra-configurational $^4F_6$ excitation lines of the Eu^{3+} ions are very weak, indicating that the excitation of Eu^{3+}

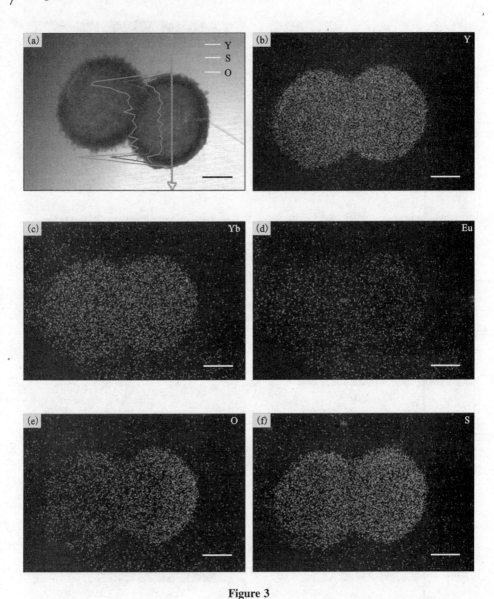

Figure 3

(a) TEM image of $Y_2O_2SO_4$: Yb, Eu hollow spheres with line-scan profile across the diameter direction. Elemental maps of (b) Y, (c) Yb, (d) Eu, (e) O and (f) S. The scale bar is 500 nm.

ions through the CT state for emission are very efficient. The emission spectrum of $Y_2O_2SO_4$: Yb, Eu under 260 nm light excitation [Figure 4(b)] demonstrates the characteristic $^5D_0 \rightarrow {}^7F_J$ (J = 1, 2, 3, 4) transitions of Eu^{3+} ions. The strongest emission peak located at 616 nm corresponds to the forced electric dipole $^5D_0 \rightarrow {}^7F_2$

transition of Eu^{3+} ions.[29-32]

Eu^{3+} as emitter ions do not favor in the cooperative UC luminescent process because of the energy level mismatch (about 5000 or 7000/cm) for the energy transfer (ET) between the $^2F_{5/2} \rightarrow {}^2F_{7/2}$ transition of Yb^{3+} and the $^7F_0 \rightarrow {}^7F_6$ transition of Eu^{3+},[15] and also the energy gap $^7F_6 \rightarrow {}^5D_0$ of Eu^{3+} compared with $^2F_{5/2} \rightarrow {}^2F_{7/2}$ transition of Yb^{3+}, as illustrated in detail in Figure 5(a). The $Y_2O_2SO_4$: Yb, Eu products are inert under illumination by a laser with wavelength of 980 nm excitation and no emission can be observed in the spectrum. DC PL spectra of $Y_2O_2SO_4$: Yb, Eu indicates that Yb^{3+} ions are ineffective in the energy transfer process. Therefore Er^{3+} ions with a ladder-like energy level structure were employed as an alternative emitter center in order to facilitate the ET from sensitizer to activator in UC processes. The emission spectra were recorded at the excitation laser wavelength of 980 nm by FluoroMax® -4 fluorescence spectrometer with a power density of about 350 mW. All the emission peaks shown in Figure 5(b) can be readily assigned to the characteristic emission peaks of Er^{3+}, indicating the successful cooperative luminescence between Yb^{3+} and Er^{3+}. The spectra are dominated by two strong f-f transitions of $^2H_{11/2} \rightarrow {}^4I_{15/2}$ and $^4F_{9/2} \rightarrow {}^4I_{15/2}$ in the green and red spectral range, with peaks centered at 523 nm (which is the strongest one), 544 nm, 651 nm, and 661 nm. Peaks around 523 nm and 544 nm are all related with transitions from thermally coupled $^2H_{11/2}$ and $^4S_{3/2}$ levels. Typically, UC can be classified by excited state absorption (ESA) or energy transfer upconversion (ETU) through the use of physically existing intermediary energy states of rare-earth ions.[33,34] The energy level and transfer scheme is shown in Figure 5(a). The detailed cooperative luminescent process can be described as following: Firstly, Yb^{3+} will be excited into $^2F_{5/2}$ state from ground state by absorbing one laser generated photon and the energy is then transferred to Er^{3+} ($^4I_{11/2}$); the Er^{3+} can be excited into the $^4F_{7/2}$ levels by absorbing another photon from Yb^{3+}. The ion decays rapidly to the $^2H_{11/2}$ or $^4S_{3/2}$ level by multiphonon relaxation. Meanwhile, the $^4I_{11/2}$ of Er^{3+} can also rapidly decay to $^4I_{13/2}$ through relaxation and then be pumped to $^4F_{9/2}$ by another photon. The $^2H_{11/2}$ and $^4F_{9/2}$ state further relax radiatively to the ground-state generating visible emissions. In the energy transfer scheme, two different cross-relaxation (CR) processes may occur: The first CR occurs from Er^{3+} ($^4S_{3/2} \rightarrow {}^4I_{13/2}$) to Yb^{3+} ($^2F_{7/2} \rightarrow {}^2F_{5/2}$). Two photons are emitted by the $^4I_{13/2} \rightarrow {}^4I_{15/2}$ transition of Er^{3+} and the $^2F_{5/2} \rightarrow {}^2F_{7/2}$ transition of Yb^{3+}. The second process is the cross-relaxation between Er^{3+} ions, $^4I_{15/2} \rightarrow {}^4I_{13/2}$ and $^4S_{3/2} \rightarrow {}^4I_{9/2}$; the energy is then redistributed between Er^{3+} and Yb^{3+} and emitted radiatively. The two processes transform a green photon into two photons at 1000 nm and 1550 nm,[35] leading to decrease in green emission intensity with increasing dopants concentration. For the red emission in the range of 650 – 660 nm, the energy transfer between Er^{3+} ions ($^4I_{13/2} + {}^4I_{11/2} \rightarrow {}^4F_{9/2} + {}^4I_{15/2}$ and $^4F_{7/2} + {}^4I_{11/2} \rightarrow {}^4F_{9/2} + {}^4F_{9/2}$) populate the $^4F_{9/2}$ level and depopulate the $^2H_{11/2}$ and $^4S_{3/2}$ levels, which will increase the red emission and decrease the green emission.[36, 37] $^4F_{7/2}, {}^2H_{11/2}$ or $^4S_{3/2} \rightarrow {}^4I_{15/2}$ and $^4F_{9/2} \rightarrow {}^4I_{15/2}$ are the energy transfer processes in which only two photons are engaged and the corresponding emissions are the dominated components in the PL spectra. Simultaneous two-photon absorption (TPA) is a well-established method for generating anti-Stokes emissions from a host of luminescent materials, such as organic dyes and semiconducting nanoparticles.[1, 2] It is also the most efficient way for energy transfer, corresponding to the most intense emission peaks in spectra. However, for the energy transfer from $^2H_{9/2}, {}^4F_{3/2}$, or $^4F_{5/2}$, to $^4I_{15/2}$, three photons are needed to pump the electrons to higher energy level, which is a more challenging and complex excitation process. The energy will be released from the excited state of $^4S_{3/2}$ by emission and it will further hinder the ESA process and lower the possibility to excite the electrons to a higher energy level. Thus, the intensity of the corresponding emission peaks ($^2H_{9/2}, {}^4F_{3/2}$, or $^4F_{5/2} \rightarrow {}^4I_{15/2}$) centered at around 409 nm, 436 nm, 455 nm was much weaker than that of the two photons processes.

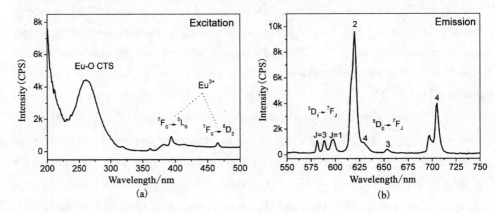

Figure 4

(a) Excitation, (b) Emission spectrum of $Y_2O_2SO_4$: Yb, Eu hollow spheres.

Energy transfer analysis indicates that an increase in Er^{3+} concentration should lead to decrease in green emission but an increase in red emission. The sensitizer and activator concentration depended PL spectra are shown in Figure 5(b). The initial ratio

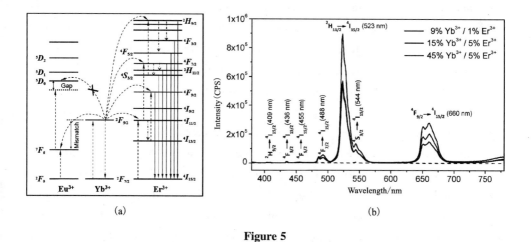

Figure 5

(a) Scheme of the energy level and energy transition of $Y_2O_2SO_4$: Yb, Eu/Er hollow spheres. (b) Emission spectrum of $Y_2O_2SO_4$: Yb, Er hollow spheres with different metal basis ratio under a laser with wavelength of 980 nm. The corresponding ratio of Y : Yb : Er calculated based on the amount of precursor is 90 : 9 : 1 (pink trace), 80 : 15 : 5 (purple trace) and 50 : 45 : 5 (blue trace), respectively.

of Y : Yb : Er was set as 90 : 9 : 1 and the corresponding product delivers the strongest emission intensity at 523 nm ($^2H_{11/2} \rightarrow {}^4I_{15/2}$). However, on increasing the Er^{3+} to 5 mol%, the emission intensity decreases sharply. This phenomenon was consistent with the CR process mentioned above, which may cause the non-radiative energy transfer or energy loss when the amount of Er^{3+} was increased. Thus, there is a stronger green emission at lower Er^{3+} concentration, but similar green emission performance when the ratio of Y : Yb : Er was 80 : 15 : 5 or 50 : 45 : 5. For the red emission, another important factor of energy back transfer (EBT) should be taken into consideration. With the increase of Yb^{3+} concentration in the hybrid material, the distance between Er^{3+} and Yb^{3+} ions is shortened, which will contribute to the increased rate of EBT processes ($^4S_{3/2}$ Er + $^2F_{7/2}$ Yb → $^4I_{13/2}$ Er + $^2F_{5/2}$ Yb),[38-41] resulting in an enhanced population of the $^4F_{9/2}$ and $^4I_{13/2}$ states. Together with the above mentioned CR process between Er^{3+} ions ($^4F_{7/2} + {}^4I_{11/2} \rightarrow {}^4F_{9/2} + {}^4F_{9/2}$), the intensity ratio of red to green emission will increase. Thus, although the overall PL intensity decreased with the increasing dopant concentration, it is worth noting in the emission spectra of calcined products, the intensity ratio of the green to the red emission PL are 2.44, 2.06, 1.14, for materials with Y : Yb : Er ratio of 90 : 9 : 1, 80 : 15 : 5, 50 : 45 : 5, respectively. A higher concentration of Yb leads to an increase in the intensity ratio of the red to the green upconversion. PL also increases due to an increased rate of EBT processes similar to what has been observed for other system.[38-41] It is commonly accepted that energy transfer of emitter ions will inevitably be affected by the local environment and its intensity depends on the symmetry of the crystal field around emitter ions.[42, 43] The amorphous hydrothermal products have a very low symmetry of crystal field around the Er^{3+} ions. In the emission spectra shown in Figure 6 (in the ESM), the emission intensity of $^4F_{9/2} \rightarrow ^4I_{15/2}$ is negligible compared to that for the well crystallized products, confirming that the efficiency of energy transfer depends on the symmetry features.

An absorption and emission model of the doped crystal structure can be proposed based on the above PL process. Figure 6 shows both activator ions of Er^{3+} (pink spheres) and sensitizer ion Yb^{3+} (light-blue spheres) doped into the host $Y_2O_2SO_4$ lattice. When an Yb^{3+} ion was excited by the pump laser, the quantized energy can be transferred to an appropriate acceptor ion. If within one ion energy gaps between three or more subsequent energy levels are very similar, sequential excitation to a highly excited state is possible with a single monochromatic light source.[2] The neighboring Er^{3+} ions are able to offer these incident energy gaps. The high energy level of Er^{3+} will require two or more photons to be excited and several surrounding Yb^{3+} ions should be coupled with a central Er^{3+} ion.

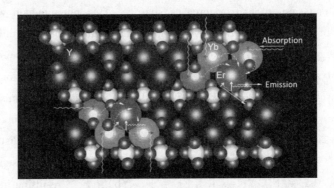

Figure 6 Absorption and emission model based the doping crystal structure

4 Conclusion

Uniform rare-earth-doped $Y_2O_2SO_4$ hollow spheres with different surface features can be successfully prepared via a biomolecule-assisted hydrothermal route followed by calcination. The formation of hollow spheres involves Ostwald ripening under hydrothermal conditions. The monoclinic $Y_2O_2SO_4$ crystal structure was shown to be a promising alternative to $NaREF_4$ as a lattice host for UC applications, since two different rare-earth ions can be easily co-doped into the $RE_2O_2SO_4$ without significantly changing the host crystal structure. Furthermore, UC emission intensity of the calcined products was significantly enhanced compared with that of the amorphous precursor. The luminescent properties were also affected by the ratio and concentration of dopant rare-earth ions due to energy transfer and symmetry of crystal field. These results are important not only in terms of providing an alternative structure for UC application, but also in giving further insights into applications of the doping strategy to yield hybrid materials with desirable physical properties.

References

[1] F. Wang, X. Liu. Recent advances in the chemistry of lanthanide-doped upconversion nanocrystals [J]. Chemical Society Reviews, 2009, 38(4): 976 – 989.

[2] M. Haase, H. Schäfer. Upconverting nanoparticles [J]. Angewandte Chemie International Edition, 2011, 50(26): 5808 – 5829.

[3] F. Auzel. Upconversion and anti-stokes processes with f and d ions in solids [J]. Chemical Reviews, 2004, 104(1): 139 – 174.

[4] H. X. Mai, Y. W. Zhang, R. Si. High-quality sodium rare-earth fluoride nanocrystals: controlled synthesis and optical properties [J]. Journal of the American Chemical Society, 2006, 128(19): 6426 – 6436.

[5] F. Wang, Y. Han, C. S Lim. Simultaneous phase and size control of upconversion nanocrystals through lanthanide doping [J]. Nature, 2010, 463(7284): 1061 – 1065.

[6] S. Jiang, Y. Zhang, K. M Lim. NIR-to-visible upconversion nanoparticles for fluorescent labeling and targeted delivery of siRNA [J]. Nanotechnology, 2009, 20(15): 155101.

[7] M. Wang, C. C. Mi, W. X. Wang. Mmunolabeling and NIR-excited fluorescent imaging of HeLa cells by using $NaYF_4$: Yb, Er upconversion nanoparticles [J]. ACS Nano, 2009, 3(6): 1580 – 1586.

[8] L. Wang, R. Yan, Z. Huo. Fluorescence resonant energy transfer biosensor based on upconversion-luminescent nanoparticles [J]. Angewandte Chemie International Edition, 2005, 44(37): 6054 – 6057.

[9] S. A Hilderbrand, F. Shao, C. Salthouse. Upconverting luminescent nanomaterials: application to in vivo bioimaging [J]. Chemical Communications, 2009, 28: 4188 – 4190.

[10] F. Wang, R. Deng, J. Wang. Tuning upconversion through energy migration in core-shell nanoparticles [J]. Nature Materials, 2011, 10(12): 968 – 973.

[11] X. Yu, M. Li, M. Xie. Dopant-controlled synthesis of water-soluble hexagonal $NaYF_4$

nanorods with efficient upconversion fluorescence for multicolor bioimaging [J]. Nano Research, 2010, 3(1): 51-60.

[12] W. Xu, Y. Zhu, X. Chen. A novel strategy for improving upconversion luminescence of $NaYF_4$: Yb, Er nanocrystals by coupling with hybrids of silver plasmon nanostructures and poly (methyl methacrylate) photonic crystals [J]. Nano Research, 2013, 6(11): 795-807.

[13] V Mahalingam, F Vetrone, R Naccache. Colloidal Tm^{3+}/Yb^{3+} - doped $LiYF_4$ nanocrystals: multiple luminescence spanning the UV to NIR regions via low-energy excitation [J]. Advanced Materials, 2009, 21(40): 4025-4028.

[14] Y. Qu, X. Kong, Y. Sun. Effect of excitation power density on the upconversion luminescence of LaF_3: Yb^{3+}, Er^{3+} nanocrystals [J]. Journal of Alloys and Compounds, 2009, 485(1): 493-496.

[15] Y. Liu, D. Tu, H. Zhu. A Strategy to achieve efficient dual-mode luminescence of Eu^{3+} in lanthanides doped multifunctional $NaGdF_4$ nanocrystals [J]. Advanced Materials, 2010, 22(30): 3266-3271.

[16] M. Machida, T. Kawano, M. Eto. Ln dependence of the large-capacity oxygen storage/release property of Ln oxysulfate/oxysulfide systems [J]. Chemistry of Materials, 2007, 19(4): 954-960.

[17] D. Zhang, F. Yoshioka, K. Ikeue. Synthesis and oxygen release/storage properties of Ce-substituted La-oxysulfates, $(La_{1-x}Ce_x)_2O_2SO_4$ [J]. Chemistry of Materials, 2008, 20(21): 6697-6703.

[18] M. Machida, K. Kawamura, T. Kawano. Layered Pr-dodecyl sulfate mesophases as precursors of $Pr_2O_2SO_4$ having a large oxygen-storage capacity [J]. Journal of Materials Chemistry, 2006, 16(30): 3084-3090.

[19] S. Kim, T. Masui, N. Imanaka. Synthesis of red-emitting phosphors based on gadolinium oxysulfate by a flux method [J]. Electrochemistry, 2009, 77(8): 611-613.

[20] T. Kijima, T. Shinbori, M. Sekita. Abnormally enhanced Eu^{3+} emission in $Y_2O_2SO_4$: Eu^{3+} inherited from their precursory dodecylsulfate-templated concentric-layered nanostructure [J]. Journal of Luminescence, 2008, 128(3): 311-316.

[21] T. Kijima, T. Isayama, M. Sekita. Emission properties of Tb^{3+} in $Y_2O_2SO_4$ derived from their precursory dodecylsulfate-templated concentric- and straight-layered nanostructures [J]. Journal of Alloys and Compounds, 2009, 485(1): 730-733.

[22] H. Hülsing, H. G. Kahle, A. Kasten. A microscopic model describing the spin-flip processes at the first-order antiferromagnetic-to-ferrimagnetic phase transition in $Dy_2O_2SO_4$ [J]. Journal of Magnetism and Magnetic Materials, 1983, 31(31): 1073-1074.

[23] J. Liang, R. Ma, F. Geng. $Ln_2(OH)_4SO_4 \cdot nH_2O$ (Ln = Pr to Tb; n ~ 2): a new family of layered rare-earth hydroxides rigidly pillared by sulfate ions [J]. Chemistry of Materials, 2010, 22(21): 6001-6007.

[24] X. W. D. Lou, L. A. Archer, Z. Yang. Hollow micro-/nanostructures: synthesis and

applications [J]. Advanced Materials, 2008, 20(21): 3987 – 4019.

[25] M. Machida, K. Kawamura, Ito K. Large-capacity oxygen storage by lanthanide oxysulfate/oxysulfide systems [J]. Chemistry of Materials, 2005, 17(6): 1487 – 1492.

[26] X. Liu, D. Zhang, J. Jiang. General synthetic strategy for high-yield and uniform rare-earth oxysulfate ($RE_2O_2SO_4$, RE = La, Pr, Nd, Sm, Eu, Gd, Tb, Dy, Y, Ho, and Yb) hollow spheres [J]. RSC Advances, 2012, 2(25): 9362 – 9365.

[27] G. Chen, W. Ma, X. Liu. Controlled fabrication and optical properties of uniform CeO_2 hollow spheres [J]. RSC Advances, 2013, 3(11): 3544 – 3547.

[28] W. Ma, Y. Guo, X. Liu. Nickel dichalcogenide hollow spheres: Controllable fabrication, structural modification, and magnetic properties [J]. Chemistry – A European Journal, 2013, 19(46): 15467 – 15471.

[29] J. Lian, X. Sun, X. Li. Synthesis. characterization and photoluminescence properties of ($Gd_{1-x}Eu_x)_2O_2SO_4$ sub-microphosphors by homogeneous precipitation method [J]. Materials Chemistry and Physics, 2011, 125(3): 479 – 484.

[30] M. N. Luwang, R. S. Ningthoujam, S. K Srivastava. Effects of Ce^{3+} codoping and annealing on phase transformation and luminescence of Eu^{3+}-doped YPO_4 nanorods: D_2O solvent effect [J]. Journal of the American Chemical Society, 2010, 132(8): 2759 – 2768.

[31] R. Yan, X. Sun, X. Wang. Crystal Structures, anisotropic growth, and optical properties: controlled synthesis of lanthanide orthophosphate one-dimensional nanomaterials [J]. Chemistry-A European Journal, 2005, 11(7): 2183 – 2195.

[32] J. Lian, X. Sun, Z. Liu. Synthesis and optical properties of ($Gd_{1-x}Eu_x)_2O_2SO_4$ nanophosphors by a novel co-precipitation method [J]. Materials Research Bulletin, 2009, 44(9): 1822 – 1827.

[33] M. F. Joubert. Photon avalanche upconversion in rare earth laser materials [J]. Optical Materials, 1999, 11(2): 181 – 203.

[34] P. A. Kurian, C. Vijayan, C. S. S. Sandeep. Two-photon-assisted excited state absorption in nanocomposite films of PbS stabilized in a synthetic glue matrix [J]. Nanotechnology, 2007, 18 (7): 075708.

[35] B. Fan, C. Chlique, O. Merdrignac-Conanec. Near-infrared quantum cutting material Er^{3+}/Yb^{3+} doped La_2O_2S with an external quantum yield higher than 100% [J]. The Journal of Physical Chemistry C, 2012, 116(21): 11652 – 11657.

[36] G. Li, M. Shang, D. Geng, Multiform La_2O_3: Yb^{3+}/Er^{3+}/Tm^{3+} submicro-/microcrystals derived by hydrothermal process: Morphology control and tunable upconversion luminescence properties [J]. CrystEngComm, 2012, 14(6): 2100 – 2111.

[37] Z. Li, W. Park, G. Zorzetto. Synthesis protocols for δ – doped $NaYF_4$: Yb, Er [J]. Chemistry of Materials, 2014, 26(5): 1770 – 1778.

[38] L. Shi, Q. Shen, Z. Qiu. Concentration-dependent upconversion emission in Er-doped and Er/Yb-codoped $LiTaO_3$ polycrystals [J]. Journal of Luminescence, 2014, 148: 94 – 97.

[39] R. Guo, B. Wang, X. Wang. Suppression of second-order cooperative up-conversion in Er/Yb silicate glass [J]. Optical Materials, 2013, 35(5): 935-939.

[40] I. R. Martin, V. D. Rodriguez, V. Lavin. Transfer and back transfer processes in Yb^{3+} - Er^{3+} codoped fluoroindate glasses [J]. Journal of Applied Physics, 1999, 86(2): 935-939.

[41] J. T. Vega-Duran. Effects of energy back transfer on the luminescence of Yb and Er ions in YAG [J]. Applied Physics Letters, 2000, 76(15): 2032-2034.

[42] D. P. Volanti, I. L. V Rosa, E. C. Paris. The role of the Eu^{3+} ions in structure and photoluminescence properties of $SrBi_2Nb_2O_9$ powders [J]. Optical Materials, 2009, 31(6): 995-999.

[43] P. Fang, H. Fan, J. Li. Lanthanum induced larger polarization and dielectric relaxation in Aurivillius phase $SrBi_{2-x}La_xNb_2O_9$ ferroelectric ceramics [J]. Journal of Applied Physics, 2010, 107(6): 064104-064104-4.

☆ G. Chen, X. H. Liu, W. Ma. H. M. Luo, J. H. Li, R. Z. Ma, G. Z. Qiu. Hollow spherical rare-earth-doped yttrium oxysulfate: a novel structure for upconversion [J]. Nano Research, 2014, 7(8): 1093-1102.

Controlled Fabrication and Optical Properties of Uniform CeO$_2$ Hollow Spheres

Abstract: Uniform CeO$_2$ hollow spheres were successfully fabricated on a large scale via the calcination of corresponding spherical Ce-organic precursor. The solution- and solid-based shape and crystal structure evolution can be consecutively controlled. Sm-doped CeO$_2$ hollow spheres have also been synthesized and their enhanced optical properties have been demonstrated.

1 Introduction

Inorganic hollow spheres with remarkable interior space have been pursued intensively over the past few decades owing to their low density, high specific surface area, good permeation and potential applications in various fields, such as in confined chemical reactions, sensors, catalysis, and drug delivery.[1-3] Numerous chemical and physicochemical strategies, such as the Kirkendall diffusion effect,[4] Ostwald ripening,[5] and template-directed techniques,[6] have been employed for the design and controlled fabrication of various micro/nanospheres with hollow interiors. The enormous developments in the synthesis of hollow spheres has greatly enhanced our ability to obtain functional materials with desirable physical properties. Cerium oxide (CeO$_2$) has stimulated considerable research interest for many years because of their outstanding properties arising from the electron transitions within the 4f shell and widespread potential applications in diverse fields, including high performance luminescence, upconversion, magnetic, electrical materials, gas sensors, and so forth.[7-10] CeO$_2$ with hollow structure may bring about a bright future for the further development of these applications. In general, doping is a widely applied technological process in materials design that involves incorporating atoms or ions of appropriate elements into host lattices to yield hybrid materials with desirable properties and functions.[11] For rare-earth compounds, doping is of fundamental importance in tuning emission properties.

Herein, we present a facile biomolecule-assisted route to prepare uniform CeO$_2$

hollow spheres via the calcination of corresponding spherical Ce-organic precursor obtained by using L-cysteine (Cys) as biomolecular template. The formation process of Ce-organic spheres involving Ostwald ripening is proposed based on the detailed experimental results. Sm(Ⅲ) ions was chosen as second emitter ions due to their similar size to the host Ce(Ⅳ) ions and strong energy transfer and emission in this doping system. Sm-doped CeO_2 has been synthesized and the crystallographic phase and optical emission properties of as-prepared products were remarkably influenced by the concentration of Sm(Ⅲ) ions.

2 Experimental Section

All chemicals were of analytical grade from Sinopharm Chemical Reagent Co., Ltd and used as starting materials without further purification.

2.1 Synthesis of CeO_2 hollow spheres. The spherical precursor compounds were synthesized by a biomolecule-assisted approach under hydrothermal condition. In a typical procedure, 1 mmol of $Ce(NO_3)_3 \cdot 6H_2O$ was loaded into a 50 mL beaker with 25 mL de-ionized water, and then 2 mmol L-cysteine and 0.3 g PVP were added into the solution under continuous stirring for 5 min. Then the solution was transferred into a 50 mL Teflon-lined stainless steel autoclave and was sealed and held at a temperature of 140 ℃ for 8 h. Subsequently, the system was allowed to cool to room temperature naturally. The resulting precipitate was collected by filtration and washed with absolute ethanol and distilled water in sequence for several times. The final product was dried in a vacuum box at 50 ℃ for 4 h. As prepared Ce-organic compound spheres were calcinated to produce CeO_2 hollow spheres in air at 600 ℃ for 2 h.

2.2 Characterization. The crystal structures of obtained specimens were characterized on a X-ray diffractometer (XRD, D/max 2550 VB+) with Cu K_α radiation ($\lambda = 1.5418$ Å). The sizes and morphologies of as-synthesized products were characterized by a field-emission scanning electron microscopy (FE-SEM, Sirion 200) with an accelerating voltage of 15 kV and transmission electron microscope (TEM, Tecnai G2 F20) with an accelerating voltage of 200 kV. The energy dispersive spectrometer (EDS) was taken on the SEM. The thermal behavior of Ce-organic compund to CeO_2 was evaluated by thermogravimetric (TG) and differential scanning calorimetry (DSC), which was obtained on a simultaneous thermal analysis (STA, NETZSCH STA 449C). The photoluminescence (PL) properties were measured on a fluorescence spectrometer (FluoroMax® -4 fluorescence spectrometer).

3 Results and Discussion

Typical TEM and SEM images of the spherical Ce-organic precursors and CeO_2 hollow spheres are shown in Figure 1. Figure 1(a) and 1(b) show spherical Ce-organic precursors with an average size of 800 nm. The inset selected area electron diffraction (SAED) pattern of Figure 1(b) demonstrates the amorphous phase of the spherical precursors. Figure 1(c) – 1(e) display the CeO_2 hollow spheres with a mean size of 600 nm after the calcination. The sharp contrast between the dark periphery and grayish center of these spheres indicates hollow feature. Correspondingly, diffraction rings in the SAED pattern shown in the inset of Figure 1(e) could be readily indexed to the cubic phase of CeO_2. The HRTEM image shown in Figure 1(f) confirms the crystalline nature of CeO_2 hollow spheres. The interplanar spacing is 0.31 nm, corresponding to the (111) plane.

Functional groups, such as —SH, —NH_2, —COOH of the Cys, have a strong tendency to coordinate with Ce(Ⅲ) ions. On the basis of metal-ligand interactions, Ce-organic coordination compound was formed with the assistance of Cys. Cys serves as a biomolecular template in the current synthetic strategy. To illustrate the formation process of the spherical Ce-organic precursor obtained via hydrothermal reaction, time-dependent experiments were carried out at 140℃ for 4, 8, and 24 h, respectively. After the reaction of 4 h [Figure 2(a)], the products were mainly of near-spherical morphology. Well-dispersed spheres with smooth surface were obtained after 8 h [Figure 2(b)]. When the reaction duration was prolonged to 24 h [Figure 2(c)], larger spheres became dominant in the final product. The morphology remained unchanged, but the size dramatically increased. Meanwhile, the generation of CeO_2 hollow structure was found to be controlled by changing the calcination duration. As shown in Figure 2(d), the interior spaces of some spheres were still occupied after annealing for 10 min. However, careful observation shows that the contrast had began to appear and some spheres have an obvious hollow structure with comparatively thick shell. After calcinating for 30 min [Figure 2(e)], the inner space of the spheres increased. The products with an apparent hollow structure obtained after 60 min [Figure 2(f)] of calcination were very similar to the typical samples displayed in Figure 1. The formation of crystals of uniform shape and size in solution reaction system commonly involves the Ostwald ripening.[12] For the spherical precursors formed in solution, we can expect that there are certain chemical equilibria established between

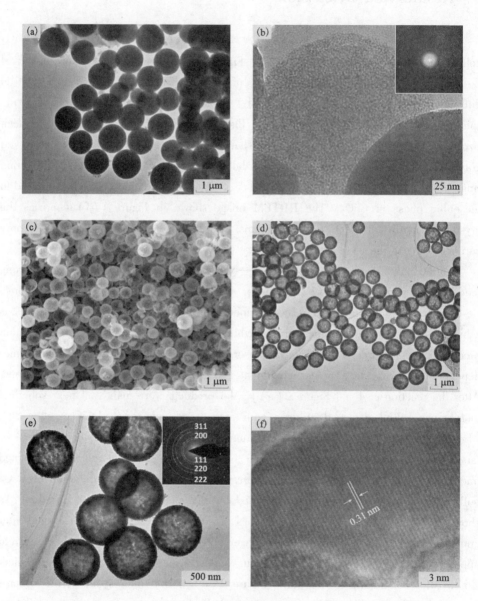

Figure 1

(a) Low and (b) high magnification TEM images of spherical Ce-organic precursors; (c) SEM, (d) low, (e) high magnification TEM and (f) HRTEM views of CeO_2 hollow spheres. Insets are corresponding SAED patterns.

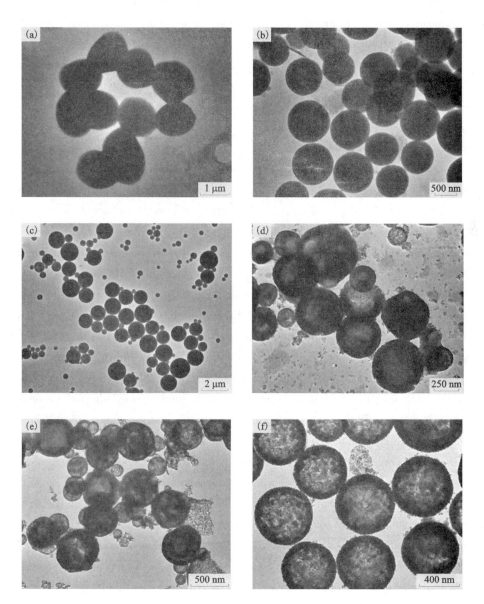

Figure 2

TEM images of Ce-organic precursors prepared at 140℃ for (a) 4 h, (b) 8 h and (c) 24 h; and CeO$_2$ prepared by calcinating the as-prepared precursor at 600℃ for (d) 10 min, (e) 30 min and (f) 60 min.

all solid-liquid interfaces. Owing to the size difference of the forming solid compounds, concentrations of growth nutrients across the solution vary. As a result, the homogenization of this concentration gradient will eventually eliminate spheres of smaller sizes whilst the growth of the large ones proceeds.[5] The morphology evolution in the solution indicates that the Ce-organic spheres originated from the ripening of precursors. On the other hand, the calcination may refer to the oxidation reaction in the surface layer of the spherical Ce-organic precursors. The organic substances were all burnt out at 600℃ and the Ce-organic compounds were gradually crystallized into CeO_2 at the peripheries of precursor spheres.

To elucidate the change and evolution of phase compositions, the crystallization behavior of the Ce-organic compound after clacination at different temperatures was checked by XRD. XRD patterns of the Ce-organic precursors on sample annealing at 400℃ and 600℃ for 2 h are shown in Figure 3(a). The initial precursor was amorphous with broad featureless peaks. The major phase of as-prepared compound at 400℃ for 2 h was cubic CeO_2 with broad feature peaks. The XRD pattern of the final product prepared at 600℃ for 2 h indicates the high crystallinity of CeO_2. The thermal behavior of Ce-organic compound oxidized to CeO_2 was investigated with thermogravimetry (TG) and differential scanning calorimetry (DSC) in the temperature range of 25 – 650℃ at a heating rate of 10℃/min in air. As shown in Figure 3(b), the gradual mass loss (6.8%) in the range of 25 – 200℃ can be attributed to evaporation of the absorbed organic residues and water species on the Ce-organic compound surfaces. With rising temperature, obvious weight loss can be seen. The first rapid mass loss (20.6%) occurred from 200℃ to 370℃, illustrating quick oxidation of the Ce-organic compound. The S element was oxidized and escaped in the form of SO_x. Subsequently, there followed a stable weight loss platform at the temperature range from 370 – 470℃, indicating that relatively stable compound was obtained. The second remarkable weight loss (13.4%) occurred between 470 – 600℃, which was attributed to the further oxidation and crystallization into CeO_2. The tremendous decrease of weight and volume was caused by the combustion or oxidation of the initial organic precursor. The DSC curve of the sample displayed two exothermal peaks in the gravimetric gain regions centered at 313℃ and 517℃ respectively, corresponding to the two remarkable mass losses.

The addition of a second emitter ion at low concentrations range had a negligible effect on nanocrystal phase and size, but induced a marked change in PL properties. In this case, Sm(Ⅲ) ions were introduced to CeO_2 nanocrystals. The addition of Sm^{3+}

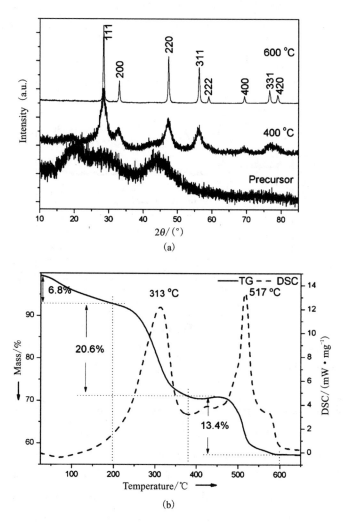

Figure 3

(a) XRD patterns of as-prepared samples obtained at different calcinating temperature;
(b) TG and DSC curve of the Ce-organic precursors annealing from 25 ℃ to 650 ℃.

with different concentrations ion into CeO_2 host lattice caused simultaneous effect on nanocrystal phase and size, as shown in Figure 4. Owing to their small amount (less that 5 mol%) and ease to dispersion, Sm(Ⅲ) ions can be readily incorporated in Ce–O to generate a cooperative luminescence system without significant crystal phase change comparing with XRD pattern of the pure CeO_2. However, with increasing the amount of Sm^{3+} to 10 mol%, some weak peaks of $CeSO_4$ could be seen in the degree range from 5° to 25°. The crystal phase of $Sm_2O_2SO_4$ occurred when 20 mol% and 50 mol% Sm^{3+} added. It is note worth that pure $Sm_2O_2SO_4$ can be prepared by the two-step biomolecule-assisted route.

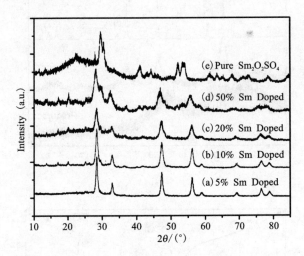

Figure 4

XRD patterns of CeO_2 with different amount of Sm^{3+} doped: (a) 5 mol%, (b) 10 mol%, (c) 20 mol%, (d) 50 mol% and (e) prue $Sm_2O_2SO_4$, respectively.

EDS analysis taken on the SEM was also used to demonstrate the composition change of CeO_2: Sm as shown in Figure 5. When 2 mol% Sm^{3+} was doped, the morphology of final products remains unchanged. When 5 mol% Sm^{3+} was doped, it can be seen that the products were still composed of spheres with some broken ones. However, the products cannot keep the spherial structure any further when 10% mol Sm^{3+} was doped. Many particles appeared. The SEM analysis agrees with the XRD results, demonstrating that the CeO_2: Sm hollow spheres with pure CeO_2 cystal structure could be obtained only in a certain Sm^{3+} doping range.

Figure 6(a) shows the PL excitation spectra of Sm-doped CeO_2. It has been demonstrated that the broad excitation bands of CeO_2 doped with lanthanide ions come from the charge transfer transition from O^{2-} to Ce^{4+}.[13-15] The excitation peak wavelength of 344 nm indicates the charge transfer energy is about 3.61 eV. The relatively weak peaks around 469 nm were caused by the electron transition of Sm^{3+} from the ground state level to the excited state level of ^4G_J and ^6P_J. PL emission spectra of as-prepared samples of Sm-doped (different concentration) CeO_2 spheres annealed at

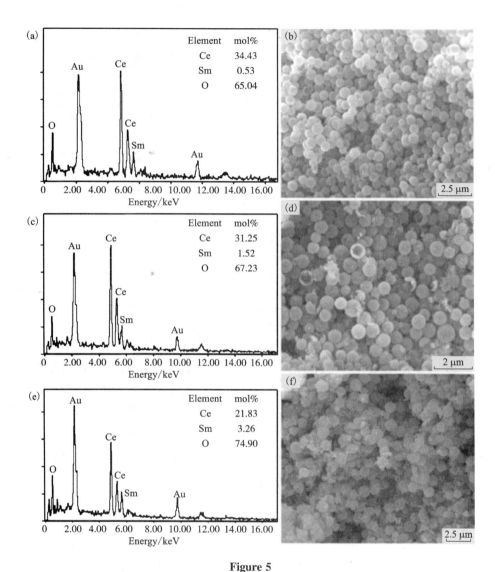

Figure 5

EDS patterns of CeO_2 with different amount of Sm^{3+} doped: (a) 2 mol%, (c) 5 mol% and (e) 10 mol%, respectively; (b), (d) and (f) are the corresponding SEM images.

600 ℃ are shown in Figure 6(b) and 6(c) recorded at excitation of 340 nm. Two strong emission bands are observed for Sm^{3+} ions in CeO_2 host lattice. The emission spectrum each contain a series of emission peaks centered at 574 nm and 617 nm. The two groups of peaks at 574 nm and 617 nm are due to the $^4G_{5/2} \rightarrow {}^6H_{5/2}$ and the $^4G_{5/2} \rightarrow {}^6H_{7/2}$ transitions, respectively.[14, 15] The point-group symmetry of the Ce sites

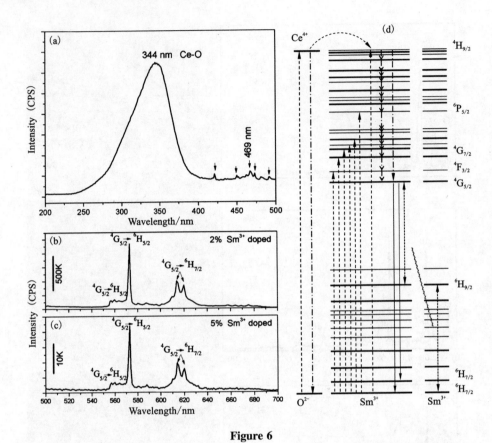

Figure 6

(a) Excitation spectrum of Sm-doped CeO_2 hollow spheres; (b) and (c) Emission spectrum of 5 mol% and 2 mol% Sm-doped CeO_2.

in the fluorite CeO_2 structure is O_h with eight-fold oxygen coordination, thereby providing an inversion symmetry.[16] The energy level scheme [Figure 6(d)] displays the probable energy transfer between different energy levels. The blue dash lines represent photon excitation, corresponding to the excitation peaks in Figure 6(a). The green dashed lines indicate the coss relaxation (CR), which usually refers to all types of energy transfer occurring between identical ions.[17] In this CR system, the sensitizer and the activator are Sm^{3+} ions, photons of the incident energy are absorbed either by the sensitizer or by the activator leading to these two ions in their excited state $^6F_{9/2}$. Then, an energy transfer promotes the activator ion in its state $^4G_{5/2}$ while the sensitizer goes down to a lower energy state. It is worth noting that emission intensity calculated from the peak area of the 2 mol% Sm-doped CeO_2 hollow spheres is approximately 25 times higher than that of 5 mol% Sm-doped CeO_2. The intensive CR process may cause the non-radiative energy transfer or energy loss so that the intensity of PL decreases sharply when a higher concentration of Sm^{3+} is doped into the solid solution.

4 Conclusion

In summary, a simple route to fabricate uniform CeO_2 hollow spheres was developed by calcination of the Ce-organic precursors. The crystal structure transformation and shape evolution of the final products could be controlled by adjusting the reaction time and temperature. Optical property studies reveal that PL intensity was dramatically influenced by the concentration of Sm(Ⅲ) ions, which provided insights of the effect of a second emitter element induced into a host crystal structure on the physical properties of doped materials. Due to its high yield and facile preparation process, this synthetic strategy has potential applications in the future large-scale synthesis of other rare-earth oxides hollow spheres.

References

[1] X. W. D. Lou, L. A. Archer, Z. Yang. Hollow micro-/nanostructures: synthesis and applications [J]. Advanced Materials, 2008, 20(21): 3987-4019.

[2] Z. R. Shen, J. G. Wang, P. C. Sun, D. T. Ding, T. H. Chen. Fabrication of lanthanide oxide microspheres and hollow spheres by thermolysis of pre-molding lanthanide coordination compounds [J]. Chemical Communications, 2009, 13: 1742-1744.

[3] X. Liu, D. Zhang, J. Jiang, N. Zhang, R. Ma, H. Zeng, B. Jia, S. Zhang, G. Qiu. General synthetic strategy for high-yield and uniform rare-earth oxysulfate ($RE_2O_2SO_4$, RE = La, Pr, Nd, Sm, Eu, Gd, Tb, Dy, Y, Ho, and Yb) hollow spheres [J]. RSC Advances, 2012, 2(25): 9362-9365.

[4] F. Zhang, Y. Shi, X. Sun, D. Zhao, G. D. Stucky. Formation of hollow upconversion rare-earth fluoride nanospheres: nanoscale kirkendall effect during ion exchange [J]. Chemistry of Materials, 2009, 21(21): 5237-5243.

[5] H. C. Zeng. Synthetic architecture of interior space for inorganic nanostructures [J]. Journal of Materials Chemistry, 2006, 16(7): 649-662.

[6] D. G. Shchukin, R. A. Caruso. Template synthesis and photocatalytic properties of porous metal oxide spheres formed by nanoparticle infiltration [J]. Chemistry of Materials, 2004, 16(11): 2287-2292.

[7] C. Laberty-Robert, J. W. Long, E. M. Lucas, K. A. Pettigrew, R. M. Stroud, M. S.

Doescher, D. R. Rolison. Sol-gel-derived ceria nanoarchitectures: synthesis, characterization, and electrical properties [J]. Chemistry of Materials, 2006, 18(1): 50 – 58.

[8] L. Liao, H. X. Mai, Q. Yuan, H. B. Lu, J. C. Li, C. Liu, C. H. Yan, Z. X. Shen, T. Yu. Single CeO_2 nanowire gas sensor supported with Pt nanocrystals: gas sensitivity, surface bond states, and chemical mechanism [J]. The Journal of Physical Chemistry C, 2008, 112(24): 9061 – 9065.

[9] X. Lu, D. Zheng, P. Zhang, C. Liang, P. Liu, Y. Tong. Facile synthesis of free-standing CeO_2 nanorods for photoelectrochemical applications [J]. Chemical Communications, 2010, 46: 7721 – 7723.

[10] Z. Wang, Z. Quan, J. Lin. Remarkable changes in the optical properties of CeO_2 nanocrystals induced by lanthanide ions doping [J]. Inorganic Chemistry, 2007, 46(13): 5237 – 5242.

[11] F. Wang, Y. Han, C. S. Lim, Y. Lu, J. Wang, J. Xu, H. Chen, C. Zhang, M. Hong, X. Liu. Simultaneous phase and size control of upconversion nanocrystals through lanthanide doping [J]. Nature, 2010, 463(7284): 1061 – 1065.

[12] W. Z. Ostwald. Studies on formation and transformation of solid materials [J]. Zeitschrift für Physikalische Chemie, 1897, 22: 289 – 330.

[13] L. Li, H. K. Yang, B. K. Moon, Z. Fu, C. Guo, J. H. Jeong, S. S. Yi, K. Jang, H. S. Lee. Photoluminescence properties of CeO_2: Eu^{3+} nanoparticles synthesized by a sol-gel method [J]. The Journal of Physical Chemistry C, 2008, 113(2): 610 – 617.

[14] H. Guo, Y. Qiao, Preparation. structural and photoluminescent properties of CeO_2: Eu^{3+} films derived by Pechini sol-gel process [J]. Applied Surface Science, 2008, 254(7): 1961 – 1965.

[15] S. Fujihara, M. Oikawa. Structure and luminescent properties of CeO_2: rare earth (RE = Eu^{3+} and Sm^{3+}) thin films [J]. Journal of Applied Physics, 2004, 95(12): 8002 – 8006.

[16] R. Srinivasan, A. C. Bose. Structural properties of Sm^{3+} doped cerium oxide nanorods synthesized by hydrolysis assisted co-precipitation method [J]. Materials Letters, 2010, 64 (18): 1954 – 1956.

[17] G. C. Liu, L. M. Chen, X. C. Duan, D. W. Liang. Synthesis and characterization of Sm^{3+}-doped CeO_2 powders [J]. Transactions of Nonferrous Metals Society of China, 2008, 18(4): 897 – 903.

[18] M. F. Joubert. Photon avalanche upconversion in rare earth laser materials [J]. Optical Materials, 1999, 11(2 – 3): 181 – 203.

☆ G. Chen, W. Ma, X. H. Liu, S. Q. Liang, G. Z. Qiu, R. Z. Ma. Controlled fabrication and optical properties of uniform CeO_2 hollowspheres [J]. RSC Advances, 2013, 3(11): 3544 – 3547.

General Synthetic Strategy for High-Yield and Uniform Rare-Earth Oxysulfates ($RE_2O_2SO_4$, RE = La, Pr, Nd, Sm, Eu, Gd, Tb, Dy, Y, Ho, and Yb) Hollow Spheres

Abstract: A general and facile synthetic strategy for fabrication of a series of high-yield and uniform rare-earth oxysulfate hollow spheres is demonstrated as the first example via calcination corresponding rare-earth coordination compounds obtained using L – cysteine as a biomolecular template. Furthermore, the hollow spheres can also be successfully doped or formed in double-shell form for property modification and technological applications.

1 Introduction

Hollow spheres have been pursued intensively owing to the fundamental requirements in basic scientific research and potential technology applications in diverse fields, including selective catalysis, controlled release and delivery, and photonic building blocks, etc. A variety of chemical and physicochemical strategies such as template-assisted synthesis, Kirkendall diffusion effects, Ostwald ripening, and chemically induced self-transformation have been employed for the design and controlled fabrication of micro/nanospheres with hollow interiors.[2] Among them, template-assisted synthesis has been demonstrated to be the most effective and versatile approach. Various templates, such as hard templates (e. g. , silica, polymer, and carbon spheres) and soft templates (e. g. , vesicles, emulsions, and surfactant), have been extensively utilized to fabricate hollow spheres.[3] In particular, biomolecules, as life's basic building blocks, have been also employed as templates for the design and synthesis of various micro/nanomaterials, including hollow spheres, due to the special structures and fascinating functions.[4,5] Despite these successes, exploiting a facile and general procedure for the fabrication of hollow spheres with narrow size distribution still remains a highly sophisticated challenge.

Rare-earth oxysulfates ($RE_2O_2SO_4$) have attracted increasing attention in recent years due to the unique luminescent and specific magnetic properties as well as significant applications in large volume oxygen storage, which offers 8 times higher oxygen storage capacity (OSC) than that of conventional cerium-based oxygen storage materials, such as $CeO_2 - ZrO_2$.[6-9] $RE_2O_2SO_4$ could be obtained through a high-temperature treatment of the corresponding hydrous sulfates [$RE_2(SO_4)_3 \cdot nH_2O$] or layered hydroxides intercalated with dodecyl sulfate (DS) ions.[10,11] Recently, layered rare-earth hydroxysulfate [$RE_2(OH)_4SO_4 \cdot nH_2O$] could also be transformed into $RE_2O_2SO_4$ by a thermal-induced dehydroxylation process.[12] However, most of $RE_2O_2SO_4$ prepared by the above methods are bulk materials with irregular morphology. There is still no report on the production of hollow spheres until now. Herein, we report a general and facile procedure for production of a series of high-yield and uniform $RE_2O_2SO_4$ hollow spheres via the calcination of corresponding rare-earth coordination compounds obtained by using L-cysteine as a biomolecular templates. Furthermore, doped and coated $RE_2O_2SO_4$ hollow spheres have also been prepared for the property modification and technological applications.

2 Experimental Section

All rare-earth nitrates used were of 99.99% purity. L-cysteine is analytical grade from the Beijing Chemical Factory, China. These chemicals were used as received without further purification. Milli-Q water was used throughout the experiments.

2.1 Synthesis of rare-earth oxysulfates and oxides hollow spheres. In a typical procedure, hydrated lanthanum nitrate [$La(NO_3)_3 \cdot 6H_2O$, 1.0 mmol] and L-cysteine (2.0 mmol) were dissolved in 40 mL deionized water under vigorous magnetic stirring. The pH of the initial mixture was about 7. The solution was then transferred into a Teflon-lined autoclave of 50 mL capacity and was sealed and maintained at 140℃ for 24 h. Subsequently, the system was allowed to cool to room temperature. The resulting La-coordination compound was collected by filtration and washed with ethanol and distilled water for several times and dried at 50℃ for 4 h, which can be transformed into $La_2O_2SO_4$ by calcination at 600℃ for 2 h. The yield of $La_2O_2SO_4$ is about 85% based on the calculation of lanthanum nitrate.

2.2 Synthesis of Eu-doped $La_2O_2SO_4$ and CeO_2 oxides hollow spheres. In a typical procedure, hydrated lanthanum nitrate [La(NO_3)$_3$·$6H_2O$, 0.95 mmol], hydrated europium nitrate [Eu(NO_3)$_3$·$6H_2O$, 0.05 mmol] and L-cysteine (2.0 mmol) were dissolved in 40 mL deionized water under vigorous magnetic stirring. The solution was then transferred into a Teflon-lined autoclave of 50 mL capacity and was sealed and maintained at 140℃ for 24 h. The resulting Eu-doped La-coordination compound was collected by filtration and washed with ethanol and distilled water for several times and dried at 50℃ for 4 h, which can be transformed into 5% mol Eu-doped $La_2O_2SO_4$ by calcination at 600℃ for 2 h.

2.3 Synthesis of ZnS-coated $La_2O_2SO_4$ and CeO_2 hollow spheres. The preparation process of ZnS-coated $La_2O_2SO_4$ and CeO_2 hollow spheres was similar to our previous work. In a typical procedure, the obtained La-coordination compounds (1 mmol) were redispersed into 100 mL thioacetamide (0.5 mmol) solution under vigorous magnetic stirring for 2 h. Subsequently, hydrated zinc nitrate [Zn(NO_3)$_2$·$6H_2O$, 0.5 mmol] was added to the above solution, which were then ultrasonicly treated for 30 min and calcinated at 600℃ for 2 h to fabricate ZnS-coated $La_2O_2SO_4$ hollow spheres.

2.4 Characterizations. XRD data were collected by a D/max2550 VB + diffractometer with Cu K_α radiation ($\lambda = 0.15405$ nm). Morphology of the synthesized products was examined using a JEOL JSM-6700F field emission scanning SEM. TEM was performed on a JEOL JEM-3100F energy-filtering (Omega type) transmission microscope. Thermogravimetric differential thermal measurements (TG-DTA) were carried out using a Rigaku TGA-8120 instrument in a temperature range of 25-1000℃ at a heating rate of 1℃/min under an air flow. The photoluminescence excitation and emission spectra were measured on a Hitachi F-4500 fluorescence spectrophotometer under the same conditions at room temperature. Magnetic measurements were conducted using a Quantum Design MPMS XP-5 superconducting quantum interference device (SQUID).

3 Results and Discussion

Figure 1(a) exhibits a typical scanning electron microscopy (SEM) image of La-coordination compound consisted of uniform spherical particles with smooth surface. The broken shell of an individual sphere confirms that the interior of the particles is hollow. This is consistent with transmission electron microscopy (TEM)

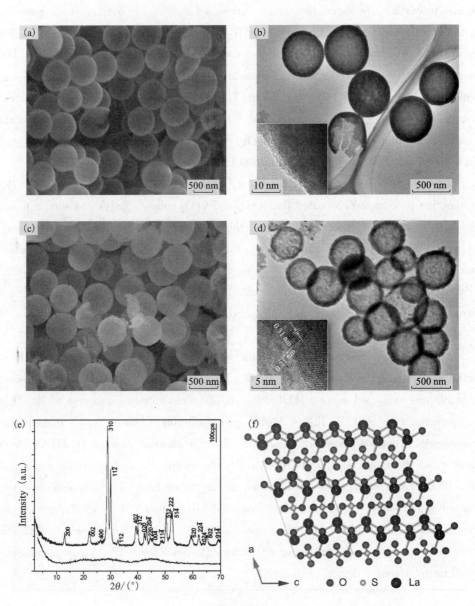

Figure 1

(a) SEM and (b) TEM images of La – coordination compound. (c) SEM and (d) TEM images of $La_2O_2SO_4$ hollow spheres. The insets in (b) and (d) depict the corresponding HRTEM images. (e) XRD patterns of La – coordination compound (blue) and $La_2O_2SO_4$ (red). (f) Crystal structure for $La_2O_2SO_4$ projected along (b)* axis. The line shows a two-dimensional unit cell. The O, S, and La species are represented by red, yellow, and violet balls, respectively.

image shown in Figure 1(b), which also reveals the hollow nature of these spherical particles. No crystal fringes is observed in the corresponding high-resolution TEM (HRTEM) image [inset of Figure 1(b)], which indicates that the hollow spheres are amorphous. The hollow spheres have a mean diameter of approximately 750 nm with a standard deviation of 70 nm and shell thickness in the range of 40 – 100 nm. Figure 1 (c) depicts a typical SEM image of $La_2O_2SO_4$ obtained by thermolysis of corresponding La – coordination compounds at 600 ℃ for 2 h. It is obvious that spherical shapes of the particles are still well retained. The surfaces, however, becomes relatively rough, which is different from the quite smooth surface of the precusory spheres. The average diameter of $La_2O_2SO_4$ hollow spheres is estimated to be approximately 550 nm with a standard deviation of 40 nm, smaller than that of the corresponding La – coordination compound, implying the tendency to shrink after calcination.[13] A typical TEM image of $La_2O_2SO_4$ hollow spheres, given in Figure 1(d), also clearly displays almost 100% hollow spheres with a mean diameter of about 550 nm and shell thickness of about 60 nm, which again suggests that the hollow nature was perfectly inherited from La – coordination compound hollow spheres. The inset of Figure 1(d) shows the corresponding HRTEM image, which provided further insight into the structure of $La_2O_2SO_4$ hollow spheres. Lattice spacing were measured to be about 0.31 nm and 0.21 nm, which are consistent with the interplanar spacings between (310) and (020) crystallographic planes in monoclinic $La_2O_2SO_4$, respectively. The evolution in crystal structure can also be interpreted from X-ray powder diffraction (XRD) results. The XRD pattern of La – coordination compound is mainly featureless, indicating the non-crystalline or amorphous nature [Figure 1(e)]. After calcination, the crystalline form of $La_2O_2SO_4$ was obtained. All the reflections can be indexed to the monoclinic structure of $La_2O_2SO_4$ [space group: C2/c (No. 15)] with lattice constant a = 14.345 Å, b = 4.259 Å, c = 8.387 Å, and β = 107°. Figure 1(f) depicts the structure of $La_2O_2SO_4$, which consists of the alternating stacking of a $La_2O_2^{2+}$ layer and a layer of sulfate (SO_4^{2-}) along a-axis.[6, 14]

It is interestingand exciting that the current strategy, converting as-prepared La – coordination compounds into $La_2O_2SO_4$ hollow spheres, can be extended to a series of rare-earth elements, such as Pr, Nd, Sm, Eu, Gd, Tb, Dy, Y, Ho, and Yb. All the reflection patterns of calcined products are found to be very similar to that of $La_2O_2SO_4$ in monoclinic symmetry (Figure 2). Representative SEM images of light rare-earth

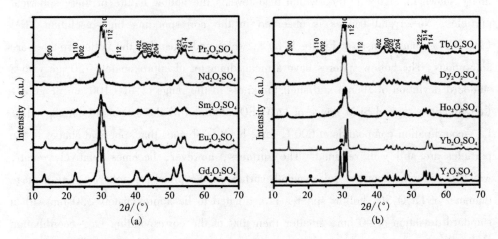

Figure 2

XRD patterns of (a) light and (b) heavy rare-earth oxysulfates hollow spheres obtained by calcination of corresponding rare-earth coordination compounds in air at 600 ℃ for 2 h. The peaks marked with asterisk, filled and empty squares are from Ho_2O_3, Yb_2O_3 and Y_2O_3, respectively.

oxysulfates such as $Pr_2O_2SO_4$, $Sm_2O_2SO_4$, and $Eu_2O_2SO_4$ are shown in Figure 3(a), 3(b) and 3(c), respectively, from which it can be seen that the products unexceptionally exhibit spherical morphologies with hollow interior. For example, the selected areas marked by a quadrangle in Figure 3(a) and 3(c) show $Pr_2O_2SO_4$ and $Eu_2O_2SO_4$ hollow spheres with broken shells. The inset clearly reveals hollow nature. For heavy rare-earth oxysulfates, the morphological features are similar with the light rare-earth counterpart. Figure 3(d) displays a typical SEM image of $Tb_2O_2SO_4$ hollow spheres. Careful observation shows that the shells of hollow spheres are composed of small nanoparticles with hollow interior [inset of Figure 3(d)].

On the other hand, the rare-earth coordination compound can also be obtained under identical conditions for the elements of Ce, Er, and Tm, but they are crystallized into oxide forms after calcination. The XRD patterns prove the formation of pure CeO_2, Er_2O_3, and Tm_2O_3 [Figure 4(a)]. Furthermore, the diffraction peaks of Y_2O_3, Ho_2O_3, and Yb_2O_3 can be found, as shown in Figure 2. It is well known that the radius of the trivalent rare-earth ions gradually decreases over the lanthanide series. In particular, the $4f^1$ electron configuration of Ce(Ⅲ) ions (1.143 Å) can easily transform into the much steadier $4f^0$ configuration of Ce(Ⅳ) ions (0.970 Å), which results in the formation of CeO_2. For Y(Ⅲ) ions, although it is much lighter in weight

Figure 3

Typical SEM and TEM images of (a) $Pr_2O_2SO_4$, (b) $Sm_2O_2SO_4$, (c) $Eu_2O_2SO_4$, and (d) $Tb_2O_2SO_4$ hollow spheres. The inset in (a), (b), and (d) depict the corresponding TEM images. The inset in (c) is a high-magnification SEM image obtained from the selected area marked by a quadrangle.

than the heavy rare-earth ions, the ionic radius (1.019 Å) is very close to that of Ho(III) ions (1.015 Å) due to the lanthanide contraction.[15] Thus, it is reasonable to believe that the formation of rare-earth oxides may be closely correlated to the structural stability and/or lanthanide contraction. Typical SEM image of CeO_2 is shown in Figure 4(b), which distinctly reveals the hollow nature of the spherical particles, consistent with TEM result [inset of Figure 4(b)]. The diameter distribution of Er_2O_3 hollow spheres is relatively uniform, as shown in Figure 4(c). Figure 4(d) depicts the TEM image of Tm_2O_3, which reveals the Tm_2O_3 hollow spheres have a rough surface with retained spherical contours. It is worth noting that all the rare-earth oxysulfates and oxides can be further functionalized through doping different rare-earth ions or coating procedure for potential applications in various fields. The doped and coated rare-earth

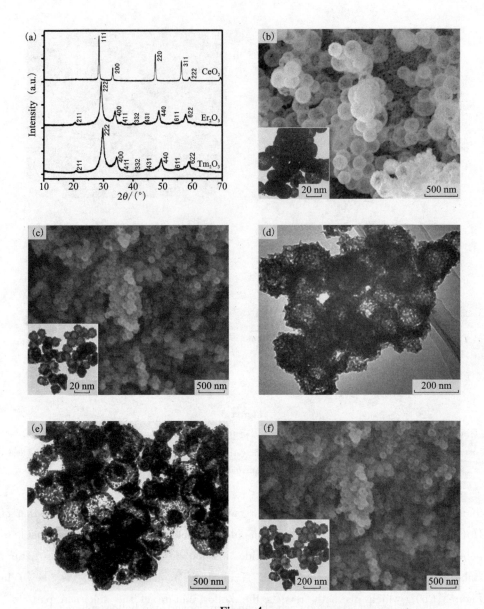

Figure 4

(a) XRD patterns of rare-earth oxides hollow spheres: face-centred cubic CeO_2 (JCPDS 04 – 0593), body-centred cubic Er_2O_3 (JCPDS 01 – 0827) and Tm_2O_3 (JCPDS 10 – 0350). Typical SEM and TEM images of rare-earth oxides hollow spheres: (b) CeO_2, (c) Er_2O_3, (d) Tm_2O_3, (e) double-shelled CeO_2/ZnS, and (f) Eu – doped CeO_2. The insets in (a), (c) and (f) are corresponding TEM images of as-prepared CeO_2, Er_2O_3 and Eu – doped CeO_2 hollow spheres, respectively.

oxysulfates and oxides maintain the morphological features as the corresponding host materials. The typical images of double-shelled $La_2O_2SO_4/ZnS$ and Eu – doped $La_2O_2SO_4$ hollow spheres are shown in Figure 4(e) and 4(f), which indicates the morphologies of Eu – doped $La_2O_2SO_4$ resemble that of $La_2O_2SO_4$ hollow spheres. Figure 4(e) indicates the typical TEM image of double-shelled CeO_2/ZnS hollow spheres. A hollow core of CeO_2 and shell of ZnS can be clearly observed.

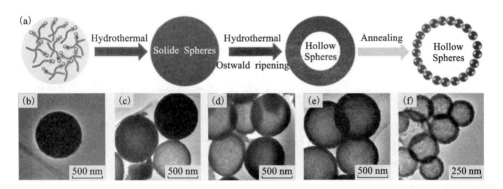

Figure 5

(a) Schematic illustration of the formation of rare-earth oxysulfates and oxides hollow spheres.
(b) – (e) TEM images exhibit the formation stages of La – coordination compounds. (f) TEM image of $La_2O_2SO_4$ hollow spheres.

Based on the experimental results, the possible scenario for the formation of hollow spheres is illustrated in Figure 5. L – cysteine plays an important role in the present synthesis procedure, which serves as both a biomolecular template and sulfide source. Firstly, L – cysteine is dispersed in deionized water and simultaneously capped by rare-earth ions due to the coordination of functional groups in the L – cysteine molecules, such as —NH_2, —COOH, and —SH, which has a strong tendency to coordinate with inorganic cations. Subsequently the freshly rare-earth coordination compound is unstable due to the high surface energy and tend to aggregate into spherical product driven by the minimization of interfacial energy, which can then convert into hollow spheres due to the Ostwald ripening process. Finally, rare-earth oxysulfates or oxides hollow spheres can be achieved via the calcination of corresponding rare-earth coordination compounds. Figure 5(b) – 5(e) indicate the typical TEM images of La – coordination compounds, in which solid spheres can gradually convert into hollow spheres due to the Ostwald ripening process.

More interestingly, Eu – doped rare-earth oxysulfate and oxide hollow spheres

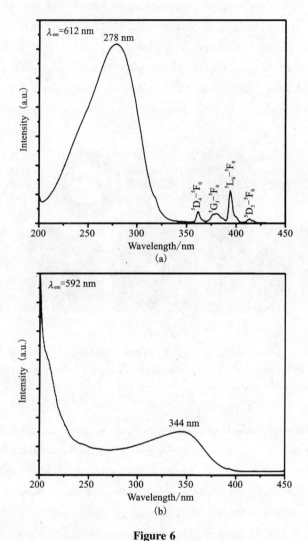

Figure 6

Room-temperature excitation spectra of Eu – doped (a) $La_2O_2SO_4$ and (b) CeO_2 hollow spheres monitored at 612 nm and 592 nm, respectively.

display apparent differences in photoluminescence properties compared with the corresponding amorphous coordination compounds. The excitation spectrum of $La_2O_2SO_4$: 0.05Eu exhibits a broad excitation band ranged from 205 nm to 350 nm with the main peak centered at about 278 nm, which can be attributed to the Eu – O charge-transfer (CT) transition [Figure 6(a)]. The relatively weak peaks in range from 350 nm to 430 nm were caused by the f-f transitions excitation lines of Eu(Ⅲ) ions, which are assigned to transitions between the ground $^7F_0$ level and the excited $^5D_4$, ^5G_J, $^5L_6$, and $^5D_3$ levels, respectively. Figure 6(b) indicates the excitation spectrum of CeO_2: 0.05Eu. It can be seen that the spectrum only exhibits one CT transition band centered at around 344 nm, while the f-f transitions excitation lines of Eu(Ⅲ) ions are not observed.

The emission spectra of the Eu – doped $La_2O_2SO_4$ and CeO_2 under the excitation wavelength of 278 nm and 344 nm originated from the excitation spectra (Figure 6) at room temperature are shown in Figure 7(a) and 7(b), respectively. It is striking that the emission maximums of the Eu – doped $La_2O_2SO_4$ and CeO_2 hollow spheres are 335 and 982 times higher than those of the amorphous precursory compounds, respectively, which exhibit a remarkable enhancement after calcination. The spectra are characteristic of the $^5D_0 - ^7F_J$ ($J = 0, 1, 2, 3, 4$) line emissions of the $4f^6$ configuration of the Eu(Ⅲ) ion. In particular, the major peaks positioned at 618 nm for Eu – doped $La_2O_2SO_4$ and nm at 592 nm for Eu – doped CeO_2 hollow spheres are attributed to $^5D_0 - ^7F_2$ and $^5D_0 - ^7F_1$ transition, respectively. Generally, the $^5D_0 - ^7F_2$ transition correlates well with the symmetry in local environments around Eu(Ⅲ) ions, while the $^5D_0 - ^7F_1$ transition is relatively insensitive to the local symmetry.[17] The ratio of the $^5D_0 - ^7F_2$ to $^5D_0 - ^7F_1$ transition, known as the asymmetric ratio, is 10.8 for the Eu – doped $La_2O_2SO_4$ and 0.8 for the Eu – doped CeO_2, suggesting that Eu – doped $La_2O_2SO_4$ hollow spheres have lower symmetry of the crystal field of the Eu(Ⅲ) ions than Eu – doped CeO_2 hollow spheres.[17b, 18] The reason may arise form the structural difference between the Eu – doped systems. As a result, the emission intensity of the Eu – doped $La_2O_2SO_4$ was higher than that of the Eu – doped CeO_2. These photoluminescent hollow spheres may be used as building blocks to construct thin film optical/display devices.

The magnetic properties of as-prepared $La_2O_2SO_4$ hollow spheres were measured with the Magnetic Property Measurement (MPMS XP – 5, SQUID). The temperature dependence of the zero-field-cooled (ZFC) and field-cooled (FC) magnetic susceptibilities of $La_2O_2SO_4$ measured under an applied field of 1000 Oe is shown in Figure 7(c). There is a signicant difference between ZFC and FC curves below about 100 K. The ZFC curve shows an obvious peak at around 60 K and a kink at about 50 K, which can be clearly observed from the inset of Figure 7(c). The feature indicates the possible presence of antiferromagnetic transitions, and such a behavior has also been observed in the case of $Gd_2O_2SO_4$ single crystals.[8a] Figure 7(d) depicts the isothermal magnetization curves measured at various temperatures. The magnetization curves are linearly increased with magnetic field between −10 kOe and 10 kOe, as shown in the inset of Figure 7(d). The magnetic moments at 2 K and 5 K depart from the linear behvior related to the magnetic field, and are close to saturation in high megnetic field, which may derive from the canted local spin array owing to edge effect. Furthermore, no hysteresise is discovered in these curves, revealing that ferromagnetic ordering component does not exist. These results further confirm that the antiferromagnetic behavior for $La_2O_2SO_4$ at low temperature with a Néel temperature $T_N = 60$ K may be a consequence of oxygen defects.

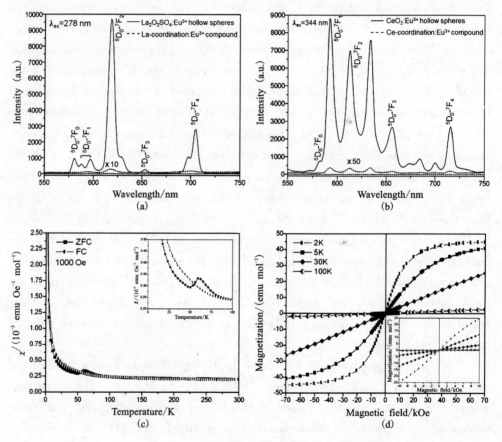

Figure 7

Room-temperature emission spectra of Eu-doped (a) $La_2O_2SO_4$, (b) CeO_2, and the corresponding coordination compounds. (c) The ZFC and FC magnetic susceptibilities of $La_2O_2SO_4$ measured under an applied field of 1000 Oe. The inset shows the enlarged plot in the low temperature region. (d) Isothermal magnetization curves of $La_2O_2SO_4$ at 2 K, 5 K, 30 K, and 100 K, respectively. The inset reveals the enlarged plot between −10 kOe and 10 kOe.

4 Conclusion

In summary, we have developed a general and facile synthetic strategy for fabrication of a series of high-yield and uniform rare-earth oxysulfates hollow spheres. Meanwhile, some oxides hollow spheres can be obtained due to the structural stability and/or lanthanide contraction. Furthermore, doped and coated rare-earth oxysulfates and oxide hollow spheres could also be successfully prepared, which will greatly endow the potential application in various fields. It is believed that the synthetic strategy may provide a general and effective route to synthesize and rationally design other inorganic micro/nanospheres with hollow interiors. These rare-earth oxysulfate and oxide hollow spheres, in particular rare-earth oxysulfates, are expected to bring new opportunities for further fundamental research, as well as for technological applications in optical/display devices, nanoscale reactors, magnetic devices, and oxygen storage materials.

References

[1] D. Mertz, P. Tan, Y. Wang, T. K. Goh, A. Blencowe, F. Caruso. Bromoisobutyramide as an intermolecular surface binder for the preparation of free-standing biopolymer assemblies [J]. Advanced Materials, 2011, 23(47): 5668 – 5673. (b) B. Wang, J. S. Chen, H. B. Wu, Z. Wang, X. W. Lou. Quasiemulsion-templated formation of α - Fe_2O_3 hollow spheres with enhanced lithium storage properties [J]. Journal of the American Chemical Society, 2011, 133 (43): 17146 – 17148. (c) D. Kim, E. Kim, J. Lee, S. Hong, W. Sung, N. Lim, C. G. Park, K, Kim. Direct synthesis of polymer nanocapsules: self-assembly of polymer hollow spheres through irreversible covalent bond formation [J]. Journal of the American Chemical Society, 2010, 132(28): 9908 – 9919. (d) Y. Zhao, L. Jiang. Hollow micro/nanomaterials with multilevel interior structures [J]. Advanced Materials, 2009, 21(36): 3621 – 3638. (e) Z. R. Shen, J. G. Wang, P. C. Sun, D. T. Ding T. H. Chen. Fabrication of lanthanide oxide microspheres and hollow spheres by thermolysis of pre-molding lanthanide coordination compounds [J]. Chemical Communications, 2009, 13: 1742 – 1744. (f) X. W. D. Lou, L. A. Archer, Z. Yang. Hollow micro-/nanostructures: synthesis and applications [J]. Advanced Materials, 2008, 20(21): 3987 – 4019. (g) X. Liang, B. Xu, S. Kuang, X. Wang. Multifunctionalized inorganic-organic rare earth hybrid microcapsules [J]. Advanced Materials, 2008, 20(19): 3739 – 3744. (h) H. Zeng, W. Cai, P. Liu, X. Xu, H. Zhou, C. Klingshirn, H. Kalt. ZnO-based hollow nanoparticles by selective etching: elimination and reconstruction of metal- semiconductor interface, improvement of blue emission and photocatalysis [J]. ACS Nano, 2008, 2(8): 1661 – 1670. (i) S. H. Im, U. Jeong, Y. Xia. Polymer hollow particles

with controllable holes in their surfaces [J]. Nature Materials, 2005, 4(9): 671-675.

[2] E. González, J. Arbiol, V. F. Puntes. Carving at the nanoscale: Sequential galvanic exchange and Kirkendall growth at room temperature [J]. Science, 2011, 334(6061): 1377-1380. (b) G. Réthoré, A. Pandit. Use of templates to fabricate nanoscale spherical structures for defined architectural control [J]. Small, 2010, 6(4): 488-498. (c) R. Yi, R. Shi, G. Gao, N. Zhang, X. Cui, Y. He, X. Liu, Hollow metallic microspheres: fabrication and characterization [J]. The Journal of Physical Chemistry C, 2009, 113(4): 1222-1226. (d) Y. Yin, R. M. Rioux, C. K. Erdonmez, S. Hughes, G. A. Somorjai, A. Paul Alivisatos. Formation of hollow nanocrystals through the nanoscale Kirkendall effect [J]. Science, 2004, 304(5671): 711-714.

[3] J. Liu, F. Liu, K. Gao, J. Wu D. Xue, Recent developments in the chemical synthesis of inorganic porous capsules [J]. Journal of Materials Chemistry, 2009, 19(34): 6073-6084. (b) H. Xu, W. Wang. Template synthesis of multishelled Cu_2O hollow spheres with a single-crystalline shell wall [J]. Angewandte Chemie International Edition, 2007, 46(9): 1489-1492. (c) X. Sun, J. Liu, Y. Li. Use of carbonaceous polysaccharide microspheres as templates for fabricating metal oxide hollow spheres [J]. Chemistry - A European Journal, 2006, 12(7): 2039-2047. (d) Z. Yang, Z. Niu, Y. Lu, Z. Hu, C. C. Han. Templated synthesis of inorganic hollow spheres with a tunable cavity size onto core-shell gel particles [J]. Angewandte Chemie, 2003, 115(17): 1987-1989. (e) S. W. Kim, M. Kim, W. Y. Lee, T. Hyeon. Fabrication of hollow palladium spheres and their successful application to the recyclable heterogeneous catalyst for Suzuki coupling reactions [J]. Journal of the American Chemical Society, 2002, 124(26): 7642-7643. (f) Z. Zhong, Y. Yin, B. Gates, Y. Xia. Preparation of mesoscale hollow spheres of TiO_2 and SnO_2 by templating against crystalline arrays of polystyrene beads [J]. Advanced Materials, 2000, 12(3): 206-209.

[4] (a) Q. Lu, F. Gao, S. Komarneni. Biomolecule-assisted synthesis of highly ordered snowflakelike structures of bismuth sulfide nanorods [J]. Journal of the American Chemical Society, 2004, 126(1): 54-55. (b) M. Knez, A. M. Bittner, F. Boes, C. Wege, H. Jeske, E. Maiβ, K. Kern. Biotemplate synthesis of 3-nm nickel and cobalt nanowires [J]. Nano Letters, 2003, 3(8): 1079-1082.

[5] B. Li, Y. Xie, Y. Xue. Controllable synthesis of CuS nanostructures from self-assembled precursors with biomolecule assistance [J]. The Journal of Physical Chemistry C, 2007, 111(33): 12181-12187. (b) P. Zhao, T. Huang, K. Huang. Fabrication of indium sulfide hollow spheres and their conversion to indium oxide hollow spheres consisting of multipore nanoflakes [J]. The Journal of Physical Chemistry C, 2007, 111(35): 12890-12897.

[6] S. Zhukov, A. Yatsenko, V. Chernyshev, E. Tserkovnaya, O. Antson, J. Hölsä, P. Baulés. Structural study of lanthanum oxysulfate $(LaO)_2SO_4$[J]. Materials Research Bulletin, 1997, 32(1): 43-50.

[7] T. Kijima, T. Shinbori, M. Sekita, M. Uota, G. Sakai. Abnormally enhanced Eu^{3+} emission in

$Y_2O_2SO_4$: Eu^{3+} inherited from their precursory dodecylsulfate-templated concentric-layered nanostructure [J]. Journal of Luminescence, 2008, 128(3): 311-316.

[8] W. Paul. Magnetism and magnetic phase diagram of $Gd_2O_2SO_4$ I. Experiments [J]. Journal of Magnetism and Magnetic Materials, 1990, 87(1): 23-28. (b) H. G. Kahle, A. Kasten. Magnetic structures and interaction constants in the two-dimensional Ising antiferromagnets $RE_2O_2SO_4$ (RE = Dy, Ho, Tb) [J]. Journal of Magnetism and Magnetic Materials, 1983, 31: 1081-1083. (c) H. Hülsing, H. G. Kahle, M. Schwab, H. J. Schwarzbauer. Specific heat of rare earth oxisulphates ($RE_2O_2SO_4$) [J]. Journal of Magnetism and Magnetic Materials, 1978, 9(1): 68-70.

[9] D. Zhang, F. Yoshioka, K. Ikeue, M. Machida. Synthesis and oxygen release/storage properties of Ce-substituted La-oxysulfates, $(La_{1-x}Ce_x)_2O_2SO_4$ [J]. Chemistry of Materials, 2008, 20(21): 6697-6703. (b) M. Machida, K. Kawamura, T. Kawano, D. Zhang, K. Ikeue. Layered Pr-dodecyl sulfate mesophases as precursors of $Pr_2O_2SO_4$ having a large oxygen-storage capacity [J]. Journal of Materials Chemistry, 2006, 16(30): 3084-3090.

[10] (a) H. G. Kahle, A. Kasten. Magnetic structures and interaction constants in the two-dimensional Ising antiferromagnets $RE_2O_2SO_4$ (RE = Dy, Ho, Tb) [J]. Journal of Magnetism and Magnetic Materials, 1983, 31: 1081-1083. (b) H. Hülsing, H. G. Kahle, A. Kasten. Magnetic ordering of $Dy_2O_2SO_4$ and $Ho_2O_2SO_4$ [J]. Journal of Magnetism and Magnetic Materials, 1983, 512: 15-18.

[11] (a) M. Machida, K. Kawamura, K. Ito, K. Ikeue. Large-capacity oxygen storage by lanthanide oxysulfate/oxysulfide systems [J]. Chemistry of Materials, 2005, 17(6): 1487-1492. (b) M. Machida, K. Kawamura, K. Ito, Novel oxygen storage mechanism based on redox of sulfur in lanthanum oxysulfate/oxysulfide [J]. Chemical Communications, 2004, 6: 662-663.

[12] J. Liang, R. Ma, F. Geng, Y. Ebina, T. Sasaki. $Ln_2(OH)_4SO_4 \cdot n H_2O$ (Ln = Pr to Tb; n ~ 2): a new family of layered rare-earth hydroxides rigidly pillared by sulfate ions [J]. Chemistry of Materials, 2010, 22(21): 6001-6007.

[13] (a) X. Liu, R. Ma, Y. Bando, T. Sasaki. Layered cobalt hydroxide nanocones: microwave-assisted synthesis, exfoliation, and structural modification [J]. Angewandte Chemie International Edition, 2010, 49(44): 8253-8256. (b) X. Liu, R. Yi, N. Zhang, R. Shi, X. Li, G. Qiu. Cobalt hydroxide nanosheets and their thermal decomposition to cobalt oxide nanorings [J]. Chemistry - An Asian Journal, 2008, 3(4): 732-738.

[14] M. Machida, T. Kawano, M. Eto, D. Zhang, K. Ikeue. Ln dependence of the large-capacity oxygen storage/release property of Ln oxysulfate/oxysulfide systems [J]. Chemistry of Materials, 2007, 19(4): 954-960.

[15] (a) M. J. Martínez-Lope, J. A. Alonso, M. Retuerto, M. T. Fernández-Díaz. Evolution of the crystal structure of RVO_3 (R = La, Ce, Pr, Nd, Tb, Ho, Er, Tm, Yb, Lu, Y) perovskites from neutron powder diffraction data [J]. Inorganic Chemistry, 2008, 47(7): 2634-2640. (b)

R. Yan, X. Sun, X. Wang, Q. Peng, Y. Li. Crystal structures, anisotropic growth, and optical properties: controlled synthesis of lanthanide orthophosphate one-dimensional nanomaterials [J]. Chemistry – A European Journal, 2005, 11(7): 2183 – 2195. (c) H. L. Sun, C. H. Ye, X. Y. Wang, J. R. Li, S. Gao, K. B. Yu. Lanthanide contraction and pH value controlled structural change in a series of rare earth complexes with p-aminobenzoic acid [J]. Journal of Molecular Structure, 2004, 702(1): 77 – 83.

[16] J. Li, H. C. Zeng. Hollowing Sn-doped TiO_2 nanospheres via Ostwald ripening [J]. Journal of the American Chemical Society, 2007, 129(51): 15839 – 15847.

[17] (a) S. Ida, C. Ogata, M. Eguchi, W. J. Youngblood, T. E. Mallouk, Y. Matsumoto. Photoluminescence of perovskite nanosheets prepared by exfoliation of layered oxides, $K_2Ln_2Ti_3O_{10}$, $KLnNb_2O_7$, and $RbLnTa_2O_7$ (Ln: lanthanide ion) [J]. Journal of the American Chemical Society, 2008, 130(22): 7052 – 7059. (b) H. X. Mai, Y. W. Zhang, R. Si, Z. G. Yan, L. D. Sun, L. P. You, C. H. Yan. High-quality sodium rare-earth fluoride nanocrystals: controlled synthesis and optical properties [J]. Journal of the American Chemical Society, 2006, 128(19): 6426 – 6436.

[18] R. Si, Y. W. Zhang, L. P. You, C. H. Yan. Rare-earth oxide nanopolyhedra, nanoplates, and nanodisks [J]. Angewandte Chemie International Edition, 2005, 117(21): 3320 – 3324.

☆ X. H. Liu, D. Zhang, J. Jiang, N. Zhang, R. Z. Ma, H. B. Zeng, B. P. Jia, S. B. Zhang, G. Z. Qiu. General synthetic strategy for high-yield and uniform rare-earth oxysulfate ($RE_2O_2SO_4$, RE = La, Pr, Nd, Sm, En, Gd, Tb, Dy, Y, Ho and Yb) hollow spheres [J]. RSC Advances, 2012, 2(25): 9362 – 9365.

Chapter V

Transition-Metal Oxide and Chalcogenides Hollow Spheres

Controllable Fabrication and Magnetic Properties of Double-shell Cobalt Oxides Hollow Particles

Abstract: Double-shell cobalt monoxide (CoO) hollow particles were successfully synthesized by a facile and effective one-pot solution-based synthetic route. The inner architecture and outer structure of the double-shell CoO hollow particles could be readily created through controlling experimental parameters. A possible formation mechanism was proposed based on the experimental results. The current synthetic strategy has good prospects for the future production of other transition-metal oxides particles with hollow interior. Furthermore, double-shell cobalt oxide (Co_3O_4) hollow particles could also be obtained through calcinating corresponding CoO hollow particles. The magnetic measurements revealed double-shell CoO and Co_3O_4 hollow particles exhibit ferromagnetic and antiferromagnetic behaviour, respectively.

1 Introduction

Multiple-shell hollow particles have attracted widely attention in recent years because of the promising properties and potential applications in various fields, such as lithium-ion batteries,[1-4] sensors,[5,6] photocatalysis,[7,8] dye-sensitized solar cells (DSSCs),[9,10] drug/gene delivery,[11] microreactors,[12] and so forth. A variety of methods including template-assisted synthesis,[13] ionic exchange reaction,[14] Ostwald ripening,[15-18] Kirkendall effects,[19] and selective etching process[20,21] have been developed to fabricate multiple-shell hollow particles. Among these methods, Ostwald ripening, a well-known physical phenomenon, which contains that the smaller size particles dissolving into the liquid phase as a nutrient supply for the growth of larger crystals and results in the formation of the hollow interior spaces, is one of the most effective approaches to the rational design of complex hollow structures. Up to now, remarkable progress has been made for the fabrication of multiple-shell hollow structures by Ostwald ripening process. In particular, Wang and co-workers demonstrated a multistep Ostwald ripening approach for the geometry-controlled fabrication of Cu_2O

particles with multilayered shell-in-shell interior structures.[22] Zeng and co-workers reported the synthesis of double-shell SiO_2 hollow spheres via Ostwald ripening process under solvothermal conditions.[23] Very recently, a family of multiple-shell structures, ($Cu_2O@$)$_n Cu_2O$ ($n = 1-4$), has been synthesized through Ostwald ripening treatment at room temperature.[24] Despite these successes, developing the Ostwald ripening methods for the fabrication of multiple-shell hollow particles remains a highly sophisticated challenge.

Cobalt oxides have drawn increasing attention in the past decades on the basis of their distinctive electronic, magnetic, and catalytic properties and wide variety of applications. It is well known that CoO and Co_3O_4 are two especially important forms among the various cobalt oxides based on their distinctive structural features and fascinating properties. Generally, CoO, crystallizing in the rocksalt (NaCl)-like structure, consists of two face-centered-cubic (fcc) sublattices of Co^{2+} and O^{2-} ions, while Co_3O_4 belongs to the spinel-like structure based on a cubic close packing array of oxide ions, in which Co^{2+} ions occupy the tetrahedral 8a sites and Co^{3+} ions occupy the octahedral 16d sites.[25] Because of the unique morphology-dependent properties, immense efforts have been dedicated to developing facile and effective approaches for the preparation of cobalt oxides with controllable morphologies, such as nanocone,[26,27] nanobelt,[28] nanoring,[29] nanocube,[30-33] nanowire,[34] nanotube,[35] hollow sphere,[36] etc. Recently, wang and co-workers have achieved a significant breakthrough in the synthesis of multiple-shell cobalt oxides hollow spheres by the use of carbonaceous microspheres (CMSs) as sacrificial templates.[37,38] Unfortunately, template-assisted synthesis generally involves tedious procedures including preparation of sacrificial templates, deposition of the designed materials, and selective removal of the templates via chemical etching or thermal decomposition. On the other hand, the removal of the CMSs templates may result in very low yield of target product based on the ion-absorption, which is unfavorable to fulfilling the application prospects of the multiple-shell hollow spheres.

Herein, we present a one-pot solvothermal method to prepare double-shell CoO hollow particles. Interestingly, the inner and outer architecture of the double-shell CoO hollow particles can be readily tuned through controlling experimental parameters. The results demonstrate that the double-shell hollow particles might form through Ostwald ripening. By using CoO hollow particles as the precursor, double-shell Co_3O_4 hollow particles can also be obtained via thermal decomposition process. Notably, the current synthetic strategy may provide an effective route for the synthesis of other transition-metal oxides particles, and is thus promising for achieving unique architectures with hollow interior for a wide range of applications.

2 Experimental Section

The synthesis use commercially available reagents: cabalt (Ⅱ) acetylacetonate [Co(acac)$_2$, chemical grade, 98%, a&k], 1 - octadecene (ODE, technical grade, 90%, ACROS), oleic acid (OA, chemical grade, SCRC), and oleylamine (OAm, technical grade, approximate C18-content 80% - 90%, ACROS) were of chemically pure and used as received.

2.1 Preparation of double-shell CoO hollow particles. Typical synthetic procedures are summarized as follows: 0.2571 g Co(acac)$_2$ (1mmol) was dissolved into 20 mL ODE with OA and OAm (in molar ratios of 4∶10) to form colloid mixture under vigorous stirring for 10 min at room temperature. Then the mixture was sealed in a Teflon-lined and maintained at 260℃ for 1 - 24 h. The autoclave was cooled to room temperature. The brown products were washed for several times with absolute ethanol and hexane. Finally, the products were dried in vacuum at 60℃ for 6 h.

2.2 Preparation of double-shell Co$_3$O$_4$ hollow particles. The double-shell Co$_3$O$_4$ hollow particles could be prepared by the thermal decomposition of as-prepared CoO hollow particles obtained at 260℃ for 2 - 12 h as precursors calcinated at 600℃ for 2 h in air.

2.3 Characterization. The structure and phase composition of the products were characterized on a X-ray diffractometer (XRD, D/max2550 VB +) with Cu K_α radiation (λ = 1.5418 Å). The morphologies and sizes of the products were characterized by a field-emission scanning electron microscopy (FE - SEM, Sirion 200) and transmission electron microscope (TEM, Tecnai G2 F20). High-resolution TEM (HRTEM) images and SAED patterns were obtained from the TEM. Magnetic measurements were conducted using a Quantum Design MPMS XP - 5 superconducting quantum interference device (SQUID).

3 Results and Discussion

Figure 1(a) displays a representative SEM image of as-prepared product obtained using cabalt (Ⅱ) acetylacetonate as cobalt source at 260℃ for 8 h, in which a large quantity of near-spherical particles with good uniformity were achieved under current conditions. The particles have a mean size of about 300 nm. There exist many broken particles, which reveals that CoO particles obviously possess hollow interiors. The inset is a high-magnification SEM image obtained from a selected area of Figure 1(a). Herein, the hollow interiors of as-prepared CoO particles can be clearly identified. Figure 1(b) presents a typical TEM image of CoO particles, which also evidently exhibits CoO particles with hollow interiors. Figure 1(c) indicates a typical TEM image of an individual CoO hollow particle. With careful observation, the double-shell structure of CoO hollow particle can be clearly observed. The outer and inner shell thicknesses of the double-shell CoO hollow particle are estimated to be about 8 nm and 100 nm, respectively. A selected area electron diffraction (SAED) pattern taken from the individual CoO hollow particles, as shown in Figure 1(d), illustrates single-crystalline structure of double-shell CoO hollow particles. Figure 1(e) depicts the HRTEM image of the individual particle. The lattice spacing is calculated to be about 0.24 nm, agreeing well with the value of {111} lattice planes of cubic CoO. The crystal structures of as-prepared products were characterized by X-ray powder diffraction (XRD). All of the reflections of the XRD pattern, as shown in Figure 1(f), can be readily indexed as a face-centered cubic phase of CoO (JCPDS 65 - 2902) with lattice constant $a = 0.426$ nm [space group: Fm - 3m (No. 225)]. No impurity peak was observed, indicating the high purity of the product obtained under such conditions. The sharp diffraction peaks also reflects the good crystallinity of as-prepared product.

For a better understanding of the growth process of double-shell hollow particles, the influences of reaction time on the morphologies of products have been investigated. Figure 2 summarizes a series of morphological observations supposed to be in different stages of forming double-shell hollow particles. When the reaction time was decreased to 1 h, the product is mostly made up of near-spherical solid particles with average size of about 150 nm, as shown in Figure 2(a). Figure 2(b) depicts a typical TEM image of product obtained for 4 h. The mean size of CoO particles is increased to about 250 nm, suggesting the elongation in size with increasing reaction time. Closer observation reveals that almost all particles possesses hollow interiors at this stage. Extending the reaction time to 12 h, the double-shell structures of the CoO hollow particles with sizes in the range 300 - 400 nm can be observed more clearly [Figure 2(c)]. It is noteworthy that if the reaction time is further extended to 24 h, almost 100% double-shell CoO hollow particles with uniform sizes about 500 nm can be obtained, as shown in Figure 2(d). In particular, the outer shell thickness of CoO hollow particle is increased to about 45 nm, comparable to that of CoO hollow particles obtained for 12 h.

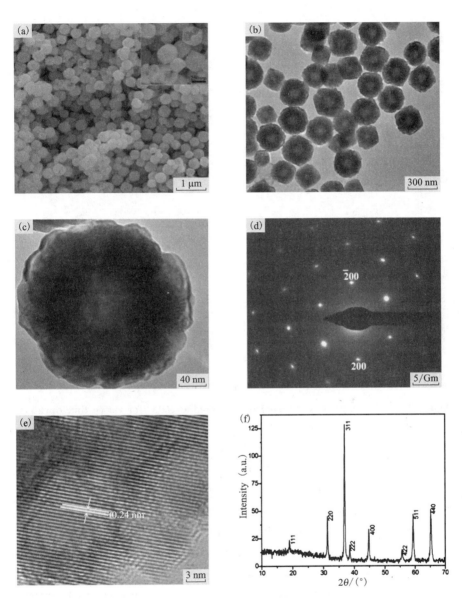

Figure 1 Solvothermal synthesis of double-shell CoO hollow particles obtained at 260 °C for 8 h

(a) SEM and (b, c) TEM images. The inset in (a) shows a higher magnification SEM image. (d) SAED pattern and (e) HRTEM image of the individual double-shell CoO hollow particle. (f) XRD pattern of as-prepared double-shell CoO hollow particles obtained at 260 °C for 8 h.

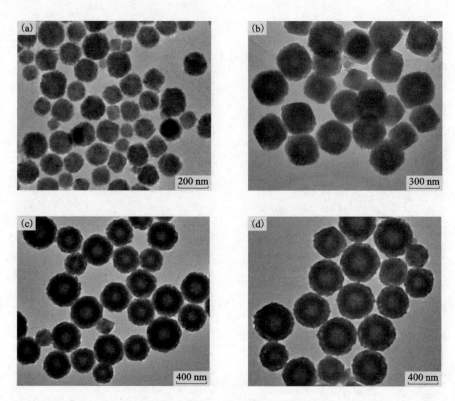

Figure 2 TEM images of CoO particles prepared at 260 ℃ for varied time durations
a) 1 h, (b) 4 h, (c) 12 h, and (d) 24 h.

Double-shelled Co_3O_4 hollow particles could be also successfully obtained via calcination method using corresponding CoO hollow particles obtained at 260 ℃ for 8 h as precursors. Figure 3(a) depicts the SEM image of as-prepared Co_3O_4 obtained by calcination of corresponding CoO hollow particles at 600 ℃ for 2 h in air. Compared with CoO hollow particles, the average size of as-prepared Co_3O_4 is estimated to be about 400 nm, which may be related to the possible oxidation of CoO to Co_3O_4. An individual particles with broken shell shown in the inset of Figure 3(a) demonstrates the ball-in-ball structure of the Co_3O_4 hollow particles. A typical TEM image of as-prepared Co_3O_4 hollow particles is shown in Figure 3(b). Herein, the Co_3O_4 hollow particles with double-shell structure can be clearly observed. In the inset, an SAED pattern was well indexed to spinel Co_3O_4, revealing a polycrystalline nature of the calcined hollow particles. The double-shell structure is also clearly revealed by TEM observation of an individual Co_3O_4 hollow particle, as shown in Figure 3(c). The inset

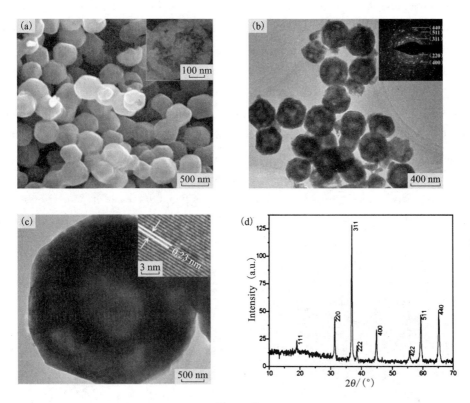

Figure 3

(a) SEM image of double-shell Co_3O_4 hollow particles obtained by calcination of as-prepared CoO hollow particles at 600 ℃ for 2 h in air. The inset shows an individual Co_3O_4 hollow particle. (b) Low-magnification and (c) high-magnification TEM images of double-shell Co_3O_4 hollow particles. The insets in (b) and (c) show SAED pattern and HRTEM image of double-shell Co_3O_4 hollow particles, respectively. (d) XRD pattern of as-prepared double-shell Co_3O_4 hollow particles.

displays the corresponding HRTEM image, which provides further insight into the structure of Co_3O_4 hollow particle. Lattice spacing is measured to be about 0.23 nm, which is consistent with the interplanar spacings of {222} for spinel Co_3O_4. Figure 3(d) shows the typical XRD pattern of double-shell Co_3O_4 hollow particles. All the reflections in the XRD pattern can be indexed as a face-centered cubic phase of spinel Co_3O_4 (JCPDS 43 – 1003) with lattice constant $a = 0.808$ nm [space group: Fd – 3m (No. 227)]. No impurity peaks were observed, indicating that cubic CoO was completely converted into spinel Co_3O_4.

The evolution of double-shell structure during the calcination of as-prepared CoO particles obtained at 260 ℃ for varied time durations as the precursor at 600 ℃ for 2 h directly mirrors a possible scenario for the formation of double-shell Co_3O_4 hollow particles. Figure 4 summarizes a series of morphological observations of forming double-shell Co_3O_4 hollow particles. Figure 4(a) shows a typical TEM image of the porous Co_3O_4 with hollow interior prepared by calcination of corresponding CoO particles obtained at 260 ℃ for 2 h. With careful observation, the shells of Co_3O_4 hollow particles were found to be composed of many pores with a mean diameter of 15 nm. It is noteworthy that if the reaction time of precursors was extended to 4 h, large interior space of Co_3O_4 hollow particles can be generated, as shown in Figure 4(b). Typical TEM image shown in Figure 4(c) indicates a significant increase of pore sizes in the shells. Furthermore, the interior space is also further improved. This suggests both the

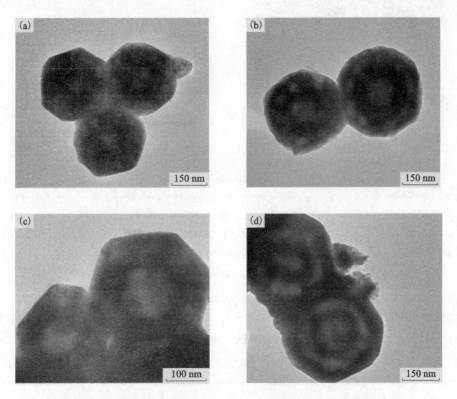

Figure 4 TEM images exhibit the formation stages of double-shell Co_3O_4 hollow particles prepared by calcination of CoO particles as the precursor obtained at 260 ℃ for varied time durations
(a) 2 h, (b) 4 h, (c) 6 h, and (d) 12 h.

expansion in interior space and the addition in pore size with increasing reaction time. More interestingly, Co_3O_4 hollow particles with apparent double-shell structure can be clearly identified using CoO hollow particles as precursors obtained at 260 ℃ for 6 h and 12 h, as shown in Figure 4(c) and 4(d), respectively.

To investigate the mechanism of double-shell CoO hollow particles, based on the experimental results, the possible scenario for the formation of double-shell CoO hollow particles is illustrated in Figure 5. Firstly, $Co(acac)_2$ dissolves in organic solvent and then undergoes thermal decomposition to form small particles under solvothermal conditions. Subsequently the fresh small particles is unstable due to the high surface energy and tends to aggregate into larger solid particles driven by the minimization of interfacial energy, which can then convert into hollow particles due to the Ostwald ripening process.[39] Finally, owing to the existence of anisotropic outer surfaces of the hollow particles, hollowing space gradually takes place at a particular region underneath the outer surface and leads to the formation of double-shelled hollow structures.[40, 41]

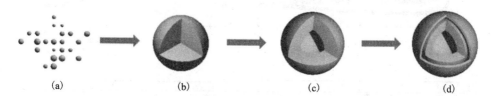

Figure 5 Schematic exhibition of the formation of double-shell CoO hollow particles

The magnetic properties of double-shell cobalt oxides hollow particles were measured on a superconducting quantum interference device (SQUID). The temperature dependences of the zero-field-cooled (ZFC) and field-cooled (FC) magnetization of the double-shell CoO hollow particles measured under an applied field of 100 Oe are shown in Figure 6(a). It is clear that there is a significant difference between ZFC and FC curves of double-shell CoO hollow particles at low temperature. Compared with FC curve, the ZFC curve depicts a distinct peak at 4.8 K, suggesting ferromagnetic (FM) behavior below 4.8 K, which may be attributed to superparamagnetic cobalt particles.[42] Figure 6(b) shows the ZFC and FC curves of double-shell Co_3O_4 hollow particles measured under an applied field of 100 Oe. The feature indicates the possible presence of antiferromagnetic (AFM) transitions. The AFM transition occurs at about 32 K (Néel temperature, T_N), being far lower than that of the bulk Co_3O_4 known at about 40 K, which is possibly resulted from the finite size and surface effect of double-shell Co_3O_4 hollow particles.[43, 44] The hysteresis loops for double-shell CoO and Co_3O_4 hollow particles at 2 K are shown in Figure 6(c) and 6(d), respectively. The coercivity value H_c for the CoO is about 1298 Oe at 2 K, indicating the presence of ferromagnetic ordering component. For Co_3O_4 hollow particles, despite of that the low temperature data show a slight curvature, the magnetization curves are nearly linear at 2 K, also indicative an antiferromagnetic ground state.

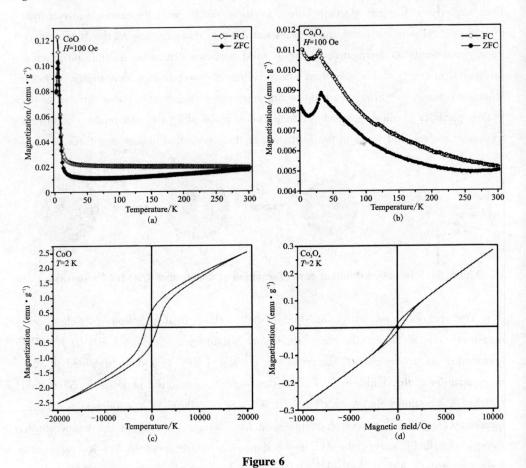

Figure 6

ZFC and FC magnetization curves for (a) CoO and (b) Co_3O_4 hollow particles measured under an applied field of 100 Oe. Isothermal magnetization curves for (c) CoO and (d) Co_3O_4 hollow particles at 2 K.

4 Conclusion

In summary, double-shell CoO hollow particles could be synthesized using a facile solvothermal method based on Ostwald ripening. The inner architecture and outer structure could be readily created through controlling experimental parameters. Double-shell Co_3O_4 hollow particles could also be obtained via the calcination of corresponding CoO hollow particles. The current synthetic strategy may provide an effective route for the synthesis of other transition-metal oxides particles with hollow interior. Owning to the excellent magnetic properties of double-shell cobalt oxides hollow particles, it is expected that the products will exhibit some important applications in, e. g. , magnetic semiconductors and other advanced materials.

References

[1] Y. Zhao, L. Jiang. Hollow micro-/nanomaterials with multilevel interior structures [J]. Advanced Materials, 2009, 21(36): 3621 – 3638.

[2] S. Xu, C. M. Hessel, H. Ren, R. B. Yu, Q. Jin, M. Yang, H. J. Zhao, D. Wang. $\alpha - Fe_2O_3$ Multi-shelled hollow microspheres for lithium ion battery anodes with superior capacity and charge retention [J]. Energy & Environmental Science, 2014, 7(2): 632 – 637.

[3] X. Wang, X. L. Wu, Y. G. Guo, Y. Zhong, X. Cao, Y. Ma, J. Yao. Synthesis and lithium storage properties of Co_3O_4 nanosheet-assembled multishelled hollow spheres [J]. Advanced Functional Materials, 2010, 20(10): 1680 – 1686.

[4] H. B. Wu, A. Pan, H. H. Hng, X. W. D. Lou. Template-assisted formation of rattle-type V_2O_5 hollow microspheres with enhanced lithium storage properties [J]. Advanced Functional Materials, 2013, 23(45): 5669 – 5674.

[5] H. Zhang, Q. Zhu, Y. Zhang, Y. Wang, L. Zhao, B. Yu. One-pot synthesis and hierarchical assembly of hollow Cu_2O microspheres with nanocrystals-composed porous multishell and their gas-sensing properties [J]. Advanced Functional Materials, 2007, 17(15): 2766 – 2771.

[6] X. Lai, J. Li, B. A. Korgel, Z. Dong, Z. Li, F. Su, J. Du, D. Wang. General synthesis and gas-sensing properties of multiple-shell metal oxide hollow microspheres [J]. Angewandte Chemie Intenational Edition, 2011, 123(12): 2790 – 2793.

[7] Y. Zeng, X. Wang, H. Wang, Y. Dong, Y. Ma, J. Yao. Multi-shelled titania hollow spheres fabricated by a hard template strategy: enhanced photocatalytic activity [J]. Chemical Communications, 2010, 46: 4312 – 4314.

[8] L. Cao, D. Chen, R. A. Caruso. Surface-metastable phase-initiated seeding and ostwald ripening: a facile fluorine-free process towards spherical fluffy core/shell, yolk/shell, and hollow

anatase nanostructures [J]. Angewandte Chemie International Edition, 2013, 52(42): 10986 – 10991.

[9] Z. Dong, H. Ren, C. M. Hessel, J. Wang, R. Yu, Q. Jin, M. Yang, Z. Hu, Y. Chen, Z. Tang, H. Zhao, D. Wang. Quintuple-shelled SnO_2 hollow microspheres with superior light scattering for high-performance dye-sensitized solar cells [J]. Advanced Materials, 2014, 26 (6): 905 – 909.

[10] Z. Dong, X. Lai, J. E. Halpert, N. Yang, L. Yi, J. Zhai, D. Wang, Z. Tang, L. Jiang. Accurate control of multishelled ZnO hollow microspheres for dye-sensitized solar cells with high efficiency [J]. Advanced Materials, 2012, 24(8): 1046 – 1049.

[11] L. Tan, T. Liu, L. Li, H. Liu, X. Wu, F. Gao, X. He, X. Meng, D. Chen, F. Tang. Uniform double-shelled silica hollow spheres: acid/base selective-etching synthesis and their drug delivery application [J]. RSC Advances, 2013, 3(16): 5649 – 5655.

[12] J. Liu, S. Z. Qiao, S. B. Hartono, G. Q. Lu. Monodisperse yolk-shell nanoparticles with a hierarchical porous structure for delivery vehicles and nanoreactors [J]. Angewandte Chemie International Edition, 2010, 49(29): 4981 – 4985.

[13] X. W. Lou, C. Yuan, L. A. Archer. Shell-by-shell synthesis of tin oxide hollow colloids with nanoarchitectured walls: cavity size tuning and functionalization [J]. Small, 2007, 3(2): 261 – 265.

[14] S. Xiong, H. C. Zeng. Serial ionic exchange for the synthesis of multishelled copper sulfide hollow spheres [J]. Angewandte Chemie International Edition, 2012, 51(4): 949 – 952.

[15] D. P. Wang, H. C. Zeng. Creation of interior space, architecture of shell structure, and encapsulation of functional materials for mesoporous SiO_2 spheres [J]. Chemistry of Materials, 2011, 23(22): 4886 – 4899.

[16] H. C. Zeng. Synthetic architecture of interior space for inorganic nanostructures [J]. Journal of Materials Chemistry, 2006, 16(7): 649 – 662.

[17] L. Cao, D. Chen, R. A. Caruso. Surface-metastable phase-initiated seeding and ostwald ripening: a facile fluorine-free process towards spherical fluffy core/shell, yolk/shell, and hollow anatase nanostructures [J]. Angewandte Chemie International Edition, 2013, 52(42): 10986 – 10991.

[18] H. C. Zeng. Ostwald ripening: a synthetic approach for hollow nanomaterials [J]. Current Nanoscience, 2007, 3(2): 177 – 181.

[19] L. Xie, J. Zheng, Y. Liu, Y. Li, X. Li. Synthesis of Li_2NH hollow nanospheres with superior hydrogen storage kinetics by plasma metal reaction [J]. Chemistry of Materials, 2007, 20(1): 282 – 286.

[20] K. An, S. G. Kwon, M. Park, H. B. Na, S. I. Baik, J. H. Yu, D. Kim, J. S. Son, Y. W. Kim, I. C. Song, W. K. Moon, H. M. Park, T. Hyeon. Synthesis of uniform hollow oxide nanoparticles through nanoscale acid etching [J]. Nano Letters, 2008, 8(12): 4252 – 4258.

[21] G. Li, Q. Shi, S. J. Yuan, K. G. Neoh, E. T. Kang, X. Yang. Alternating silica/polymer multilayer hybrid microspheres templates for double-shelled polymer and inorganic hollow microstructures [J]. Chemistry of Materials, 2010, 22(4): 1309 – 1317.

[22] L. Zhang, H. Wang. Interior structural tailoring of Cu_2O shell-in-shell nanostructures through multistep ostwald ripening [J]. The Journal of Physical Chemistry C, 2011, 115(38): 18479 – 18485.

[23] D. P. Wang, H. C. Zeng. Creation of interior space, architecture of shell structure, and encapsulation of functional materials for mesoporous SiO_2 spheres [J]. Chemistry of Materials, 2011, 23(22): 4886 – 4899.

[24] C. C. Yec, H. C. Zeng. Synthetic architecture of multiple core-shell and yolk-shell structures of $(Cu_2O@)_nCu_2O$ ($n = 1 - 4$) with centricity and eccentricity [J]. Chemistry of Materials, 2012, 24(10): 1917 – 1929.

[25] R. R. Shi, G. Chen, W. Ma, D. Zhang, G. Z. Qiu, X. H. Liu. Shape-controlled synthesis and characterization of cobalt oxides hollow spheres and octahedra [J]. Dalton Transactions, 2012, 41(19): 5981 – 5987.

[26] X. H. Liu, R. Z. Ma, Y. Bando, T. Sasaki. Layered cobalt hydroxide nanocones: microwave-assisted synthesis, exfoliation, and structural modification [J]. Angewandte Chemie International Edition, 2010, 49(44): 8253 – 8256.

[27] X. H. Liu, R. Z. Ma, Y. Bando, T. Sasaki. High-yield preparation, versatile structural modification, and properties of layered cobalt hydroxide nanocones [J]. Advanced Functional Materials, 2014, 24(27): 4292 – 4302.

[28] L. Tian, H. L. Zou, J. X. Fu, X. F. Yang, Y Wang, H. L. Guo, X. H. Fu, C. L. Liang, M. M. Wu, P. K. Shen, Q. M. Gao. Topotactic conversion route to mesoporous quasi-single-crystalline Co_3O_4 nanobelts with optimizable electrochemical performance [J]. Advanced Functional Materials, 2010, 20(4): 617 – 623.

[29] X. H. Liu, R. Yi, N. Zhang, R. R. Shi, X. Li, G. Z. Qiu. Cobalt hydroxide nanosheets and their thermal decomposition to cobalt oxide nanorings [J]. Chemistry-An Asian Journal, 2008, 3(4): 732 – 738.

[30] X. H. Liu, G. Z. Qiu, X. G. Li. Shape-controlled synthesis and properties of uniform spinel cobalt oxide nanocubes [J]. Nanotechnology, 2005, 16(12): 3035.

[31] R. Xu, H. C. Zeng. Self-generation of tiered surfactant superstructures for one-pot synthesis of Co_3O_4 nanocubes and their close- and non-close-packed organizations [J]. Langmuir, 2004, 20(22): 9780 – 9790.

[32] R. Xu, H. C. Zeng. Mechanistic investigation on salt-mediated formation of free-standing Co_3O_4 nanocubes at 95℃ [J]. The Journal of Physical Chemistry B, 2003, 107(4): 926 – 930.

[33] J. Feng, H. C. Zeng. Size-controlled growth of Co_3O_4 nanocubes [J]. Chemistry of Materials, 2003, 15(14): 2829 – 2835.

[34] P. Y. Keng, B. Y. Kim, I. -B. Shim, R. Sahoo, P. E. Veneman, N. R. Armstrong, H. Yoo, J. E. Pemberton, M. M. Bull, J. J. Griebel, E. L. Ratcliff, K. G. Nebesny, J. Pyun. Colloidal polymerization of polymer-coated ferromagnetic nanoparticles into cobalt oxide nanowires [J]. ACS Nano, 2009, 3(10): 3143-3157.

[35] L. Zhuo, J. Ge, L. Cao, B. Tang. Solvothermal synthesis of CoO, Co_3O_4, $Ni(OH)_2$ and $Mg(OH)_2$ nanotubes [J]. Crystal Growth & Design, 2008, 9(1): 1-6.

[36] X. Wang, X. L. Wu, Y. G. Guo, Y. Zhong, X. Cao. Y. Ma, J. Yao, Synthesis and lithium storage properties of Co_3O_4 nanosheet-assembled multishelled hollow spheres [J]. Advanced Functional Materials, 2010, 20(10): 1680-1686.

[37] J. Wang, N. Yang, H. Tang, Z. Dong, Q. Jin, M. Yang, D. Kisailus, H. Zhao, Z. Tang, D. Wang. Accurate control of multishelled Co_3O_4 hollow microspheres as high-performance anode materials in lithium-ion batteries [J]. Angewandte Chemie International Edition, 2013, 125(25): 6545-6548..

[38] X. Lai, J. Li, B. A. Korgel, Z. Dong, Z. Li, F. Su, J. Du, D. Wang. General synthesis and gas-sensing properties of multiple-shell metal oxide hollow microspheres [J]. Angewandte Chemie International Edition, 2011, 123(12): 2790-2793.

[39] J. Li, H. C. Zeng. Hollowing Sn-doped TiO_2 nanospheres via ostwald ripening [J]. Journal of the American Chemical Society, 2007, 129(51): 15839-15847.

[40] Q. Xie, F. Li, H. Guo, L. Wang, Y. Chen, G. Yue, D. L. Peng. Template-free synthesis of amorphous double-shelled zinc-cobalt citrate hollow microspheres and their transformation to crystalline $ZnCo_2O_4$ microspheres [J]. ACS Applied Materials & Interfaces, 2013, 5(12): 5508-5517.

[41] J. Liu, H. Xia, D. Xue, L. Lu. Double-shelled nanocapsules of V_2O_5-based composites as high-performance anode and cathode materials for Li ion batteries [J]. Journal of the American Chemical Society, 2009, 131(34): 12086-12087.

[42] S. Kundu, A. J. Nelson, S. K. McCall, T. van Buuren, H. Liang. Shape-influenced magnetic properties of CoO nanoparticles [J]. Journal of Nanoparticle Research, 2013, 15(5): 1-13.

[43] W. L. Roth. The magnetic structure of Co_3O_4 [J]. Journal of Physics and Chemistry of Solids, 1964, 25(1): 1-10.

[44] L. He, C. P. Chen, N. Wang, W. Zhou, L. Guo. Finite size effect on Neel temperature with Co_3O_4 nanoparticles [J]. Journal of Applied Physics, 2007, 102(10): 103911-103911-4.

☆ D. Zhang, J. Y. Zhu, N. Zhang, T. Liu. L. M. Chen, X. H. Liu, R. Z. Ma, H. T. Zhang, G. Z. Qiu. Controllable fabrication and magnetic properties of double-shell cobalt oxides hollow particles [J]. Scientific Reports, 2015, 5: 8737.

Shape-Controlled Synthesis and Characterization of Cobalt Oxides Hollow Spheres and Octahedra

Abstract: We demonstrate that single-crystalline cobalt monoxide (CoO) hollow spheres and octahedra could be selectively synthesized via thermal decomposition of cobalt(II) acetylacetonate in 1-octadecene solvent in the presence of oleic acid and oleylamine. The morphologies and sizes of as-prepared CoO nanocrystals could be controlled by adjusting the reaction parameters. Cobalt oxide (Co_3O_4) hollow spheres and octahedra could also be selectively obtained via calcination method using corresponding CoO hollow spheres and octahedra as precursors. The morphology, size and structure of the final products were investigated in detail by XRD, SEM, TEM, HRTEM, DSC, TG, and XPS. The results revealed that the electrochemical performance of cobalt oxide hollow spheres is much better than that of cobalt oxide octahedra, which may be related to the degree of crystallinity, size, and morphology of cobalt oxides.

1 Introduction

Inorganic nanocrystals with novel morphologies and desired compositions have drawn immense attention due to their unique morphology-dependent and composition-dependent physicochemical properties and their importance in basic scientific research and potential technology applications.[1-3] Cobalt oxides have aroused much more attention in recent years on the basis of their distinctive electronic, magnetic, and catalytic properties and wide variety of practical and potential applications. It is well known that CoO and Co_3O_4 are two especially important forms among the various cobalt oxides based on their distinctive structural features and fascinating properties. Especially, CoO, crystallizing in the rock salt structure, is promising functional materials owing to their potential applications based on magnetic, catalytic and gas-sensing properties.[4-6] Co_3O_4 belongs to the normal spinel crystal structure based on a cubic close packing array of oxide ions, in which Co(II) ions occupy the tetrahedral

8a sites and Co(Ⅲ) ions occupy the octahedral 16d sites. Co_3O_4 is an important magnetic p-type semiconductor that has been demonstrated to have considerable application in, e. g., heterogeneous catalysts, gas sensors, electrochromic devices, solar energy absorbers, pigments, etc.[7-11]

In recent years, considerable effort has been devoted to preparing cobalt oxides nanocrystals with controlled morphology and desired compositions. Several chemical and physicochemical methods have been employed to prepare cobalt oxides nanocrystals, for example, spray pyrolysis,[12] chemical vapor deposition,[13] sputtering,[14] thermal decomposition,[15] electrospinning technique,[16] electrochemical and sonochemical synthesis.[17] Recently a variety of novel shapes such as cobalt oxides nanocone,[18] nanobelt,[19] nanoring,[20] nanocube,[21] nanofiber,[22] nanorods,[23] nanotube,[24] multishelled hollow sphere[25,26] have been reported. However, a few routes have been proposed for the synthesis of pure CoO nanocrystals. As far as we know, CoO nanocrystals have been obtained mainly based on the following two methods: thermal decomposition of metal-surfactant complexes in noncoordinating solvents[27-29] and controlled oxidization of $Co_2(CO)_8$ or metallic cobalt nanocrystals.[30,31]

Herein we demonstrated that single-crystalline CoO hollow spheres and octahedra could be selectively synthesized in large quantities by thermal decomposition of cobalt(Ⅱ) acetylacetonate [$Co(acac)_2$] in 1-octadecene solvent in the presence of oleic acid and oleylamine. The morphologies and sizes of as-prepared CoO nanocrystals could be controlled by adjusting the reaction parameters, such as surfactants and reaction atmosphere. In particular, we could obtain CoO nanocrystals via a facile thermolysis method without protected atmosphere. Co_3O_4 hollow spheres and octahedra could also be obtained via calcination method using corresponding CoO as precursor. The results revealed that the electrochemical performance of hollow spheres is much better to compare with that of octahedra, which may be related to the degree of crystallinity, morphology, and particle size of cobalt oxides. It is worthy to note that the current synthetic strategy can be used to synthesize other metal oxides nanocrystals, and it will have a good prospect in the future large-scale application due to its high yields, simple reaction apparatus.

2　Experimental Section

The synthesis use commercially available reagents: cobalt(Ⅱ) acetylacetonate [$Co(acac)_2$, analytical grade, 99%, ACROS], trioctylphosphine oxide (TOPO, analytical grade, 99%, ACROS), 1 - octadecene (ODE, technical grade, 90%, ACROS), oleic acid (OA, chemical grade, SCRC), oleylamine (OAm, technical grade, approximate C_{18}-content 80 - 90%, ACROS), hexadecyl trimethyl ammonium bromide (CTAB), polyvinyl pyrrolidone (PVP) and sodium dodecyl sulfate (SDS) were of analytical grade and used without any further purification.

All the reactions were conducted in a three-neck flask equipped with a stirring and heating attachment. In a typical reaction, 0.2571 g Co(acac)$_2$ (1 mmol) and 0.0364 g CTAB (0.1 mmol) were dissolved into 20 mL ODE with 4 mmol OA and 10 mmol OAm. The reaction mixture was degassed for 20 min at room temperature using high purity gas of nitrogen. The solution was heated up to 260℃ with approximately 8℃/min under vigorous stirring. At the beginning of reaction, a balloon was used to seal the system and buffer high pressure generated during the reaction. Reflux for 2 h, the solution was then cooled to room temperature, and a mixture of alcohol and hexane was added to the solution to yield a waxy precipitate, which was separated by centrifugation. At the end, the products were dried in an oven at room temperature. The resulting precipitate was found to be re-dispersible in many organic solvents, such as n-hexane and chloroform. Shape control of the products can be achieved by adjusting some of parameters, such as the surfactants and reaction atmosphere.

The obtained products were characterized on a D/max2550 VB + X-ray powder diffractometer (XRD) with Cu-K_α radiation ($\lambda = 1.54178$ Å). The operation voltage and current were kept at 40 kV and 40 mA, respectively. The size and morphology of as-prepared products were determined at 20 kV by a XL30 S-FEG scanning electron microscope (SEM) and at 160 kV by a JEM - 200CX transmission electron microscope (TEM) and a JEOL JEM - 2010F high-resolution transmission electron microscope (HRTEM). Thermogravimetric analysis and differential scanning calorimetry (TGA/DSC) were carried out with a NETZSCH STA - 449C simultaneous TG-DTA/DSC apparatus at a heating rate of 10 K/min in flowing air. X-ray photoelectron spectroscopy (XPS) data were acquired on a VG ECA-LAB MK2. Sintering process was conducted in box-type resistance furnace (SX - 4 - 10).

In order to evaluate the electrochemical characteristics, electrodes were fabricated using the powder by mixing 80 wt.% active materials, 10 wt.% carbon black and 10 wt.% polyvinylidene fluoride (PVDF) dissolved in N-methyl-2-pyrrolidone. The resultant slurries were spread on aluminum foil substrates. After coating, the electrodes were pressed and dried at 120℃ under vacuum for 12 h and then pressed between two stainless steel plates at 1 MPa. Prior to cell assembling, the electrodes with area of

0.64 cm^2 were dried at 120℃ for 4 h under vacuum. The testing cells had a typical two-electrode construction using a polypropylene microporous sheet as the separator, and 1 M LiPF$_6$ dissolved in ethylene carbonate (EC) and dimethyl carbonate (DMC) (1:1, v/v) are used as the electrolyte. A pure lithium foil was used as the counter electrode and the samples under test were used as the working electrode during electrochemical measurements. All cells were assembled in an argon-filled glove box. The electrode capacity was measured by a galvanostatic charge/discharge experiment with a current density of 200 mA/g at a potential between 0 V and 2.5 V.

3 Results and Discussion

Figure 1(a) shows the low-magnification TEM image of CoO obtained by using 0.1 mmol CTAB in the presence of oleic acid and oleylamine at 260℃ for 2 h under a nitrogen environment. We found that the product mainly consisted of hollow spheres with an outer-diameter of about 250 nm and wall thickness of about 50 nm. The remarkable contrast between the shells and centers indicates the hollow nature of the spheres. More details of structure can be obtained from a further magnified image, as shown in Figure 1(b). The surfaces of hollow spheres are rough with some flocculates. High-resolution TEM provided further insight into the structures of as-prepared products. Figure 1(c) shows the HRTEM image of the area marked in Figure 1(b). The lattice spacing is calculated to be 0.21 nm, corresponding to d-spacing of (200) crystal plane of cubic CoO, which further confirmed its single-crystalline structure. The inset of Figure 1(c) is the fast Fourier Transform (FFT) image of HRTEM result. Figure 1(d) shows the XRD pattern of as-prepared CoO hollow spheres. All diffraction peaks can be readily indexed to a pure cubic structure with lattice constants $a = 4.263$ Å, well consistent well with the standard PDF database (JCPDS file No.65-2902). Diffraction peaks of the sample are sharp and narrow, indicative of a relatively good crystallinity.

Generally, the chemical and physical properties of inorganic materials are directly related to their morphology, size, and phase composition. Therefore, it is very important to tailor their electronic, magnetic and optical properties through the morphology and shape control of the final products. Figure 2(a) shows the low-magnification SEM image of CoO nanocrystals prepared by using 0.1 mmol CTAB in the presence of oleic acid and oleylamine at 260℃ for 2 h without a protected atmosphere. The products comprise large quantities of CoO octahedra. More details of the structure

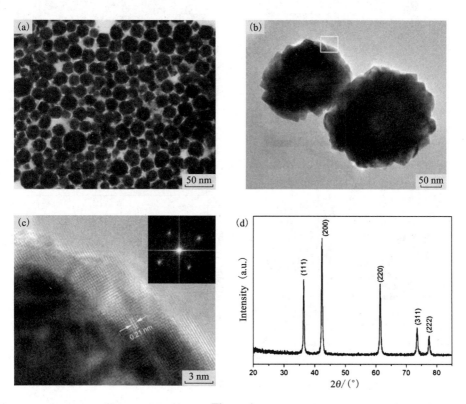

Figure 1

(a) Low-magnification TEM image of CoO hollow spheres obtained by using 0.1 mmol CTAB in the presence of oleic acid and oleylamine at 260℃ for 2 h under nitrogen environment; (b) High-magnification TEM image of CoO hollow spheres; (c) HRTEM image of the area marked by the square in Figure (b); (d) XRD pattern of CoO hollow spheres. The inset of (c) is the fast Fourier transform image.

can be obtained from a further magnified image, as shown in Figure 2(b), from which uniform CoO octahedra with the size of about 150 nm and edge length of 80 – 120 nm can be seen clearly. Figure 2(c) is the representative TEM image of CoO octahedra, which clearly shows those octahedral structures from different viewing angles. Inset of Figure 2(c) is the HRTEM image taken from individual octahedra. The lattice spacing was calculated to be 0.21 nm, corresponding to the d-spacing of (200) crystal plane of cubic CoO. Figure 2(d) shows the XRD pattern of as-prepared octahedral CoO nanocrystals. All diffraction peaks can be readily indexed to a pure cubic structure with lattice constants $a = 4.274$ Å (JCPDS file No. 65 – 2902), which indicate that cubic CoO phase can be obtained without protected atmosphere.

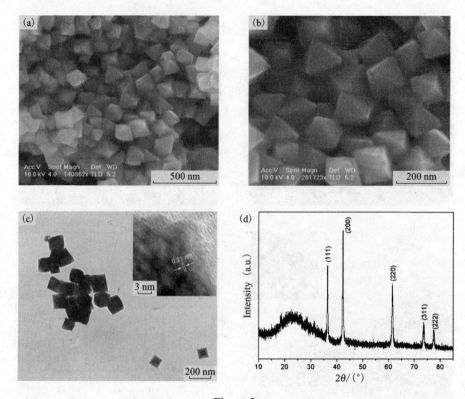

Figure 2

(a) Low-magnification SEM image of as-prepared octahedral-like CoO nanocrystals prepared by using 0.1 mmol CTAB in the presence of oleic acid and oleylamine at 260 ℃ for 2 h without a protected atmosphere; (b) High-magnification SEM image of octahedral CoO nanocrystals selected from (a); (c) TEM image of octahedral CoO nanocrystals; (d) XRD pattern of octahedral CoO nanocrystals. Inset of (c) is the corresponding HRTEM image.

The surfactant is also a very important factor influencing the morphology of the final products.[32] Figure 3(a) and 3(b) show the typical SEM and TEM images of as-prepared product obtained by using 0.03 g PVP in the presence of oleic acid and oleylamine at 260 ℃ for 2 h under nitrogen environment. The product is nearly monodisperse and the heterogeneous contrast effect from its TEM image [Figure 3(b)] makes a guess that nanocrystals might be unsmooth surface. The SEM image confirms this point. On the whole, these quasi-spherical particles have the trends of polyhedron. However, it is worth noting that in our previous research with similar reaction system, PVP is a more favorable surfactant for the dispersity of nanoparticles.[33] In particular, when 0.1 mmol TOPO was employed as surfactant to control morphology of product, CoO nanocrystals with octahedral structures can be obtained, as shown in Figure 3(c).

SEM image shows that the majority morphology of as-prepared CoO nanocrystals is octahedra. The edge-length of octahedral CoO is about 80 nm. Figure 3(d) is the TEM image of CoO nanocrystals prepared by using 0.1 mmol SDS in the presence of oleic acid and oleylamine at 260 ℃ for 2 h under nitrogen environment, from which it can be seen that products are composed of irregular monodisperse nanoparticles in the size range of 20 – 30 nm.

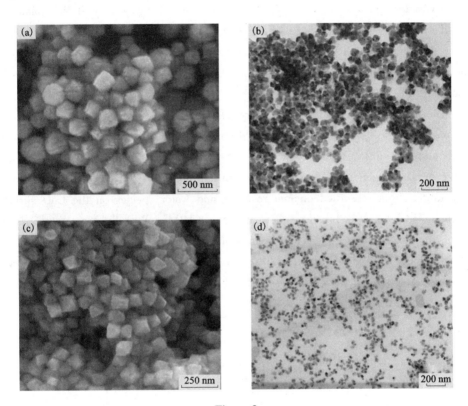

Figure 3

Morphologies of CoO nanocrystals prepared obtained by using different surfactants in the presence of oleic acid and oleylamine at 260 ℃ for 2 h under nitrogen environment: SEM image (a) and TEM image (b) of CoO nanocrystals prepared with 0.03 g PVP; SEM image (c) of CoO nanocrystals prepared with 0.1 mmol TOPO; TEM image (d) of CoO nanocrystals prepared with 0.1 mmol SDS.

It should be mentioned that it is failure to control the morphology of the products if the reaction was adopted in air atmosphere. Ammonia released accompanied as the reaction provides an anaerobic atmosphere. All of these experimental condition and phenomenon ensured the formation of cobalt monoxide, preventing it from further oxidation. Experimental results suggest that reaction atmosphere and surfactant CTAB may play the important roles in the formation of CoO hollow spheres. As is well-known, most of the methods for preparation of hollow spheres are template-based ones, in which preparation of sacrificial templates, either hard or soft ones.[34,35] We considered that nitrogen gas could help surfactant CTAB to serve as template, and CoO crystallites deposited along the template surface, resulting in formation of the shells around the CTAB vesicles. Based on the hypothesis above, it is suggestive that the amount of CTAB surfactant used should be effective on the formation of CoO hollow spheres. When the amount of CTAB increased from 0.5 mmol to 1.0 mmol, CoO spheres with hollow interior and floccular structures can be clearly observed due to the excessive amount of CTAB.

The thermal behavior of CoO nanocrystals oxidized to Co_3O_4 was investigated with thermogravimetric analysis (TGA) and differential scanning calorimetry (DSC) in the temperature range 25 – 750 ℃ with the heating rate of 10 ℃/min in air, as shown in Figure 4. The gradual mass loss (3.39%) in the range 25 – 290 ℃ can be attributed to evaporation of the absorbed organic residues and water species on the CoO surfaces accompanied by a weak broad exothermal peak at 240 ℃. With the temperature rising, the obvious gravimetric gain can be seen from Figure 4. The first rapid mass increase (2.78%) occurred from 290 ℃ to 320 ℃, illustrating quickly oxidation of the surface of products. The second mass increase (1.81%) occurred before 445 ℃, which was

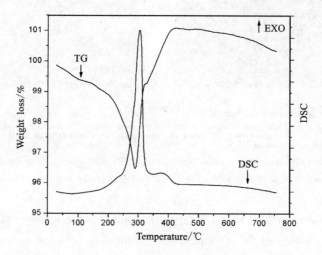

Figure 4   DSC and TGA curves of CoO hollow spheres

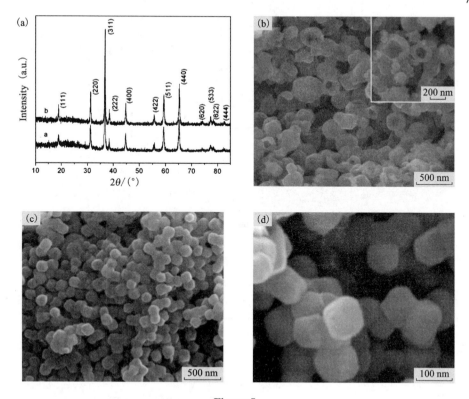

Figure 5

(a) XRD patterns of inner hollow [curve (a)] and octahedral [curve (b)] Co_3O_4 nanocrystals calcined at 600 ℃ for 2 h using the corresponding CoO nanocrystals as precursor, respectively. (b) Low-magnification SEM image of Co_3O_4 nanocrystals with inner hollow structures calcined at 600 ℃ for 2 h using CoO hollow spheres as precursor. (c) Low-magnification SEM image of octahedral Co_3O_4 nanocrystals calcined at 600 ℃ for 2 h using octahedral CoO as precursor; (d) High-magnification SEM image of octahedral Co_3O_4 nanocrystals; Inset of (b) is the corresponding high-magnification SEM image of Co_3O_4 hollow spheres.

attributed to the further oxidation of CoO. The DSC curve of the sample displayed two exothermal peaks in the gravimetric gain region: one is centered at 290 ℃ and another at about 380 ℃. The totally mass increase on the curve was smaller than the theoretical value (7.1 %), which may be associated with partially oxidation of CoO in early stage and organic residues.

Co_3O_4 hollow spheres and octahedra could be selectively obtained via calcination method using corresponding CoO hollow spheres and octahedra as precursor. Figure 5(a) shows XRD patterns of Co_3O_4 hollow spheres [curve (a)] and octahedra [curve (b)] obtained by the calcination of corresponding CoO nanocrystals at 600 ℃

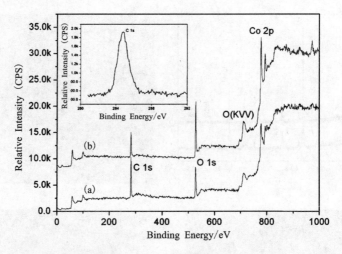

Figure 6 Full survey XPS spectra of CoO [curve (a)] and Co_3O_4 [curve (b)] hollow spheres. The inserted profile is the C 1s survey curve.

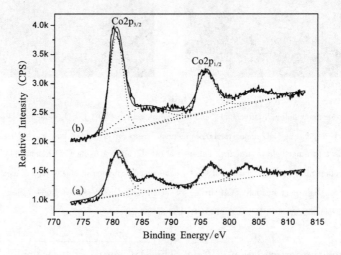

Figure 7

Co 2p electron XPS spectra for (a) CoO hollow spheres and (b) Co_3O_4 hollow spheres. Data were acquired with Mg K_α radiation and were fitted after removal of a linear background.

for 2 h in air, respectively. All the reflections in the XRD patterns can be indexed to the pure face-centered-cubic phase [space group: Fd3m (No. 227)] of spinel cobalt oxide with lattice constant $a = 8.085$ Å (JCPDS file no. 78 – 1969). Without impurity peaks were observed, which indicates that cubic CoO was completely converted into the spinel structure Co_3O_4. Low-magnification SEM image of Co_3O_4 hollow spheres is showed in Figure 5(b). And high-magnification SEM image inset in Figure 5(b) further shows the spheres with hollow interior. Figure 5(c) and 5(d) are low- and high-magnification SEM images of Co_3O_4 octahedra, respectively. The morphology analysis of Co_3O_4 hollow spheres and octahedra reveals the perfect inheritation from corresponding CoO nanocrystals.

X-ray photoelectron spectroscopy (XPS) is a reliable method for intensive investigation of samples, especially efficient for studying the statues of atoms less than 10 nm deeply from the surface of metastable materials with partially filled valence band. It was introduced here to evaluate the surface of as-prepared CoO and Co_3O_4 nanocrystals. The full survey spectra of the two samples are just in coherence with each other, which are shown in Figure 6, where each of the main peaks are indexed to O 1s, C 1s and Co 2p regions, confirming the cobalt oxides with nonexistence of impurities. The inserted profile is the binding energy (BE) spectrum for C 1s electrons of CoO sample. A main peak emerged at BE of 284.8 eV, which is the standard BE for the amorphous carbon as an inert reference. The surface of the nanomaterials is very vulnerable to being affected by circumstance because of its special immense surface area and high activity. The cubic 3d transition metal Co has a partially filled valence band within which strong electron correlations among the 3d electrons were localized. It is of great importance to study the stability and status of atoms special requirements rising from electrochemical application of cobalt oxides.

Table 1 XPS peak positions (BE in eV) obtained for CoO and Co_3O_4 hollow spheres compared with the literature values

Samples	Co $2p_{3/2}$/eV	Satellite/eV	Co $2p_{1/2}$	Satellite/eV
CoO (this work)	780.890	786.278	796.773	802.796
Co_3O_4 (this work)	780.591	784.929	795.935	803.621
CoO	780.0 – 780.919,[38, 41]	785.339	796.0 – 796.519,[39-41]	802.639
Co_3O_4	779.5, 780.743	787.945	794.5, 795.038	803.045

CoO and Co_3O_4 have different coupling into the two possible final states giving rise to the main and satellite peaks. It is well known that cobalt series occupy the identical BE region belonging to Co $2p_{1/2}$ and Co $2p_{3/2}$ electrons, but the distinction among many of the cobalt compounds is still vague, as chemical shift of main peaks in XPS spectra of

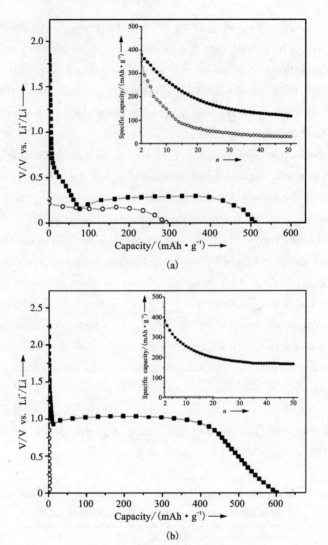

Figure 8

The first discharge curves of the Li − CoO (a) and Li − Co_3O_4 (b) cells made by CoO and Co_3O_4 with different morphologies, hollow spheres (filled squares) and octahedral structures (empty circles), at a current density of 200 mA/g. Inset: cycle performance of Li − profiles of the cells at a current density of 200 mA/g at room temperature.

Co 2p electrons in Co^{2+} and Co^{3+} is not obvious enough. The survey spectra for Co 2p region of inner hollow CoO nanocrystals [curve (a)] prepared using 0.1 mmol CTAB with the molar ration of OA to OAm 4 : 10 at 260 ℃ for 2 h in nitrogen atmosphere and Co_3O_4 nanocrystals [curve (b)] sintered at 600 ℃ for 2 h using inner hollow CoO nanocrystals as precursor are shown in Figure 7. The curves are fitted by pseudo-Voigt function; the binding energy (BE) values emerged on the fitting lines are listed in Table 1. It was in fact observed that the BE values of the most intense Co photoelectronic peak (Co 2p) did not allow a clear distinction between CoO pure Co(II) and Co_3O_4 $Co^{II}Co_2^{III}O_4$. The two major peaks appeared on the curve a shown in Figure 7 are at 780.890 eV and 796.773 eV, separated by 15.9 eV, corresponding to the Co $2p_{3/2}$ and Co $2p_{1/2}$ spin-orbit peaks, respectively, of the CoO phase, which are consistent with the supposed BE values of earlier reported literature.[36, 37] Furthermore, two satellites located at approximately 6 eV above the primary binding energy peaks were detected at 786.278 eV and 802.796 eV, which were used as a fingerprint for the recognition of high-spin Co (II) species in CoO.[38, 39] In addition, the Co 2p spectrum of Co_3O_4 emerged in Figure 7b yields a Co $2p_{3/2}$ peak at 780.591 eV and Co $2p_{1/2}$ peaks at 795.935 eV with the corresponding satellites at 784.929 eV and 803.621 eV, respectively, are just in the same shape and are the identical BE of Co $2p_{3/2}$ and Co $2p_{1/2}$ electrons, respectively, which is in agreement with the literature for Co_3O_4.[38, 40-42] Unlike in Co (II) compounds, in the low-spin Co (III) compounds, the satellite structure is weak or missing.[41] Co_3O_4, a mixed-valance oxide shows a weak satellite structure symptomatic of shake-up from the minor Co (II) component.[42] The (Co $2p_{1/2}$- Co $2p_{3/2}$) energy separation is approximately 15.3 eV, which is also identical to that of the pure phase Co_3O_4 in the literature.[43, 44]

The electrochemical performances of as-prepared CoO and Co_3O_4 with different morphologies, hollow spheres and octahedral structures, were investigated. Figure 8 gives the typical first discharge curves of Li – CoO and Li – Co_3O_4 cells made by CoO and Co_3O_4 with inner-hollow (filled squares) and octahedral structures (empty circles), between 0 V and 2.5 V at a current density of 200 mA/g. Generally speaking, nanocrystals with an inner-hollow structure have higher capacity than octahedra, whether CoO or Co_3O_4. Figure 8 (a) shows that inner-hollow CoO nanocrystals have an initial discharge capacity as high as 510.8 mAh/g, and 116.6 mAh/g after 50 cycles. Octahedral ones have an initial discharge of at 291.7 mAh/g and 29.7 mAh/g after 50 cycles. The discharge curves of Co_3O_4 hollow spheres electrode shown in Figure 8(b) indicated that the discharge voltage decreased sharply from 2.5 V to the discharge plateau located in the range of 1.0 – 0.8 V at the first discharge cycle. The initial capacity of as-prepared Co_3O_4 hollow spheres reaches 601.2 mAh/g. The Co_3O_4 hollow spheres electrode demonstrates a stable reversible lithium storage capacity of 165.9 mAh/g within 50 cycles. Obviously, metal oxides nanocrystals with different morphologies have different lithium storage capacity, and the reason of this difference is complicated, just as many literature reports have mentioned.[45, 46] The higher discharge capacity of hollow spheres may be attributed to its unique morphology. The inner hollow structure is favorable for increasing the interface area between electrode and electrolyte, which can result in a higher diffusion rate and faster electrode kinetics, and also can increase the usage factor of the active component of the electrode.[47] However, the discharge capacity of octahedral Co_3O_4 electrode was closed to zero, indicating its poor electrode kinetics.

4 Conclusion

In summary, CoO hollow spheres and octahedra can be selectively synthesized via thermal decomposition of cobalt(Ⅱ) acetylacetonate in 1 - octadecene solution using oleic acid and oleylamine as capping ligands. The morphologies and sizes of as-prepared CoO nanocrystals can be controlled by adjusting the reaction parameters. This simple and reliable synthetic strategy may be carried out to synthesize other metal oxides nanocrystals. Furthermore, Co_3O_4 hollow spheres and octahedra can be obtained by the calcination of corresponding CoO hollow spheres and octahedra at 600℃ for 2 h. The electrochemical performance of cobalt oxides hollow spheres is much better than that of cobalt oxides octahedra. These cobalt oxides hollow spheres and octahedra can also be expected to bring new opportunities for further fundamental research, as well as for technological applications in catalysts, solid-state sensors, and as anode materials in Li - ion rechargeable batteries.

References

[1] A. P. Alivisatos. Semiconductor clusters, nanocrystals, and quantum dots [J]. Science, 1996, 271(5251): 933 - 937.

[2] M. V. Kovalenko, M. Scheele, D. V. Talapin. Colloidal nanocrystals with molecular metal chalcogenide surface ligands [J]. Science, 2009, 324(5933): 1417 - 1420.

[3] N. Tessler, V. Medvedev, M. Kazes, S. H. Kan, U. Banin. Efficient near-infrared polymer nanocrystal light-emitting diodes [J]. Science, 2002, 295(5559): 1506 - 1508.

[4] V. Skumryev, S. Stoyanov, Y. Zhang, G. Hadjipanayis, D. Givord, J. Nogués. Beating the superparamagnetic limit with exchange bias [J]. Nature, 2003, 423(6942): 850 - 853.

[5] W. S. Seo, J. H. Shim, S. J. Oh, E. K. Lee, N. H. Hur, J. T. Park. Phase-and size-controlled synthesis of hexagonal and cubic CoO nanocrystals [J]. Journal of the American

Chemical Society, 2005, 127(17): 6188 –6189.

[6] A. Lagunas, A. M. I Payeras, C. Jimeno, M. A. Pericàs. A doubly folded spacer in a self-assembled hybrid material [J]. Chemical Communications, 2006, 12: 1304 – 1306.

[7] W. Y. Li, L. N. Xu, J. Chen. Co_3O_4 nanomaterials in Lithium-ion batteries and gas sensors [J]. Advanced Functional Materials, 2005, 15(5): 851 – 857.

[8] S. Weichel, P. Møller, J. Chen. Annealing-induced microfaceting of the CoO (100) surface investigated by LEED and STM [J]. Surface Science, 1998, 399(2): 219 – 224..

[9] X. W. Xie, W. J. Shen. Morphology control of cobalt oxide nanocrystals for promoting their catalytic performance [J]. Nanoscale, 2009, 1(1): 50 – 60.

[10] M. Ando, T. Kobayashi, S. Lijima, M. Haruta. Optical recognition of CO and H_2 by use of gas-sensitive Au – Co_3O_4 composite films [J]. Journal of Materials Chemistry, 1997, 7(9): 1779 – 1783.

[11] R. J. Wu, C. H. Hu, C. T. Yeh, P. G. Su. Nanogold on powdered cobalt oxide for carbon monoxide sensor [J]. Sensors and Actuators B: Chemical, 2003, 96(3): 596 – 601.

[12] V. R. Shinde, S. B. Mahadik, T. P. Gujar, C. D. Lokhande. Supercapacitive cobalt oxide (Co_3O_4) thin films by spray pyrolysis [J]. Applied Surface Science, 2006, 252(20): 7487 – 7492.

[13] N. Bahlawane, P. H. T. Ngamou, V. Vannier, T. kottke, J. Heberle, K. Kohse-Höinghaus. Tailoring the properties and the reactivity of the spinel cobalt oxide [J]. Physical Chemistry Chemical Physics, 2009, 11(40): 9224 – 9232.

[14] R. R. Owings, G. J. Exarhos, C. F. Windisch, P. H. Holloway, J. G. Wen. Process enhanced polaron conductivity of infrared transparent nickel-cobalt oxide [J]. Thin Solid Films, 2005, 483 (1), 175 – 184.

[15] C. K. Xu, Y. K. Liu, G. D. Xu, G. H. Wang. Fabrication of CoO nanorods via thermal decomposition of CoC_2O_4 precursor [J]. Chemical Physics Letters, 2002, 366(5): 567 – 571.

[16] N. A. M. Barakat, M. S. Khil, F. A. Sheikh. Synthesis and optical properties of two cobalt oxides (CoO and Co_3O_4) nanofibers produced by electrospinning process [J]. The Journal of Physical Chemistry C, 2008, 112(32): 12225 – 12233.

[17] D. P. Dutta, G. Sharma, P. K. Manna, A. K. Tyagi, S. M. Yusuf. Room temperature ferromagnetism in CoO nanoparticles obtained from sonochemically synthesized precursors [J]. Nanotechnology, 2008, 19(24): 245609 – 245615.

[18] X. H. Liu, R. Ma, Y. Bando, T. Sasaki. Layered cobalt hydroxide nanocones: microwave-assisted synthesis, exfoliation, and structural modification [J]. Angewandte Chemie Intenational Edition, 2010, 49(44): 8253 – 8256.

[19] G. B. Sun, X. Q. Zhang, M. H. Cao, B. Q. Wei, C. W. Hu. Facile synthesis. characterization, and microwave absorbability of CoO nanobelts and submicrometer spheres [J]. The Journal of Physical Chemistry C, 2009, 113(17): 6948 – 6954.

[20] X. H. Liu, R. Yi, N. Zhang, R. R. Shi, X. G. Li, G. Z. Qiu. Cobalt hydroxide nanosheets

and their thermal decomposition to cobalt oxide [J]. Chemistry – An Asian Journal, 2008, 3 (4): 732 – 738.

[21] X. H. Liu, G. Z. Qiu, X. G. Li. Shape-controlled synthesis and properties of uniform spinel cobalt oxide nanocubes [J]. Nanotechnology, 2005, 16(12): 3035 – 3040.

[22] H. Y. Guan, C. L. Shao, S. B. Wen, B. Chen, J. Gong, X. H. Yang. A novel method for preparing Co_3O_4 nanofibers by using electrospun PVA/cobalt acetate composite fibers as precursor [J]. Materials Chemistry and Physics, 2003, 82(3): 1002 – 1006.

[23] K. An, N. Lee, J. Park, S. C. Kim, Y. Hwang, J. G. Park, J. Y. Kim, J. H. Park, M. J. Han, J. Yu, T. Hyeon. Synthesis, characterization, and self-assembly of pencil-shaped CoO nanorods [J]. Journal of the American Chemical Society, 2006, 128(30): 9753 – 9760.

[24] L. H. Zhuo, J. C. Ge, L. H. Cao. Solvothermal Synthesis of CoO, Co_3O_4, $Ni(OH)_2$ and $Mg(OH)_2$ nanotubes [J]. Journal of the American Chemical Society, 2006, 128(30): 9753 – 9760.

[25] X. Wang, X. L. Wu, Y. G. Guo, Y. T. Zhong, X. Q. Cao, Y. Ma, J. N. Yao. Synthesis and lithium storage properties of Co_3O_4 nanosheet-assembled multishelled hollow spheres [J]. Advanced Functional Materials, 2010, 20(10): 1680 – 1686.

[26] X. Y. Lai, J. Li, B. A. Korgel, Z. H. Dong, Z. M. Li, F. B. Su, J. A. Du, D. Wang. General synthesis and gas-sensing properties of multiple-shell metal oxide hollow microspheres [J]. Angewandte Chemie International Edition, 2011, 50(12): 2738 – 2741.

[27] N. R. Jana, Y. F. Chen, X. G. Peng. Size- and shape-controlled magnetic (Cr, Mn, Fe, Co, Ni) oxide nanocrystals via a simple and general approach [J]. Chemistry of Materials, 2004, 16(20): 3931 – 3935.

[28] M. Ghosh, E. V. Sampathkumaran, C. N. R. Rao. Synthesis and magnetic properties of CoO nanoparticles [J]. Chemistry of Materials, 2005, 17(9): 2348 – 2352.

[29] Y. L. Zhang, J. Zhu, X. Song. X. H. Zhong. Controlling the synthesis of CoO nanocrystals with various morphologies [J]. The Journal of Physical Chemistry C, 2008, 112(14): 5322 – 5327.

[30] J. S. Yin, Z. L. Wang. Ordered self-assembling of tetrahedral oxide nanocrystals [J]. Physical Review Letters, 1997, 79(13): 2570 – 2572.

[31] Y. L. Zhang, X. H. Zhong, J. Zhu, X. Song. Alcoholysis route to monodisperse CoO nanotetrapods with tunable size. Nanotechnology, 2007, 18(19), 195605 – 195610.

[32] Y. G. Sun, Y. N. Xia. Shape-controlled synthesis of gold and silver nanoparticles [J]. Science, 2002, 298(5601): 2176 – 2179.

[33] R. R. Shi, G. H. Gao, R. Yi, K. C. Zhou, G. Z. Qiu, X. H. Liu. Controlled synthesis and caracterization of monodisperse Fe_3O_4 nanoparticles [J]. Chinese Journal of Chemistry, 2009, 27(4): 739 – 744.

[34] X. Xu, S. A. Asher. Synthesis and utilization of monodisperse hollow polymeric particles in photonic crystals [J]. Journal of the American Chemical Society, 2004, 126(25): 7940 –

7945.

[35] Z. Y. Zhong, Y. D. Yin, B. Gates, Y. N. Xia. Preparation of mesoscale hollow spheres of TiO_2 and SnO_2 by templating against crystalline arrays of polystyrene beads [J]. Advanced Materials, 2000, 12(3): 206-209.

[36] V. V. Nemoshalenko, V. V. Didyk, V. P. Krivitskii, A. I. Senekevich. Zhurnal Organicheskoi Khimii, 1983, 28: 2182.

[37] J. P. Bonnelle, J. Grimblot, A. D'huysser. Influence de la polarisation des liaisons sur les spectres esca des oxydes de cobalt [J]. Journal of Electron Spectroscopy and Related Phenomena, 1975, 7(2): 151-162.

[38] H. M. Yang. J. Ouyang, A. D. Tang, Single step synthesis of high-purity CoO nanocrystals [J]. The Journal of Physical Chemistry B, 2007, 111(28): 8006-8013.

[39] M. Burriel, G. Garcia, J. Santiso, A. Abrutis, Z. Saltyte, A. Figueras. Growth kinetics, composition, and morphology of Co_3O_4 thin films prepared by pulsed liquid-injection MOCVD [J]. Chemical Vapor Deposition, 2005, 11(2): 106-111.

[40] T. J. Chuang, C. R. Brundle, D. W. Rice. Interpretation of the X-ray photoemission spectra of cobalt oxides and cobalt oxide surfaces [J]. Surface Science, 1976, 59(2): 413-429.

[41] C. V. Chenck, J. G. Dillard, J. W. Murrray. Surface analysis and the adsorption of Co(II) on goethite [J]. Journal of Colloid and Interface Science, 1983, 95(2): 398-409.

[42] N. S. McIntyre, M. G. Cook. X-ray photoelectron studies on some oxides and hydroxides of cobalt, nickel, and copper [J]. Analytical Chemistry, 1975, 47(13): 2208-2213.

[43] M. Salavati-Niasari, N. Mir, F. Davar. Synthesis and characterization of Co_3O_4 nanorods by thermal decomposition of cobalt oxalate [J]. Journal of Physics and Chemistry of Solids, 2009, 70(5): 847-852.

[44] G. A. Carson, M. A. Nassir, M. A. Langell. Epitaxial growth of Co_3O_4 on CoO (100) [J]. Journal of Vacuum Science & Technology A, 1996, 14(3): 1637-1642.

[45] W. Li, F. Cheng, Z. Tao, J. Chen. Vapor-transportation preparation and reversible lithium intercalation/deintercalation of $\alpha-MoO_3$ microrods [J]. The Journal of Physical Chemistry B, 2006, 110(1): 119-124.

[46] P. Poizot, S. Laruelle, S. Grugeon, L. Dupont, J. M. Tarascon. Nano-sized transition-metal oxides as negative-electrode materials for lithium-ion batteries [J]. Nature, 2000, 6803(407): 496.

[47] F. F. Tao, C. L. Gao, Z. H. Wen, Q. Wang, J. H. Li, Z. Xu. Cobalt oxide hollow microspheres with micro- and nano-scale composite structure: Fabrication and electrochemical performance [J]. Journal of Solid State Chemistry, 2009, 182(5): 1055-1060.

☆ R. R. Shi, G. Chen, W. Ma, D. Zhang, G. Z. Qiu, X. H. Liu. Shape-controlled Synthesis and Characterization of Cobalt Oxides Hollow Spheres and Octahedra [J]. Dalton Transactions, 2012, 41(19): 5981-5987.

Shape-Controlled Synthesis and Properties of Dandelion-like Manganese Sulfide Hollow Spheres

Abstract: Dandelion-like gamma-manganese (Ⅱ) sulfide (MnS) hollow spheres assembled with nanorods have been prepared via a hydrothermal process in the presence of L-cysteine and polyvinylpyrrolidone (PVP). L-cysteine was employed as not only sulfur source, but also coordinating reagent for the synthesis of dandelion-like MnS hollow spheres. The morphology, structure and properties of as-prepared products have been investigated in detail by X-ray diffraction (XRD), scanning electron microscopy (SEM), transmission electron microscopy (TEM), Energy dispersive X-ray spectroscopy (EDS), selected area electron diffraction (SAED), high-resolution transmission electron microscopy (HRTEM) and photoluminescence spectra (PL). The probable formation mechanism of as-prepared MnS hollow spheres was discussed on the basis of the experimental results. This strategy may provide an effective method for the fabrication of other metal sulfides hollow spheres.

1 Introduction

Over the past decades, the design and controlled synthesis of inorganic materials with well-controlled morphology has drawn increasing attentions owing to the chemical and physical properties associated with not only the chemical compositions, but also the sizes and morphologies.[1-5] Recently, tremendous efforts have been devoted to the exploration of hierarchical architectures consisting of nano- or microstructured units, and they exhibit unique properties and potential technological applications in various fields, such as electronics, optoelectronics, catalysts, sensors, etc.[6-8] Furthermore, to meet the requirements of applications, the novel architectures with hollow interior may be very useful.[9-11] Although considerable achievements have been made, it remains a highly sophisticated challenge to fabricate the novel hierarchical architectures with hollow interior.

As an important magnetic p-type semiconductor with a wide gap (band gap energy, $E_g(T=0)$ about 3.7 eV), MnS has become the focus in recent years due to its novel optical, electric and magnetic properties and the important potential application in solar cells as a window/buffer material, optoelectronic devices, electrode material.[12-14] As a diluted magnetic semiconductor, MnS has three polymorphs: α - MnS, β - MnS and γ - MnS. The pink β - MnS with zinc blende crystal structure and γ - MnS with wurtzite crystal structure were synthesized in the low substrate temperature and they can transform into green α - MnS with rock salt structure at 100 - 400 ℃ or at high pressure.[15] Inspired by these excellent properties, the study for the fabrication of MnS with desired shapes and architectures is continually being intensified.[16-18] Nevertheless, up to now there are few reports about the synthesis of dandelion-like MnS hollow spheres. Herein we have successfully developed a simple one-pot hydrothermal route to synthesize the dandelion-like MnS hollow spheres assembled with nanorods in the presence of L - cysteine and PVP. L - cysteine and PVP play a significant role in the synthesis of dandelion-like MnS hollow spheres assembled with nanorods.[19] The morphologies and sizes of as-prepared products have been affected by surfactants, reaction time and reaction temperature. The photoluminescence (PL) properties of as-prepared products were investigated. The probable formation mechanism of dandelion-like MnS hollow spheres was proposed.

2 Experiment Section

All chemicals used in this work, such $MnCl_2 \cdot 4H_2O$, polyvinylpyrrolidone (PVP) and L - cysteine were of analytical grade and were used without further purification. $MnCl_2 \cdot 4H_2O$, PVP and L - cysteine were purchased from China National Medicines Corporation Ltd.

2.1　Synthesis of dandelion-like MnS hollow spheres. $MnCl_2 \cdot 4H_2O$ (0.5 mmol) and 0.15 g PVP were put into Teflon-lined autoclave of 50 mL capacity with 20 mL deionized water at room temperature. Then, L - cysteine (2 mmol) was added to the above-mention solution under an ultrasonic treatment. The autoclave was heated and maintained 180 ℃ for 8 h in an electric oven without any agitating and then was allowed to gradually cool down to room temperature after heat treatment. The final product was collected by sedimentation, washed with absolute alcohol several times, and then dried in a vacuum box at 60 ℃ for 4 h.

2.2　Characterization. The chemical composition and crystal structure of as-synthesized samples was determined by X-ray powder diffraction (XRD) using a D/max2550 VB +X-ray diffractometer with Cu K_α radiation ($\lambda = 1.5418$ Å) at a scan rate of 4°/min for 2θ ranging from 10° to 85°. The morphology and size of as-prepared products were characterized at 20 kV by a XL30 S-FEG SEM and at 160 kV by a JEM - 200CX TEM. EDS was collected on the SEM whereas TEM images and SAED patterns were obtained on the TEM. The room temperature PL measurement was carried out on a Hitachi F - 4500 fluorescence spectrophotometer using the 245 nm excitation line of Xe light.

3 Results and Discussion

The morphology and size of as-synthesized products were characterized by using SEM and TEM. Figure 1(a) and 1(b) shows the typical different magnification SEM images of as-produced MnS microspheres, indicating good uniformity. A large amount of MnS microspheres assembled with nanorods were achieved using this approach. The length of nanorods are about several micrometers, growing radially from the center into dandelion-like structures. The diameters of microspheres are in the range from 9 μm to 12 μm. Figure 1 (c) shows the reverse side of as-prepared MnS microspheres, confirming the hollow structure. And the hollow structure of as-synthesized product is also further confirmed by TEM shown in Figure 1 (d). The internal diameter of dandelion-like MnS hollow spheres is in the range of 3 - 4.5 μm. As shown in Figure 1(e), the SAED pattern of single nanorod can be indexed to be hexagonal MnS growing along the [001] direction. The HRTEM image taken on the tip of an individual nanorod shows that the interlayer spacing is calculated to be 0.32 nm, corresponding to the (002) crystal plane of hexagonal MnS.

The crystal structures of products were characterized by X-ray powder diffraction (XRD). Figure 1(f) shows the representative XRD pattern of as-prepared dandelion-like MnS hollow spheres obtained at 180℃ for 8 h. All the diffraction peaks agree well with the hexagonal MnS with a = 3.979 Å, and c = 6.447 Å, respectively [space group: P63mc (186), JCPDS 40 - 1289]. No peak of other impurities can be detected under the current experimental conditions, indicating the high purity of as-prepared products.

Further studies suggested that the surfactants have great influence on the size and morphologies of as-prepared products. Figure 2 (a) shows the SEM images of MnS crystals in the absence of PVP. The morphology of as-prepared MnS was dandelion-like with hollow structure, but the nanorods on the surface of as-prepared products were shorter and irregular. Figure 2(b) shows the SEM image of final product with adding

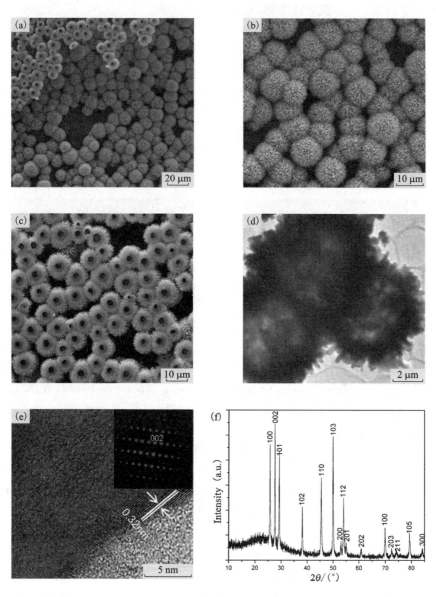

Figure 1

(a), (b) The SEM images of dandelion-like MnS hollow spheres with different magnifications. (c) The image of reverse side of MnS hollow sphere. (d) TEM image of MnS hollow spheres. (e) HRTEM image of single nanorod (inset: the corresponding SAED pattern). (f) XRD pattern of as-prepared dandelion-like MnS hollow spheres at 180 ℃ for 8 h.

Figure 2 The SEM images of as-synthesized products with different surfactants at 180 ℃ for 8 h

(a) without PVP; (b) 0.3 g PVP; (c) 0.15 g PEG; (d) 0.5 mmol CTAB.

0.3 g PVP. The morphology of as-prepared MnS hollow spheres has little difference with MnS obtained using 0.15 g PVP. However, the products obtained at this condition are mingled with block of irregular shape. When 0.15 g polyethylene glycol (PEG), instead of PVP, was used [Figure 2(c)], the morphology of as-prepared product is spherical structure assembled with solid particle and the diameter of as-prepared MnS is in the range of 10 – 15 μm. If hexadecyl trimethyl ammonium bromide (CTAB) was added [Figure 2(d)], the morphology of product was observed as anomalous spheres. The results suggest that CTAB has smaller size in comparison with PVP. Thus the binding energy between CTAB and MnS may be weaker. As a polymer, PVP has been proved to exert positive effect in the synthesis of 1D nanomaterials because PVP acts as a protective agent against the particles agglomeration and face-inhibited functional surfactant, favoring the 1D growth.[20-23]

The influences of reaction temperature on the morphology of final products are also explored. Figure 3(a) shows the SEM image of as-prepared dandelion-like MnS hollow sphere obtained at 160℃ for 8 h. Compared with the products obtained at 180℃, the nanorods of dandelion-like MnS are slenderer and pell-mell at 160℃. When the reaction temperature was raised up to 200℃, the nanorods of dandelion-like MnS hollow spheres are shorter than those obtained at 180℃ [Figure 3(b)]. It is thus suggested that the release rate of S^{2-} anions from L – cysteine becomes faster with the elevation of reaction temperature, a probable reason for various morphologies. Other reaction temperatures are found unfavorable for the growth of dandelion-like MnS hollow spheres even other conditions are unaltered.[20]

Figure 3　The images of as-prepared products obtained at different temperature for 8 h
(a) 160℃; (b) 200℃

To understand the formation mechanism of dandelion-like MnS hollow spheres assembled with nanorods, time-dependent experiments were carried out. As shown in SEM images in Figure 4, four obvious evolution stages could be identified. Figure 4(a) shows the rough MnS spheres obtained for 1 h and the diameter of as-prepared spheres is ranged from 3.5 μm to 5.5 μm. Prolonging the reaction time to 2 h, the morphology of as-obtained MnS changed into spheres with core-shell structure and the obtained sphere assembled with rough nanorods [Figure 5(b)]. The rough nanorods became longer with prolonging the reaction time to 4 h [Figure 4(c)]. With further lengthening of reaction time to 12 h, the rough rods turned into smooth nanorods due to the ripening process [Figure 4(d)]. The diameter of as-prepared products is generally increased with the lengthening of reaction time.

Based on the experimental results, the formation process of dandelion-like MnS hollow spheres assembled with nanorods may be divided into several steps, as shown in

Figure 4 The images of as-synthesized products obtained at 180 °C different reaction time
(a) 1 h; (b) 2 h; (c) 4 h; (d) 12 h

Scheme 1. We assume that L – cysteine played an undoubtedly significant role in the synthesis process. In the L – cysteine molecule, there are many functional groups, such as —NH_2, —COOH, and —SH, which have a strong tendency to coordinate with inorganic cations and metals.[22-24] In the present synthesis produce, L – cysteine could coordinated with Mn^{2+} cations to form Mn-cysteine complex,[22] which may subsequently form irregular structures owing to hydrogen-bond interactions. With the elevating of reaction temperature, the C = S bond of L – cysteine began to break and release S^{2-} anions gradually, the S^{2-} anions combined with Mn^{2+} cations and formed MnS monomers. The freshly MnS crystalline particles were unstable due to the high surface energy[25] and they tend to aggregate into spherical product driven by the minimization of interfacial energy. With the release of S^{2-}, the irregular structure of Mn-cysteine complex shrank gradually into irregular solid microspheres. With the prolonging of reaction time, many spots and similarly subuliform nanostructure may form on the surface of microspheres under the effect of stochastic diffusive force. Then radially growing nanorods are formed due to the preferred 1D growth of the spots and subuliform nanostructures under the influence of PVP.[26, 27] At the same time, the solid microspheres converted into hollow spheres due to the Ostwald ripening process the outward migration of crystals would result in continuing expansion of interior space within the original aggregate, the core region would transferred to out section, which can be reconfirmed by gradually increasing diameter along nanorods.[28] Finally dandelion-like gamma-manganese sulfide hollow spheres assembled with nanorods were achieved for longer reaction time.

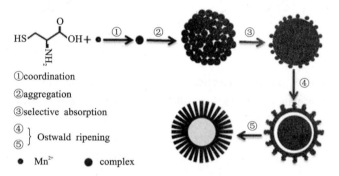

Scheme 1. Proposed growth process of dandelion-like MnS hollow spheres

Figure 5 presents the room-temperature PL spectra of as-prepared products obtained at 160 ℃ [Figure 5(a)], 180 ℃ [Figure 5(c)], and 200 ℃ [Figure 5(d)] for 8 h and 180 ℃ for 2 h [Figure 5(b)], respectively, which were measured on a fluorescence spectrophotometer using a Xe lamp with an excitation wavelength of 245 nm. It is obvious that the as-prepared MnS exhibited a maximum intensity emission at 393 nm, which is consistent with the calculated bandgap of the bulk counterpart.[29] With the elevation of reaction temperature from 160 ℃ to 200 ℃, the PL intensity is rising gradually. The intensity of the emission peaks of dandelion-like MnS hollow spheres obtained 200 ℃ for 8 h is the strongest, indicting its optical property is superior to MnS obtained with other conditions. PL intensity of as-prepared MnS obtained at 180 ℃ for 2 h and 8 h is also slightly different, indicating the reaction time also has an effect on the optical property. Although the PL emission mechanism of MnS is not fully understood, the emission peaks may be attributed to the recombination of charge carriers in deep traps of surface localized states and a photogenerated holes caused by surface defects. The optical properties are directly influenced by the shape and size of as-prepared products.[30-32]

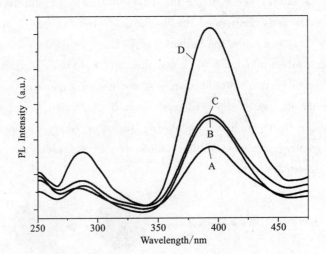

Figure 5 Room-temperature PL of as-prepared MnS obtained at different temperature for 8 h with an excitation wavelength of 245 nm
(a) 160 ℃; (c) 180 ℃; (d) 200 ℃, and (b) 180 ℃ for 2 h

4 Conclusion

In summary, dandelion-like MnS hollow spheres assembled with nanorods were successfully prepared via a simple and convenient hydrothermal route. In this synthetic strategy, L-cysteine, as a coordinating reagent and shape modifier, plays a significant role in synthesizing MnS hollow spheres. On the other hand, PVP exhibits a great influence on the synthesis of dandelion-like MnS hollow spheres assembled with nanorods. MnS microcrystals with different morphologies could be synthesized by adjusting surfactant, reaction temperature and reaction time. This strategy may provide more opportunities for the synthesis of hollow spheres composed of other metal sulfides.

References

[1] Z. L. Xiao, C. Y. Han, W. K. Kwok, H. H. Wang, U. Welp, J. Wang and G. W. Crabtree. Tuning the architecture of mesostructures by electrodeposition [J]. Journal of the American Chemical Society, 2004, 126(8): 2316-2317.

[2] B. C. Wang, Y. T. Dung. Structural and optical properties of passivated silicon nanoclusters with different shapes: A theoretical investigation [J]. The Journal of Physical Chemistry A,

2008, 112(28): 6351-6357.

[3] Y. W. Jun, J. Cheon. Shape control of semiconductor and metal oxide nanocrystals through nonhydrolytic colloidal routes [J]. Angewandte Chemie International Edition, 2006, 45(21): 3414-3439.

[4] J. B. Zhao, L. L. Wu, K. Zou. Fabrication of hollow mesoporous NiO hexagonal microspheres via hydrothermal process in ionic liquid [J]. Materials Research Bulletin, 2011, 46(12): 2427-2432.

[5] J. H. L. Beal, P. G. Etchegoin. Transition metal polysulfide complexes as single-source precursors for metal sulfide nanocrystals [J]. The Journal of Physical Chemistry C, 2010, 114(9): 3817-3821.

[6] F. Cao, L. Zhao, H. Zhang. Hydrothermal synthesis and high photocatalytic activity of 3D wurtzite ZnSe hierarchical nanostructures [J]. The Journal of Physical Chemistry C, 2008, 112(44): 17095-17101.

[7] Z. Li, Y. Ding, Y. Xiong, Y. Xie. One-step solution-based catalytic route to fabricate novel alpha-MnO_2 hierarchical structures on a large scale [J]. Chemical Communications, 2005, 7: 918-920.

[8] H. Li, L. Chai, Y. Liu, Y. Qian. Hydrothermal growth and morphology modification of β-NiS three-dimensional flowerlike architectures [J]. Crystal Growth & Design, 2007, 7(9): 1918-1922.

[9] B. Liu and H. C. Zeng. Fabrication of ZnO "dandelions" via a modified Kirkendall process [J]. Journal of the American Chemical Society, 2004, 126(51): 16744-16746.

[10] B. Z. Liu, H. Chun. Mesoscale organization of CuO nanoribbons: formation of "dandelions" [J]. Journal of the American Chemical Society, 2004, 126(26): 8124-8125.

[11] H. C. Zeng. Synthetic architecture of interior space for inorganic nanostructures [J]. Journal of Materials Chemistry, 2006, 16(7): 649-662.

[12] Y. H. Zheng, Y. Cheng, C. Jia. Metastable γ-MnS hierarchical architectures: synthesis, characterization, and growth mechanism [J]. The Journal of Physical Chemistry B, 2006, 110(16): 8284-8288.

[13] S. S. Aplesnin, L. I. Ryabinkina, D. A. Velikanov, A. D. Balaev. Conductivity, weak ferromagnetism, and charge instability in an α-MnS single crystal [J]. Physical Review B, 2005, 71(12): 125204.

[14] N. Zhang, R. Yi, X. H. Liu. Hydrothermal synthesis and electrochemical properties of alpha-manganese sulfide submicrocrystals as an attractive electrode material for lithium-ion batteries [J]. Materials Chemistry and Physics, 2008, 111(1): 13-16.

[15] F. Tao, Z. J. Wang, X. G. Li. Hydrothennal synthesis of 3D alpha-MnS flowerlike nanoarchitectures [J]. Materials Letters, 2007, 61(28): 4973-4975.

[16] F. Zuo, B. Zhang, X. Z. Tang, Y. Xie. Porous metastable gamma-MnS networks: biomolecule-assisted synthesis and optical properties [J]. Nanotechnology, 2007, 18(21):

215608.

[17] B. Peng, Z. T. Deng, F. Q. Tang, D. Chen, X. L. Ren, J. Ren. Self-healing self-assembly of aspect-ratio-tunable chloroplast-shaped architectures [J]. Crystal Growth & Design, 2009, 9 (11): 4745-4751.

[18] J. Joo, H. B. Na, T. Yu, J. H. Yu, Y. W. Kim, F. X. Wu, J. Z. Zhang, T. Hyeon. Generalized and facile synthesis of semiconducting metal sulfide nanocrystals [J]. Journal of the American Chemical Society, 2003, 125(36): 11100-11105.

[19] F. Zuo, S. Yan, Y. Zhao, Y. Xie. L-cysteine-assisted synthesis of PbS nanocube-based pagoda-like hierarchical architectures [J]. The Journal of Physical Chemistry C, 2008, 112 (8): 2831-2835.

[20] P. T. Zhao, Q. M. Zeng, K. X. Huang. Fabrication of beta-NiS hollow sphere consisting of nanoflakes via a hydrothermal process [J]. Materials Letters, 2009, 63(2): 313-315.

[21] Y. Xia, Y. J. Xiong, B. Lim, S. E. Skrabalak. Shape-controlled synthesis of metal nanocrystals: simple chemistry meets complex physics? [J]. Angewandte Chemie International Edition, 2009, 48(1): 60-103.

[22] B. X. Li, Y. Xie, Y. Xue. Controllable synthesis of CuS nanostructures from self-assembled precursors with biomolecule assistance [J]. The Journal of Physical Chemistry C, 2007, 111 (33): 12181-12187.

[23] F. Zuo, S. Yan, B. Zhang, Y. Zhao, Y. Xie. L-cysteine-assisted synthesis of PbS nanocube-based pagoda-like hierarchical architectures [J]. The Journal of Physical Chemistry C, 2008, 112(8): 2831-2835.

[24] B. Zhang, X. C. Ye, Y. Xie. Biomolecule-assisted synthesis and electrochemical hydrogen storage of Bi_2S_3 flowerlike patterns with well-aligned nanorods [J]. The Journal of Physical Chemistry B, 2006, 110(18): 8978-8985.

[25] Y. Zhao, X. Zhu, Y. Y. Huang, Y. Xie. Synthesis, growth mechanism, and work function at highly oriented {001} surfaces of bismuth sulfide microbelts [J]. The Journal of Physical Chemistry C, 2007, 111(33): 12145-12148.

[26] Y. H. Zheng, Y. Cheng, L. H. Zhou, F. Bao, C. Jia. Metastable γ-MnS hierarchical architectures: synthesis, characterization, and growth mechanism [J]. The Journal of Physical Chemistry B, 2006, 110(16): 8284-8288.

[27] X. H. Liu, X. D. Liang. Selective synthesis and characterization of sea urchin-like metallic nickel nanocrystals [J]. Materials Science and Engineering: B, 2006, 132(3): 272-277.

[28] Q. R. Zhao, Y. Xie, T. Dong, Z. G. Zhang. Oxidation-crystallization process of colloids: An effective approach for the morphology controllable synthesis of SnO_2 hollow spheres and rod bundles [J]. The Journal of Physical Chemistry C, 2007, 111(31): 11598-11603.

[29] S. H. Kan, I. Felner and U. Banin. Synthesis, characterization, and magnetic properties of α-MnS nanocrystals [J]. Israel Journal of Chemistry, 2001, 41(1): 55-62.

[30] J. Mu, Z. F. Gu, L. Wang, Z. Q. Zhang, H. Sun, S. Z. Kang. Phase and shape controlling

of MnS nanocrystals in the solvothermal process [J]. Journal of Nanoparticle Research, 2008, 10(1): 197-201.

[31] X. L. Yu, C. B. Cao, C. L. Liu, Q. H. Gong. Nanometer-sized copper sulfide hollow spheres with strong optical-limiting properties [J]. Advanced Functional Materials, 2007, 17(8): 1397-1401.

[32] P. T. Zhao and K. X. Huang. Preparation and characterization of netted sphere like CdS nanostructures [J]. Crystal Growth and Design, 2007, 8(2): 717-722.

☆ W. Ma, G. Chen, D. Zhang, J. Y. Zhu, G. Z. Qiu, X. H. Liu. Shape-controlled synthesis and properties of dandelion-like manganese sulfide hollow spheres [J]. Materials Research Bulletin, 2012, 47(9): 2182-2187.

Nickel Dichalcogenides Hollow Spheres: Controllable Fabrication, Structural Modification and Magnetic Properties

Abstract: A facile hydrothermal synthetic route was developed for the synthesis of uniform NiS_2 hollow spheres, which could be respectively transformed into $NiSe_2$ and $NiTe_2$ hollow spheres through chemical conversion process. The magnetic measurements reveal an antiferromagnetic behavior for NiS_2 and $NiTe_2$ hollow spheres while a paramagnetic behavior for $NiSe_2$ hollow spheres.

1 Introduction

Morphology-dependent properties of nano-/micro-spheres with hollow interiors have attracted immense attention because of the unique properties and potential technological applications in various fields, such as electronics, optoelectronics, selective catalysis, controlled release and delivery, etc.[1] Inspired by the excellent properties, the study on the fabrication of transition-metal dichalcogenides with desired morphologies is continually being intensified.[2] In particular, nickel dichalcogenides (such as NiS_2, $NiSe_2$ and $NiTe_2$) have become the focus in recent years due to their novel electric and magnetic properties and important technological applications in various fields. It is well known that NiS_2 is of particular interest as a Mott-Hubbard insulator with anomalous antiferromagnetic long-range order and a metal-insulator transitional body in $NiS_{2-x}Se_x$, whereas $NiSe_2$ is a typical Pauli paramagnet with metallic conductivity and, therefore, is suitable as a storage energy material for rechargeable lithium batteries.[3] More importantly, $NiSe_2$ can also be applied as the prominent Pt-free electrocatalyst for dye-sensitized solar cells (DSSCs), which shows better catalytic activity in the reduction of triiodide than Pt electrode.[4] Furthermore, $NiTe_2$ is a promising back contact to CdTe/CdS solar cells with exceeding 10% conversion efficiencies.[5] A variety of chemical and physicochemical methods, such as solid-state synthesis, elemental direct reaction, and mechanical alloying have been used to fabricate nickel dichalcogenides; nevertheless, there is barely a report about the synthesis of NiS_2, $NiSe_2$ and $NiTe_2$

hollow spheres up to now, partly due to the difficulties in fabricating transition-metal dichalcogenides with regular morphologies.[6]

Herein, we have successfully devised a simple and general hydrothermal synthetic route to prepare the uniform NiS_2 hollow spheres in the presence of L – cysteine and urea. Especially, $NiSe_2$ and $NiTe_2$ hollow spheres could be obtained through a chemical conversion method under hydrothermal conditions by using NiS_2 hollow spheres as the precursor, respectively. Meanwhile, NiS and NiO hollow spheres could also be successfully synthesized by the calcinations of NiS_2 hollow spheres at different temperatures. A probable formation mechanism of the resulting products has been proposed, which offers a designable route to prepare transition-metal dichalcogenides as well as transformations between sulfides and oxides. The magnetic properties of as-prepared nickel dichalcogenides have been investigated in detail and indicate an antiferromagnetic behavior for NiS_2 and $NiTe_2$ hollow spheres.

2 Experimental Section

All chemicals were of analytical grade form China National Medicines Corporation Ltd and used as starting materials without further purification.

2.1 Synthesis of NiS_2, NiS and NiO hollow spheres. In a typical procedure, $Ni(NO_3)_2 \cdot 6H_2O$ (0.5 mmol) was put into Teflon-lined autoclave of 50 mL capacity with 20 mL of de-ionized water at room temperature. Then, L – cysteine (2 mmol) and urea (0.5 mmol) were added to the above-mention solution under an ultrasonic treatment. The autoclave was sealed and maintained 140 ℃ for 24 h in an electric oven without any agitating and then was allowed to gradually cool down to room temperature after heat treatment. The resulting precipitate was collected by free sedimentation and washed with absolute ethyl alcohol several times. The final product was dried in a vacuum box at 60 ℃ for 4 h, which could be transformed into NiS and NiO hollow spheres by calcination in air at 400 ℃ and 770 ℃ for 1 h, respectively.

2.2 Synthesis of $NiSe_2$ and $NiTe_2$ hollow spheres. In a typical procedure, NiS_2 (0.5 mmol) hollow spheres obtained under hydrothermal conditions was put into Teflon-lined autoclave of 50 mL capacity with 15 mL of de-ionized water at room temperature. Then, H_2SeO_3 (1 mmol) or H_2TeO_3 (1 mmol) and 15 mL of $N_2H_4 \cdot H_2O$ were added to the above solution, respectively. The autoclave was sealed and maintained 140 ℃ for 24 h in an electric oven without any agitating and then was allowed to gradually cool down to room temperature after heat treatment. The final product was collected by free sedimentation, washed with absolute ethyl alcohol several times, and then dried in a vacuum box at 60 ℃ for 4 h.

2.3 Characterization. The chemical composition and crystal structure of the as-synthesized samples was characterized by XRD using a D/max2550 VB + X-ray diffractometer with Cu K_α radiation ($\lambda = 1.5418$ Å). The morphologies and sizes of as-prepared products were characterized at 20 kV by a XL30 S – FEG SEM and at 160 kV by a JEM – 200CX TEM. HRTEM images and SAED patterns were obtained on the TEM. Thermogravimetric (TG) measurements were carried out using a NETZSCH STA 449C instrument in a temperature range of 35 – 900 ℃ at a heating rate of 10 ℃/min under an air flow. The magnetic properties of as-prepared products were measured using a Quantum Design MPMS XP – 5 superconducting quantum interference device.

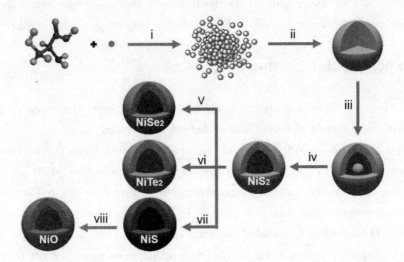

Scheme 1. Proposed growth process of NiS_2, $NiSe_2$, $NiTe_2$, NiS and NiO hollow spheres

(i) coordination and selective absorption of L – cysteine with Ni ions; (ii) formation of NiS_2 solid spheres; (iii, iv) an Ostwald ripening process; (v, vi) a chemical conversion process from NiS_2 to $NiSe_2$ and $NiTe_2$ hollow spheres; (vii, viii) calcinations of NiS_2 into NiS and NiO hollow spheres.

3 Results and Discussion

Scheme 1 illustrates the designe process and possible scenario for the formation of nickel dichalcogenides hollow spheres. Firstly, L – cysteine is dispersed in deionized water and simultaneously capped by Ni ions due to the coordination of functional groups in the L-cysteine molecules, such as —NH_2, —COOH, and —SH, which has a strong tendency to coordinate with inorganic cations. [7] Urea can be used to provide an alkaline environment and facilitate the formation of NiS_2 based the hydrolysis of L-cysteine under the hydrothermal conditions. Subsequently the freshly prepared NiS_2 tend to aggregate into spherical products driven by the minimization of interfacial energy. The initial NiS_2 solid spheres might not be well crystallized owing to rapid nucleation and growth, containing parts of the Ni-coordination compound. Therefore, the Ni-coordination compound from inside would dissolve and recrystallize to form a new particle. Meanwhile, the organic residues would transfer out to form inner space, producing core/shell-structured NiS_2 hollow spheres. Finally, the resulting NiS_2 hollow spheres could be obtained with the prolonging of hydrothermal time based on the Ostwald ripening process. [8] Meanwhile, $NiSe_2$ or $NiTe_2$ hollow spheres can be obtained due to the inequivalent exchange of metallic atoms: Se^{2-} or Te^{2-} could react with Ni^{2+} slowly dissolved from the surface of NiS_2 hollow spheres to form spherical $NiSe_2$ or $NiTe_2$ with hollow interiors, which are more thermodynamically stable due to their lower solubility. Furthermore, NiS and NiO hollow spheres could be obtained through calcination of NiS_2 hollow spheres at different temperatures. According to the above discussion, the possible formation of various nickel chalcogenides and oxides can be summarized as shown in Scheme 1.

Figure 1(a) exhibits a typical scanning electron microscopy (SEM) image of as-prepared NiS_2 obtained by using L-cysteine (2 mmol) and urea (0.5 mmol) at 140 ℃ for 24 h and indicates a large quantity of uniform spherical particles with the diameter of about 2 μm were achieved by using this approach. The high-magnification SEM image shown in Figure 1(b) reveals that the interior of the spheres with broken shell is hollow, and the hollow nature of these spherical particles is also further confirmed by transmission electron microscopy (TEM) image shown in Figure 1(c) based on the sharp contrast between the dark periphery and grayish center of these spheres with a shell thickness in the range of 60 – 150 nm. The selected area electron diffraction (SAED) pattern of NiS_2 hollow spheres is given in the inset of Figure 1(c), which can be readily indexed to the cubic phase of NiS_2. Figure 1(d) shows the high-resolution TEM (HRTEM) image taken on the surface of an individual hollow sphere, which provides further insight into the structure of NiS_2 hollow spheres. Lattice spacing is measured to be about 0.28 nm, which is consistent with the interplanar spacing of the (200) crystal plane in cubic NiS_2.

To understand the formation mechanism of NiS_2 hollow spheres, time-dependent

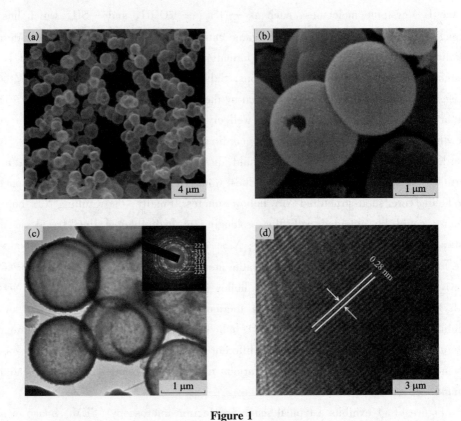

Figure 1

(a) Low- and (b) high-magnification SEM images of NiS$_2$ hollow spheres. (c) TEM image of a NiS$_2$ hollow sphere. The inset in C shows an SAED pattern taken on hollow spheres. (d) HR-TEM image of NiS$_2$ hollow sphere.

experiments were carried out. As shown in Figure 2, two obvious evolution stages of NiS$_2$ can be identified. Figure 2(a) and 2(b) display the typical SEM and TEM images of NiS$_2$ solid spheres obtained at 140 ℃ for 1 h and the mean diameter of solid spheres is about 1.0 μm. Prolonging the hydrothermal time to 8 h, the core/shell structured NiS$_2$ hollow spheres can be obtained, as shown in Figure 2(c) and 2(d). The results further reveal NiS$_2$ solid spheres can gradually transform into hollow spheres with the prolonging of hydrothermal time due to the Ostwald ripening process. Figure 2(e) depicts the evolution of the X-ray diffraction (XRD) patterns of as-prepared products obtained at 140 ℃ for 1 h and 8 h, respectively. From this Figure, it is clear that all the reflections of the XRD patterns can be finely indexed to the cubic NiS$_2$ with lattice constants $a = 5.678$ Å [Space group: Pa-3 (205), JCPDS 65-3325].

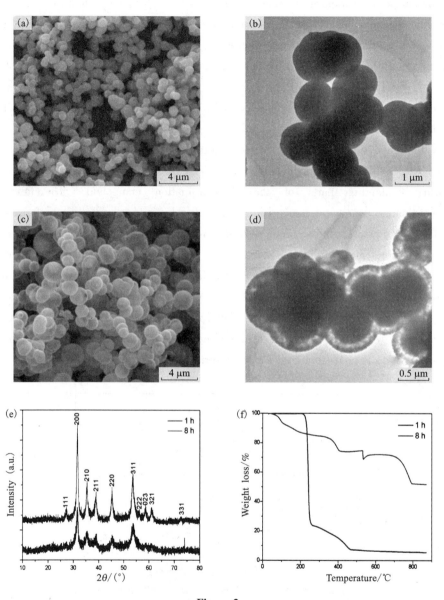

Figure 2

The SEM and TEM images of NiS2 at 140 ℃ for different reaction duration: (A, B) 1 h; (C, D) 8 h; (E) XRD patterns of as-synthesized products obtained at 140 ℃ for 1 and 8 h, respectively; (F) TG analysis curves of resulting products obtained at 140 ℃ for 1 and 8 h under ambient condition in the temperature rang of 35 ℃ to 900 ℃ with a heating rate of 10 ℃ min−1, respectively.

Diffraction peaks of impurities cannot be observed, which indicates that the products are pure phase compounds. With the elevation of hydrothermal time, the crystallinity of the products is continuously improved. Figure 2(f) indicates the TG curves of NiS_2 spheres obtained at 140 ℃ for 1 h and 8 h. The weight loss of as-prepared NiS_2 spheres obtained at 140 ℃ for 1 h in the range 150 –300 ℃ was 77.5%, which showed a sharp decrease compared to other NiS_2 spheres obtained at 140 ℃ for 8 h. The TG curves demonstrated that the organic contents of as-prepared products decrease gradually with the prolonging of hydrothermal time. According to the results presented above, we assume that the initial NiS_2 solid spheres might not be well crystallized owing to rapid nucleation and growth, containing parts of the Ni-coordination compound. Therefore, the Ni-coordination compound from inside would dissolve and recrystallize to form a new particle. Meanwhile, the organic residues would transfer out to form inner space, producing core/shell structured NiS_2 hollow spheres. In particular, nanoparticles located in the inner cores have higher solubilities than outer nanoparticles due to the higher surface energies associated with their larger curvatures. Wherefore the shells grew at the expense of the inner cores. Finally, the resulting NiS_2 hollow spheres could be obtained with the prolonging of hydrothermal time based on Ostwald ripening process, which is consistent with earlier reports.[8]

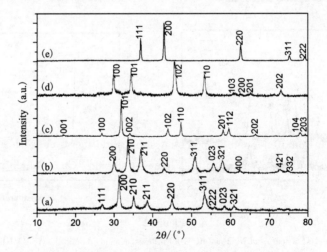

Figure 3　XRD patterns of as-prepared products
(a) NiS_2, (b) $NiSe_2$, (c) $NiTe_2$, (d) NiS, and (e) NiO.

More importantly, various nickel dichalcogenides could be readily obtained through the chemical conversion process by using NiS_2 hollow spheres as a precursor, and the evolution of the X-ray diffraction (XRD) patterns from different products is shown in Figure 3. Figure 3(a) shows the representative XRD pattern of as-prepared NiS_2 hollow spheres. All the diffraction peaks in this pattern can be indexed to the cubic NiS_2 with lattice constants $a = 5.678$ Å [Space group: Pa-3 (205)], which are consistent with the values in the literature (JCPDS 65-3325). Through the chemical conversion process, pure $NiSe_2$ and $NiTe_2$ could be successfully obtained under hydrothermal conditions in the presence of $N_2H_4 \cdot H_2O$ at 140℃ for 24 h, separately, as shown in Figure 3(b) and 3(c). The diffraction peaks in the patterns can be indexed as the cubic $NiSe_2$ with lattice parameters $a = 5.960$ Å [Space group: Pa-3 (205), JCPDS 65-1843], and hexagonal $NiTe_2$ with lattice constants $a = 3.843$ Å, $c = 5.265$ Å [Space group: P-3m1 (164), JCPDS 08-0004]. As the substitution of Se for S is random without a valence change, NiS_2 and $NiSe_2$ keep the same crystal structure. However, with the increasing of anion radius, the original crystal structure changed from cubic phase (NiS_2 and $NiSe_2$) to hexagonal phase ($NiTe_2$). Moreover, the single-phase NiS and NiO were also selectively obtained by calcination of as-prepared NiS_2 in air at 400℃ and 770℃ for 1 h, on the basis of the thermogravimetric analysis (TGA). XRD patterns of NiS and NiO are shown in Figure 3(d) and 3(e), respectively, from which it can be seen that all the diffraction peaks can be indexed to the hexagonal NiS with lattice constant $a = 3.439$ Å, $c = 5.352$ Å [Space group: P63/mmc (194), JCPDS 65-5762], and cubic NiO with lattice constant $a = 4.195$ Å [Space group: Fm-3m (225), JCPDS 65-2901]. No peak of other impurities is detected, which indicates the high purity of NiS_2, $NiSe_2$, $NiTe_2$, NiS and NiO were selectively obtained under the current conditions.

It is interesting and exciting that all of the resulting products obtained by the chemical conversion of NiS_2 hollow spheres could maintain the original morphological features, which would endow their potential application in various fields. Typical TEM images of $NiSe_2$ and $NiTe_2$ are given in Figure 4(a) and 4(b), which reveals that the $NiSe_2$ and $NiTe_2$ hollow spheres hold a rough surface with retained spherical contours, respectively. The average diameter of $NiSe_2$ and $NiTe_2$ hollow spheres ranges from 1-2 μm. SEM and TEM images of as-prepared NiS are shown in Figure 4(c), from which it can be seen that the products exhibit spherical morphologies with hollow interiors. Figure 4(d) depicts a representative TEM image of NiO, which distinctly reveals the hollow nature of the spherical particles. The average diameter of NiO hollow spheres is estimated to be about 1 μm, smaller than that of the NiS_2 hollow spheres, which implies the tendency to shrink after calcination. The SAED patterns of as-prepared $NiSe_2$,

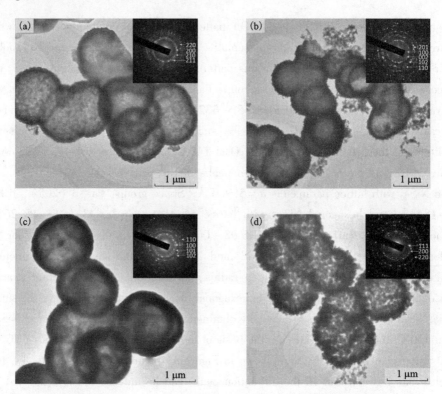

Figure 4

Typical TEM images of (a) NiSe$_2$, (b) NiTe$_2$, (c) NiS and (d) NiO hollow spheres. Insets depict the corresponding SAED patterns.

NiTe$_2$, NiS and NiO hollow spheres are shown in the insets of Figure 4(a) – 4(d), respectively, consistent with the corresponding XRD results. The HRTEM images taken on the surface of individual NiSe$_2$, NiTe$_2$, NiS and NiO hollow sphere are shown in Figure 5. Figure 5(a) – 5(d) show the HRTEM images of (a) NiSe$_2$, (b) NiTe$_2$, (c) NiS and (d) NiO hollow spheres taken on surface of the individual hollow sphere, and the interlayer spacings are calculated to be 0.30 nm, 0.29 nm, 0.27 nm and 0.25 nm, corresponding to the (200), (101), (002) and (111) crystal plane of cubic NiSe$_2$, hexagonal NiTe$_2$, hexagonal NiS and cubic NiO.

The magnetic properties of as-prepared products were measured with the Magnetic property measurement (MPMS XP-5, SQUID). The representative hysteresis loops of NiS$_2$, NiSe$_2$ and NiTe$_2$ hollow spheres at 2 K are shown in Figure 6. The magnetization (M_s) values of NiS$_2$, NiSe$_2$ and NiTe$_2$ hollow spheres at 50 kOe are 27.79 emu/g, 4.29 emu/g and 6.03 emu/g, respectively. However, no hysteresises are observed in

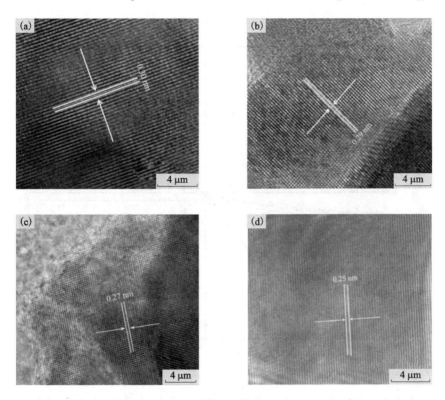

Figure 5

The HRTEM images of as-prepared (a) $NiSe_2$, (b) $NiTe_2$, (c) NiS and (d) NiO hollow spheres.

the curves, as shown in the lower inset of Figure 6, revealing that ferromagnetic ordering component does not exist. The temperature dependence of the zero-field-cooled (ZFC) and field-cooled (FC) magnetic susceptibilities of NiS_2 hollow spheres measured under an applied field of 1 kOe is shown in the upper inset of Figure 6. The magnetization of NiS_2 hollow spheres gradually increases with decreasing temperature in both ZFC and FC curves. The isothermal magnetizations measured at 300 K for nickel dichalcogenides do not show any hysteresis loops and evolve nearly linearly within the magnetic field range, which suggests the absence of ferromagnetic orderings. In particular, the Curie-Weiss plots to the magnetic susceptibilities of NiS_2 and $NiTe_2$ hollow spheres yield the Weiss temperatures of −23.5 K and −101 K, respectively, which suggests the dominated antiferromagnetic ground state, whereas the plot to the data of $NiSe_2$ hollow spheres gives a very large Weiss temperature with an unphysical large effective moment, implying the plot is in fact not effective due to the paramagnetic nature of $NiSe_2$ hollow spheres.[9]

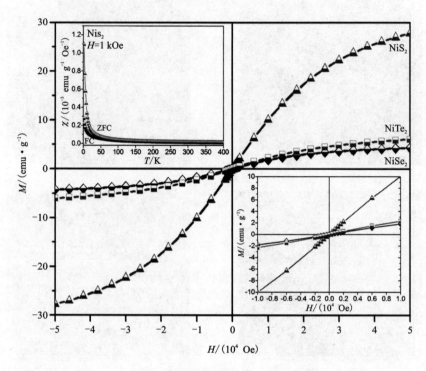

Figure 6 Hysteresis loops of NiS$_2$, NiSe$_2$ and NiTe$_2$ hollow spheres at 2 K

The lower inset is the enlarged plot between −10 kOe and 10 kOe; the upper inset is the temperature dependence of the magnetic susceptibility of NiS$_2$ hollow spheres measured under an applied field of 1 kOe.

4　Conclusion

In summary, we have developed a convenient and reliable synthetic strategy for the fabrication of uniform NiS$_2$ hollow spheres, which could then be transformed into NiSe$_2$ and NiTe$_2$, respectively, by a chemical conversion process. Meanwhile, NiS and NiO hollow spheres could also be successfully prepared by the calcination of NiS$_2$ hollow spheres at different temperature in air, which will greatly endow the potential application in various fields. The formation mechanism of as-prepared products has been proposed. The synthetic strategy presented here may provide an effective route to synthesize other transition-metal dichalcogenides hollow spheres. These nickel dichalcogenides hollow spheres are expected to bring new opportunities for further fundamental research, as well as for technological applications in magnetic devices, solar cells and other advanced materials.

References

[1] (a) X. H. Liu, D. Zhang, J. H. Jiang, N. Zhang, R. Z. Ma, H. B. Zeng, B. P. Jia, S. B. Zhang, G. Z. Qiu. General synthetic strategy for high-yield and uniform rare-earth oxysulfate ($RE_2O_2SO_4$, RE = La, Pr, Nd, Sm, Eu, Gd, Tb, Dy, Y, Ho, and Yb) hollow spheres [J]. RSC Advances, 2012, 2(25): 9362 – 9365. (b) J. G. Guan, F. Z. Mou, Z. G. Sun, W. D. Shi. Preparation of hollow spheres with controllable interior structures by heterogeneous contraction [J]. Chemical Communications, 2010, 46: 6605 – 6607. (c) H. B. Li, L. L. Chai, X. Q. Wang, X. Y. Wu, Y. T. Qian. Hydrothermal growth and morphology modification of beta – NiS three-dimensional flowerlike architectures [J]. Crystal Growth & Design, 2007, 7 (9): 1918 – 1922. (d) Y. Zhao, L. Jiang. Hollow micro-/nanomaterials with multilevel interior structures [J]. Advanced Materials, 2009, 21(36): 3621 – 3638. (e) R. Yi, R. R. Shi, G. H. Gao, N. Zhang, X. M. Cui, Y. H. He, X. H. Liu. Hollow metallic microspheres: fabrication and characterization [J]. The Journal of Physical Chemistry C, 2009, 113(4): 1222 – 1226. (f) X. W. Lou, L. A. Archer, Z. Yang. Hollow micro-/nanostructures: synthesis and applications [J]. Advanced Materials, 2008, 20(21): 3987 – 4019. (g) Y. Hu, J. F. Chen, W. M. Chen, X. L. Li. Synthesis of nickel sulfide submicrometer-sized hollow spheres using a gamma-irradiation route [J]. Advanced Functional Materials, 2004, 14(4): 383 – 386. (h) X. L. Yu, C. B. Cao, H. S. Zhu, Q. S. Li, C. L. Liu, Q. H. Gong. Nanometer-sized copper sulfide hollow spheres with strong optical-limiting properties [J]. Advanced Functional Materials, 2007, 17(8): 1397 – 1401. (i) C. Z. Wu, Y, Xie, L. Y. Lei, S. Q. Hu, C. Z. Ouyang. Synthesis of new-phased VOOH hollow "dandelions" and their application in lithium-ion batteries [J]. Advanced Materials, 2006, 18 (13): 1727 – 1732.

[2] (a) W. M. Du, X. F. Qian, X. S. Niu, Q. Gong. Symmetrical six-horn nickel diselenide nanostars growth from oriented attachment mechanism [J]. Crystal Growth & Design, 2007, 7 (12): 2733 – 2737. (b) X. H. Chen, R. Fan. Low-temperature hydrothermal synthesis of transition metal dichalcogenides [J]. Chemistry of Materials, 2001, 13(3): 802 – 805. (c) J. H. Wang, Z. Cheng, J. L. Brédas, M. Liu. Electronic and vibrational properties of nickel sulfides from first principles [J]. The Journal of Chemical Physics, 2007, 127(21): 214705. (d) Z. Y. Zhong, Y. D. Yin, B. Gates, Y. N. Xia. Preparation of mesoscale hollow spheres of TiO_2 and SnO_2 by templating against crystalline arrays of polystyrene beads [J]. Advanced Materials, 2000, 12(3): 206 – 209. (e) X. J. Zhang, G. F. Wang, A. X. Gu, Y. Wei, B. Fang. CuS nanotubes for ultrasensitive nonenzymatic glucose sensors [J]. Chemical Communications, 2008, 45: 5945 – 5947. (f) X. X. Yu, J. G. Yu, B. Cheng, B. B. Huang. One-Pot template-free synthesis of monodisperse zinc sulfide hollow spheres and their photocatalytic properties [J]. Chemistry – A European Journal, 2009, 15(27): 6731 – 6739.

(g) N. Wang, X. Cao, L. Guo, S. H. Yang, Z. Y. Wu. Facile synthesis of PbS truncated octahedron crystals with high symmetry and their large-scale assembly into regular patterns by a simple solution route [J]. ACS Nano, 2008, 2(2): 184 – 190.

[3] (a) J. Hubbard. Electron correlations in narrow energy bands. II. The degenerate band case [C]. Proceedings of the Royal Society of London A: Mathematical, Physical and Engineering Sciences. The Royal Society, 1964, 277(1369): 237 – 259. (b) M. Z. Xue, Z. W. Fu. Lithium electrochemistry of $NiSe_2$: A new kind of storage energy material [J]. Electrochemistry Communications, 2006, 8(12): 1855 – 1862. (c) C. Marini, A. Perucchi, D. Chermisi, P. Dore, M. Valentini, D. Topwal, D. D. Sarma, S. Lupi, P. Postorino. Combined Raman and infrared investigation of the insulator-to-metal transition in $NiS_{2-x}Se_x$ compounds [J]. Physical Review B, 2011, 84(23): 235134.

[4] F. Gong, X. Xu, Z. Q. Li, G. Zhou, Z. S. Wang. $NiSe_2$ as an efficient electrocatalyst for a Pt-free counter electrode of dye-sensitized solar cells [J]. Chemical Communications, 2013, 49: 1437 – 1439.

[5] (a) O. Rotlevi, K. D. Dobson, D. Rose, G. Hodes. Electroless Ni and $NiTe_2$ ohmic contacts for CdTe/CdS PV cells [J]. Thin Solid Films, 2001, 387(1): 155 – 157. (b) K. D. Dobson, O. Rotlevi, D. Rose, G. Hodes. Formation and characterization of electroless-deposited $NiTe_2$ back contacts to CdTe/CdS thin-film solar cells [J]. Journal of the Electrochemical Society, 2002, 149(2): 147 – 152.

[6] (a) S. H. Yu, M. Yoshimura. Fabrication of powders and thin films of various nickel sulfides by soft solution-processing routes [J]. Advanced Functional Materials, 2002, 12(4): 277 – 285. (b) Y. Wang, Q. S. Zhu, L. Tao, X. W. Su. Controlled-synthesis of NiS hierarchical hollow microspheres with different building blocks and their application in lithium batteries [J]. Journal of Materials Chemistry, 2011, 21(25): 9248 – 9254. (c) G. Q. Zhang, X. L. Lu, W. Wang, X. G. Li. Facile one-pot synthesis of PbSe and $NiSe_2$ hollow spheres: Kirkendall-effect-induced growth and related properties [J]. Chemistry of Materials, 2009, 21(5): 969 – 974. (d) S. L. Yang, H. B. Yao, M. R. Gao, S. H. Yu. Monodisperse cubic pyrite NiS_2 dodecahedrons and microspheres synthesized by a solvothermal process in a mixed solvent: thermal stability and magnetic properties [J]. CrystEngComm, 2009, 11(7): 1383 – 1390. (e) P. R. Bonneau, R. K. Shibao, R. B. Kaner. Low-temperature precursor synthesis of crystalline nickel disulfide [J]. Inorganic Chemistry, 1990, 29(13): 2511 – 2514. (f) R. Luo, X. Sun, L. F. Yan, W. M. Chen, Synthesis and optical properties of novel nickel disulfide dendritic nanostructures [J]. Chemistry Letters, 2004, 33(7): 830 – 831.

[7] (a) G. Chen, W. Ma, D. Zhang, J. Y. Zhu, X. H. Liu. Shape evolution and electrochemical properties of cobalt sulfide via a biomolecule-assisted solvothermal route [J]. Solid State Sciences, 2013, 17: 102 – 106. (b) G. Chen, W. Ma, X. H. Liu, S. Q. Liang, G. Z. Qiu, R. Z. Ma. Controlled fabrication and optical properties of uniform CeO_2 hollow spheres [J]. RSC Advances, 2013, 3(11): 3544 – 3547.

[8] (a) X. W. Lou, Y. Wang, C. Yuan, J. Y. Lee, L. A. Archer. Template-free synthesis of SnO_2 hollow nanostructures with high lithium storage capacity [J]. Advanced Materials, 2006, 18(17): 2325 - 2329. (b) J. Li, H. C. Zeng. Hollowing Sn-doped TiO_2 nanospheres via Ostwald ripening [J]. Journal of the American Chemical Society, 2007, 129(51): 15839 - 15847. (c) Y. Cheng, Y. S. Wang, C. Jia, F. Bao. MnS hierarchical hollow spheres with novel shell structure [J]. The Journal of Physical Chemistry B, 2006, 110(48): 24399 - 24402. (d) Y. H. Zheng, Y. Cheng, Y. S. Wang, L. H. Zhou, F. Bao C. Jia. Metastable gamma-MnS hierarchical architectures: Synthesis, characterization, and growth mechanism [J]. The Journal of Physical Chemistry B, 2006, 110(16): 8284 - 8288. (e) B. Liu, H. C. Zeng. Symmetric and asymmetric Ostwald ripening in the fabrication of homogeneous core-shell semiconductors [J]. Small, 2005, 1(5): 566 - 571. (f) G. F. Lin, J. W. Zheng, R. Xu, Template-free synthesis of uniform CdS hollow nanospheres and their photocatalytic activities [J]. The Journal of Physical Chemistry C, 2008, 112(19): 7363 - 7370.

[9] (a) S. Ogawa. Magnetic properties of 3d transition-metal dichalcogenides with the pyrite structure [J]. Journal of Applied Physics, 1979, 50(B3): 2308 - 2311. (b) N. Inoue, H. Yasuoka, S. Ogawa. Paramagnetism in $NiSe_2$ [J]. Journal of the Physical Society of Japan, 1980, 48(3): 850 - 856. (c) S. Waki, N. Kasai, S. Ogawa. Thermal expansion of $CoSe_2$ [J]. Solid State Communications, 1982, 41(11): 835 - 837. (d) J. Kuneš, L. Baldassarre, B. Schächner, K. Rabia, C. A. Kuntscher, D. M. Korotin, V. I. Anisimov, J. A. McLeod, E. Z. Kurmaev, A. Moewes. Metal-insulator transition in $NiS_{2-x}Se_x$ [J]. Physical Review B, 2010, 81(3): 035122 - 035122 - 6. (e) C. Schuster, M. Gatti, A. Rubio. Electronic and magnetic properties of NiS_2, NiSSe and $NiSe_2$ by a combination of theoretical methods [J]. The European Physical Journal B-Condensed Matter and Complex Systems, 2012, 85(9): 1 - 10.

☆ W. Ma, Y. F. Guo, X. H. Liu, D. Zhang, T. Liu, R. Z. Ma, K. C. Zhou, G. Z. Qiu. Nickel Dichalcogenides Hollow Spheres: Controllable Fabrication, Structural Modification and Magnetic Properties [J]. Chemistry-A European Journal, 2013, 19(46): 15467 - 15471.

Biomolecule-Assisted Hydrothermal Synthesis and Properties of Manganese Sulfide Hollow Microspheres

Abstract: Gamma-manganese sulfide (γ – MnS) hollow microspheres have been successfully synthesized via a biomolecule-assisted hydrothermal process in the presence of L – cysteine and urea at 180 ℃ for 24 h. In the synthesis system, L – cysteine was employed as not only a sulfur source, but also a coordination agent. The structure, morphology and optical properties of as-prepared products have been investigated by X-ray diffraction (XRD), scanning electron microscopy (SEM), energy-dispersive X-ray spectroscopy (EDS) and photoluminescence (PL) spectrum. Reaction parameters such as ratio of L – cysteine to urea, surfactants and reaction time played a significant role in controlling the morphology of as-prepared products. The probable formation mechanism of the γ – MnS hollow microsphere was proposed on the basis of the experimental results.

1 Introduction

Over the past decades, inorganic micro-/nano-materials with well-controlled morphologies have drawn great attention and interests owing to the unique size-dependent and shape-dependent physicochemical properties.[1-5] Many kinds of inorganic materials with novel morphologies and controllable sizes have been successfully prepared by different approaches, e. g. , solid-state reaction, chemical vapor deposition, solvothermal/hydrothermal synthesis, microwave-assisted synthesis, and so forth.[6-12] Recently, a biomolecule-assisted synthesis approach has attracted increasing attention owing to the unique structure and fascinating self-assembling function of the biomolecules. Many kinds of biomolecules such as DNA, protein, amino acid have been utilized to synthesize various inorganic materials.[13, 14] Although considerable achievements have been made, it remains a highly sophisticated challenge to fabricate inorganic materials with novel hierarchical architectures.

Manganese sulfide (MnS), as an important magnetic p-type semiconductor with a wide gap [band gap energy, Eg($T=0$) about 3.7 eV], has become the focus in recent years due to its novel optical, electric and magnetic properties and the important potential applications in solar cells as a window/buffer material, optoelectronic devices, electrode material.[6, 15-18] It is well known that MnS has three different polymorphs: stable phase α - MnS, metastable tetrahedral structures β - MnS, and γ - MnS. The pink β - MnS with zinc blende structure and γ - MnS with wurtzite structure can be synthesized at a low substrate temperature, and they can transform into green α - MnS with rock salt structure at 100 - 400℃ or at high pressure.[15] Inspired by these excellent properties, many efforts have been devoted to the preparation of MnS with various shapes. Yu et al.[10] have synthesized novel flower-like α - MnS via a precipitation reaction of manganese dichloride and thiourea (Tu) in ethylene glycol (EG) solvent at 180℃ for 24 h. Michel et al.[16] have used the inorganic reagents to synthesize pure α - MnS via hydrothermal route. Joo et al.[17] have developed a precursor strategy to synthesize MnS with various shapes such as wires, spheres and cubes. Due to the large specific surface area, unique structure, optical and surface properties, a great number of groups have focused their attention on the fabrication of nano-/micro-crystals with hollow interior.[18]

Herein we have successfully developed a facile biomolecule-assisted hydrothermal method to synthesize the γ - MnS hollow microspheres. The morphologies of the final products can be effectively controlled by adjusting the experimental parameters such as the ratio of L - cysteine to urea, surfactants and reaction time. The photoluminescence (PL) properties of as-prepared γ - MnS hollow microspheres were also investigated and the probable formation mechanism of γ - MnS hollow microspheres was proposed on the basis of the experimental results.

2 Experimental Section

All the chemical reagents used in the experiment were of analytical grade and used without further purification.

2.1 Synthesis. In a typical procedure for fabrication of γ - MnS hollow microspheres, $MnCl_2 \cdot 4H_2O$ (0.5 mmol) was put into Teflon-lined autoclave of 50 mL capacity with 20 mL de-ionized water at room temperature. Then L - cysteine (2 mmol) and urea (0.5 mmol) were added to the above-mentioned solution under ultrasonic treatment. The autoclave was sealed and maintained at 180℃ for 24 h without any agitating and then was allowed to gradually cool down to room temperature. The final product was collected by free sedimentation, washed with absolute ethyl alcohol and deionized water several times, and then dried at 60℃ for 4 h.

2.2 Characterization. The chemical composition and the crystal structure of as-synthesized products were determined by X-ray diffraction (XRD) using a D/max2550 VB + X-ray diffractometer with Cu K_α radiation ($\lambda = 1.5418$ Å). The morphology and size of as-prepared products were characterized at 20 kV by a XL30 S – FEG scanning electron microscope (SEM). Energy-dispersive X-ray spectroscopy (EDS) was taken on the SEM. The room-temperature photoluminescence (PL) measurement was carried out on a Hitachi F – 4500 fluorescence spectrophotometer using the 245 nm excitation line of Xe light.

3 Results and Discussion

The phase and purity of as-obtained products were firstly characterized by X-ray diffraction (XRD). Figure 1(a) and 1(b) show the representative XRD patterns of as-synthesized γ – MnS obtained at 180℃ for 2 h and 24 h, respectively. All diffraction peaks in the patterns can be indexed to the hexagonal phase γ – MnS with lattice constants $a = 3.979$ Å, $b = 3.979$ Å and $c = 6.447$ Å [space group: P63mc (186), JCPDS 40 – 1289]. Diffraction peak of α – MnS, β – MnS or impurities were not detected, indicating the high purity of the products, which could be successfully synthesized under the current conditions. With the prolonging of reaction time, the XRD patterns show that the crystallinity of the products is greatly improved.

The size and morphology of as-prepared products were characterized by scanning electron microscopy (SEM). The typical SEM images of the product obtained at 180℃ for 24 h are shown in Figure 2(a) – 2(c). Figure 2(a) indicates a low-magnification SEM image of as-produced γ – MnS microspheres with the diameter in the range from 3 μm to 6 μm. Figure 2(b) and 2(c) shows the typical high-magnification SEM images of as-produced γ – MnS microspheres, in which we can clearly observe the γ – MnS microspheres with hollow interior.

The chemical composition of γ – MnS hollow microspheres was also further investigated by energy-dispersive X-ray spectroscopy (EDS). The EDS image shown in Figure 2(d) demonstrates that the as-prepared product is composed of Mn and S elements, and the atomic ratio is about 49.1∶50.8, which is very close to the stoichiometry of MnS. Furthermore, no oxygen peak was detected, indicating that there was no manganese oxide layer on the surface of the microspheres.

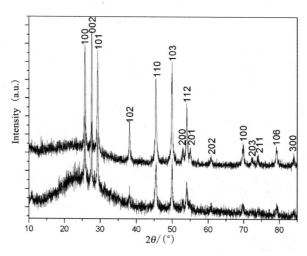

Figure 1 XRD pattern of as-prepared product obtained at 180 ℃ for different reaction time
(a) 2 h; (b) 24 h

To the best of our knowledge, the experimental parameters such asmolar ratio of L-cysteine to urea, surfactant and reaction time have great influences on the morphologies and sizes of as-synthesized products. In order to explore the influence of molar ratio of L-cysteine to urea, the experiments were carried out at 180 ℃ for 24 h by using different molar ratio of L-cysteine to urea. Figure 3 shows the images of as-prepared γ-MnS with different ratios of L-cysteine to urea. Figure 3(a) displays the SEM image of γ-MnS prepared in the present of 2 mmol L-cysteine and without urea. The diameter of γ-MnS spheres is about 10 μm, and the surface of spheres is rugged. When 1 mmol urea is added in above solution, γ-MnS hollow spheres could be obtained, as shown in Figure 3(b). The diameter of as-prepared γ-MnS hollow spheres is about 15 μm and the surface of hollow spheres is smooth. When the amount of urea is kept unchanged, a tremendous change in the morphologies of products can be observed by adjusting the amount of L-cysteine. For example, when the amount of L-cysteine was reduced to 1 mmol, the product was mostly made up of hollow spheres with smooth surface, as shown in Figure 3(c). When the amount of L-cysteine was increased to 3 mmol, the γ-MnS hollow spheres with rugged surface were obtained [Figure 3(d)]. Based on the experimental results, we assume that L-cysteine could be utilized as template and urea played a significant part in determining the morphology of γ-MnS microspheres. As a coordination agent, urea served as not only as inducement to the decomposition of L-cysteine, but also a structure directing

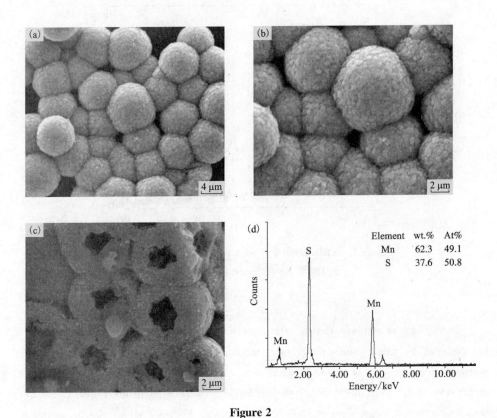

Figure 2

(a) Low and (b) high magnification SEM images, (c) SEM image of reverse side, and (d) EDS image of as-prepared γ - MnS hollow microspheres.

agent.[19] And raising the amount of L - cysteine, the diameter of γ - MnS was correspondingly increasing. It may be a result based on the increasing of sulfur source.

Further studies suggested that the surfactants have great influence on the morphologies of γ - MnS microcrystals. Figure 4 displays the SEM images of γ - MnS microcrystals obtained by using different surfactants. When 0.15 g PVP [Figure 4(a)] or 0.5 mmol CTAB [Figure 4(b)] was added, nearly cubic γ - MnS microcrystals with the mean diameter of about 3.5 μm or 6.5 μm could be obtained, respectively. We thought that PVP and CTAB, as the capping agents, can be added to a solution-phase synthesis to control the morphology of microcrystals. The capping agent selective adsorption on specific crystal facets results in the formation of cubic MnS crystals.[4]

To understand the formation mechanism of γ - MnS hollow spheres, time-dependent experiments were carried out. Figure 5(a) shows the SEM image of γ - MnS

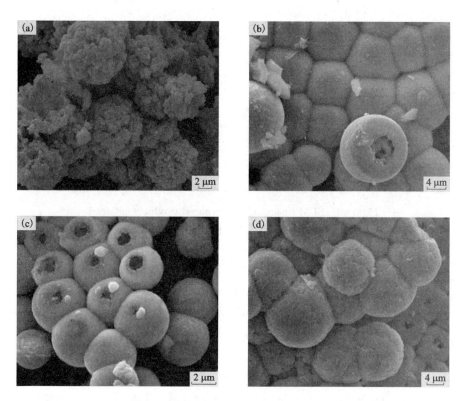

Figure 3 SEM images of as-prepared products obtained at 180 ℃ for 24 h using different ratios of L - cystenine to urea

(a) 2 mmol L - cysteine without urea; (b) 2 mmol L - cysteine and 1 mmol urea; (c) 1 mmol L - cysteine and 0.5 mmol urea; (d) 3 mmol and 0.5 mmol urea.

microspheres with nanorods obtained at 180 ℃ for 0.5 h and the diameter ranging from 12 μm to 16 μm. On prolonging the reaction duration to 1 h, the γ - MnS transformed into solid microspheres with smooth surfaces [Figure 5(b)]. Figure 5(c) shows the SEM image of γ - MnS microspheres core-shell structures obtained at 180 ℃ for 4 h. Prolonging the reaction time to 24 h, γ - MnS hollow spheres with diameter approximately 7 μm could be obtained, as shown in Figure 5(d). The shell thickness of hollow spheres is about 1.8 μm.

Based on the above experimental results, the chemical reactions we employed can be expressed as follows:

$$Mn^{2+} + (L - cysteine) \longrightarrow Mn(L - cysteine)^{2+}$$
$$Mn(L - cysteine)^{2+} \longrightarrow MnS$$

Figure 4 SEM images of as-prepared products obtained at 180 ℃ for 24 h with different surfactants
(a) 0.15 g PVP and (b) 0.5 mmol CTAB

Furthermore, the probable mechanism of MnS hollow microspheres can also be deduced, although the exact formation mechanism of the MnS hollow microspheres is still unclear. The formation process of MnS hollow microspheres may be divided into several steps, as shown in Scheme 1. L－cysteine and urea played undoubtedly significant roles in the synthesis process. The functional groups in the L－cysteine molecule such as —NH_2, —COOH and —SH have a tendency to interact with inorganic cations. Urea was used as a coordinate agent to adjust the morphology of as-prepared MnS and provided alkaline environment to induce the decomposition of L－cysteine. In the present synthesis procedure, L－cysteine could coordination with Mn^{2+} cations to form Mn-cysteine complexes through the ultrasonic treatment. Meanwhile, the complexes aggregated into cluster-like structure due to intermolecular absorption forces such as the Van der Waals force. With the elevating of reaction temperature, urea molecules began to hydrolyze and released ammonia, the C=S bond of L－cysteine began to break and release S^{2-} with the promoting of pH value and reaction temperature.[19, 20] The S^{2-} anions released from L－cysteine combined with Mn^{2+} and formed MnS nanocrystals. The initial MnS nanocrystals were unstable due to the surface energy. The increasing MnS nanocrystals deposited on the surface of complexes and formed solid microspheres. With prolonging reaction time, the solid microspheres converted into core-shell structure due to the Ostwald ripening process. The outward migration of crystals would result in continuous expansion of interior space within the original aggregate, and the core region would get transferred to a outer

Figure 5 SEM images of as-prepared products obtained at 180 ℃ for different reaction duration

(a) 0.5 h; (b) 1 h; (c) 4 h; (d) 24 h

section[6]. MnS microspheres with hollow interior were obtained for a longer reaction time.

Figure 6 presents the room-temperature photoluminescence (PL) spectrum of as-prepared γ – MnS hollow spheres obtained at 180 ℃ for 2 h and 24 h. As the spectrum shows, a very weak band edge emission peak appears at 289 nm and the double-peaked trap state emission peaks emerge at 392 nm and 396 nm. Although the photoluminescence emission mechanism of MnS microcrystals is not fully understood, the different emission peaks are attributed to the recombination of charge carriers in deep traps of surface localized states and photogenerated hole caused by surface defects of MnS hollow microspheres.[21]

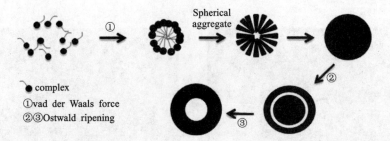

Scheme 1 Proposed growth process of as-prepared γ-MnS hollow microspheres

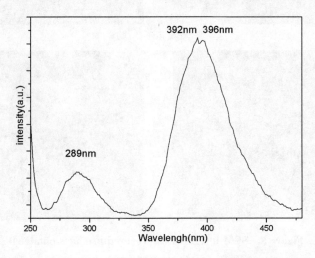

Figure 6 Room temperature PL of as-prepared γ-MnS hollow microspheres obtained at 180℃ for 24 h with an excitation wavelength of 245 nm

4 Conclusion

In summary, the hexagonal phase γ - MnS hollow microspheres were successfully synthesized via a simple and convenient biomolecule-assisted hydrothermal approach. In the synthetic process, L - cysteine and urea were used, as coordinating reagents and shape modifiers, which play a significant role in the synthesis of γ - MnS hollow microspheres. γ - MnS microcrystals with different morphologies could be obtained by adjusting the experimental parameters such as reaction time, surfactant, and the ratio of L - cysteine to urea. With prolonging reaction duration, γ - MnS hollow microspheres were obtained via Ostwald ripening process. Cubic MnS microcrystals could also be synthesized in the presence of CTAB or PVP. This synthetic strategy may provide more opportunities to synthesize other metal sulfides hollow spheres.

References

[1] X. H. Liu, R. Ma, Y. Bando, and T. Sasaki. A general strategy to layered transition-metal hydroxide nanocones: Tuning the composition for high electrochemical performance [J]. Advanced Materials, 2012, 24(16): 2148–2153.

[2] H. J. Cho, P. G. Hwang, D. Jung. Preparation and photocatalytic activity of nitrogen-doped TiO_2 hollow nanospheres [J]. Journal of Physics and Chemistry of Solids, 2011, 72(12): 1462–1466.

[3] X. H. Liu, R. Ma, Y. Bando, T. Sasaki. Layered cobalt hydroxide nanocones: Microwave-assisted synthesis, exfoliation, and structural modification [J]. Angewandte Chemie International Edition, 2010, 49(44): 8253–8256.

[4] Y. Xia, Y. J. Xiong, B. Lim, S. E. Skrabalak. Shape-controlled synthesis of metal nanocrystals: Simple chemistry meets complex physics? [J]. Angewandte Chemie International Edition, 2009, 48(1): 60–103.

[5] B. C. Wang, Y. M. Chou, J. P. Deng, Y. T. Dung. Structural and optical properties of passivated silicon nanoclusters with different shapes: a theoretical investigation [J]. The Journal of Physical Chemistry A, 2008, 112(28): 6351–6357.

[6] Y. Cheng, Y. S. Wang, C. Jia, F. Bao. MnS hierarchical hollow spheres with novel shell structure [J]. The Journal of Physical Chemistry B, 2006, 110(48): 24399–24402.

[7] N. D. Sankir, B. Dogan. Investigation of structural and optical properties of the CdS and CdS/PPy nanowires [J]. Journal of Materials Science, 2010, 45(23): 6424–6432.

[8] X. H. Liu, L. B. Zhou, R. Yi. Single-crystalline indium hydroxide and indium oxide microcubes: Synthesis and characterization [J]. The Journal of Physical Chemistry C, 2008, 112(47): 18426–18430.

[9] C. O'Sullivan, R. D. Gunning, A. Sanyal, C. A. Barrett, H. Geaney, F. R. Laffir, S. Ahmed, K. M. Ryan. Spontaneous room temperature elongation of CdS and Ag_2S nanorods via oriented attachment [J]. Journal of the American Chemical Society, 2009, 131(34): 12250–12257.

[10] J. G. Yu, H. Tang. Solvothermal synthesis of novel flower-like manganese sulfide particles [J]. Journal of Physics and Chemistry of Solids, 2008, 69(5): 1342–1345.

[11] S. S. Wu, H. Q. Cao, S. F. Yin, X. W. Liu, X. R. Zhang. Amino acid-assisted hydrothermal synthesis and photocatalysis of SnO_2 nanocrystals [J]. The Journal of Physical Chemistry C, 2009, 113(41): 17893–17898.

[12] S. B. Wang, K. W. Li, R. Zhai, H. Wang, Y. D. Hou, H. Yan. Synthesis of metastable gamma-manganese sulfide crystallites by microwave irradiation [J]. Materials Chemistry and Physics, 2005, 91(2): 298–300.

[13] A. Jatsch, E. K. Schillinger, S. Schmid, P. Bauerle. Biomolecule assisted self-assembly of

pi-conjugated oligomers [J]. Journal of Materials Chemistry, 2010, 20(18): 3563 – 3578.

[14] A. P. Alivisatos, K. P. Johnsson, X. G. Peng, T. E. Wilson, C. J. Loweth, M. P. Bruchez, P. G. Schultz. Organization of 'nanocrystal molecules' using DNA [J]. Nature, 1996, 382(6592): 609 – 611.

[15] N. Zhang, R. Yi, Z. Wang, R. R. Shi, H. D. Wang, G. Z. Qiu, X. H. Liu. Hydrothermal synthesis and electrochemical properties of alpha-manganese sulfide submicrocrystals as an attractive electrode material for lithium-ion batteries [J]. Materials Chemistry and Physics, 2008, 111(1): 13 – 16.

[16] F. M. Michel, M. A. A. Schoonen, X. V. Zhang, S. T. Martin, J. B. Parise. Hydrothermal synthesis of pure alpha-phase manganese(II) sulfide without the use of organic reagents [J]. Chemistry of Materials, 2006, 18(7): 1726 – 1736.

[17] J. Joo, H. B. Na, T. Yu, J. H. Yu, Y. W. Kim, F. X. Wu, J. Z. Zhang, T. Hyeon. Generalized and facile synthesis of semiconducting metal sulfide nanocrystals [J]. Journal of the American Chemical Society, 2003, 125(36): 11100 – 11105.

[18] X. L. Yu, C. B. Cao, H. S. Zhu, Q. S. Li, C. L. Liu, Q. H. Gong. Nanometer-sized copper sulfide hollow spheres with strong optical-limiting properties [J]. Advanced Functional Materials, 2007, 17(8): 1397 – 1401.

[19] H. Zhu, X. Wang, F. Yang, X. Yang. Template-free, surfactantless route to fabricate $In(OH)_3$ monocrystalline nanoarchitectures and their conversion to In_2O_3 [J]. Crystal Growth & Design, 2008, 8(3): 950 – 956.

[20] H. Zhu, X. L. Wang, W. Yang, F. Yang, X. R. Yang. Indium sulfide microflowers: Fabrication and optical properties [J]. Materials Research Bulletin, 2009, 44(10): 2033 – 2039.

[21] J. J. Hui, R. N. Yu, X. H. Liu. Shape-controlled synthesis and properties of manganese sulfide microcrystals via a biomolecule-assisted hydrothermal process [J]. Materials Chemistry and Physics, 2009, 115(2): 502 – 506.

☆ W. Ma, G. Chen, D. Zhang, K. C. Zhou, G. Z. Qiu, X. H. Liu. Biomolecule-assisted hydrothermal synthesis and properties of manganese sulfide hollow microspheres [J]. Journal of Physics and Chemistry of Solids, 2012, 73(11): 1385 – 1389.

Biomolecule-Assisted Hydrothermal Synthesis and Electrochemical Properties of Copper Sulfide Hollow Spheres

Abstract: Copper Sulfide (CuS) hollow spheres could be successfully synthesized in large quantities through a facile biomolecule-assisted hydrothermal synthetic method. The as-fabricated CuS hollow spheres showed specific capacitances and excellent cycling stability in supercapacitors.

1 Introduction

Hollow spheres with nanometer or micrometer dimensions are attracting considerable attention because of their distinctive properties and wide variety of practical and potential applications in, e. g., selective catalysis, controlled delivery, and biomedical diagnosis and therapy, etc.[1,2] A variety of chemical and physicochemical strategies have been employed for the design and controlled fabrication of micro/nanospheres with hollow interiors over the past decades. Especially, biomolecule-assisted synthesis route has draw the increasing attention in recent years owing to the unique structure and fascinating self-assembling functions of the biomolecules and the important technological applications in the synthesis of inorganic micro/nanostructure materials, including hollow spheres.[3] DNA, protein, amino acid have been utilized to synthesize various nanomaterials.[4] At the meantime, a great effort has been devoted to the various transition-metal sulfide because of their novel magnetic, optical and electrical properties and their wide variety of potential applications.[5] However, the great challenge and difficulties are presented in fabricating desired architecture-structure micro/nanomaterials owing to their extremely small size and complexity of the influencing factors.

As an important magnetic p-type semiconductor with a wide gaps of 1.2 – 2.0eV, copper sulfide (CuS) has gained extensive attraction in recent years due to its different unique properties and the important potential application in catalyst, solar cells, optical filters, superconductors, gas sensors, cathode materials for lithium-ion batteries and nonlinear optical materials.[6] In particular, the properties of CuS sensitively depend on their morphologies. Thus it is a significant challenge to fabricate CuS with novel morphologies. Up to now, CuS nanomaterials with various morphologies, for example, nanowires, nanotubes, nanorods and nanoplates, have been successfully synthesized through various methods, such as spray pyrolysis, hydrothermal, solvothermal, chemical deposition, ion exchange reaction, template technology, and so on.[7] Due to the large surface area, unique structure, optical and surface properties, more and more efforts have been devoted to the preparation of CuS hollow spheres. Recently, Zhu et al. reported the synthesis of CuS hollow spheres via the chemical transformation of in situ formed Cu_2O spheres as sacrificial templates.[8] Lu et al. also synthesized hollow CuS nanospheres based on a template-assisted approach.[9] However, sacrificial template strategies normally involves intricate processing: (i) the selection and preparation of templates, (ii) coating the templates with the desired materials to form a core@shell structure, and (iii) removal of templates by calcination or wet chemical etching to achieve the desired interiors.[10] In the present work, we demonstrated that hollow CuS spheres could be successfully synthesized in large quantities through a simple one-pot synthetic method. The effects of synthetic parameters on the morphologies of as-prepared products are also investigated. The hollow CuS microspheres showed specific capacitances and excellent cycling stability in supercapacitors.

2 Experimental Section

All chemicals were of analytical grade from Beijing Chemical Reagents Factory and used as starting materials without further purification.

2.1 Synthesis of CuS hollow spheres. In a typical procedure, $Cu(NO_3)_2 \cdot 3H_2O$ (0.5 mmol) was put into Teflon-lined autoclave of 50 ml capacity with 30 ml deionized water at room temperature. And L-cysteine (2 mmol) was added to the above-mentioned solution, the milk white solution was formed with the stirring, then thiourea (0.5 mmol) was added to the solution under an ultrasonic treatment for a few minutes. The autoclave was heated and maintained at 180 ℃ for 24 h without any agitating, and then was allowed to gradually cool to room temperature. The final product was collected by centrifugation, washed with absolute ethyl alcohol several times, and then dried at 60 ℃ for 4 h.

2.2 Characterization. The crystal structures of obtained specimens were characterized on a X-ray diffractometer (XRD, D/max2550 VB +) with $Cu_{K\alpha}$ radiation (λ = 1.5418 Å). The sizes and morphologies of as-synthesized products were characterized by a field-emission scanning electron microscopy (FE-SEM, Sirion 200) with an accelerating voltage of 15 kV and transmission electron microscope (TEM, Tecnai G2 F20) with an accelerating voltage of 200 kV. Cyclic voltammetry and galvanic charge-discharge measurements were carried out on a Solartron electrochemistry workstation using a three electrode cell with 1 M KOH as the electrolyte, a Ag/AgCl electrode as reference electrode, and platinum wire as counter electrode, respectively.

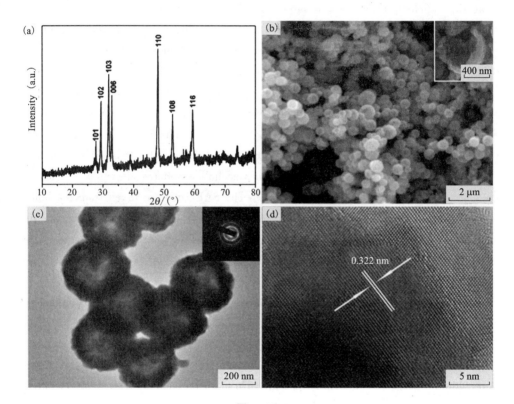

Figure 1

(a) XRD pattern of as-prepared product obtained at 180℃ for 24 h. (b) SEM images of as-prepared CuS hollow sphere. The inset in (b) is a high-magnification SEM image of an individual CuS hollow sphere with broken shells. c) TEM images of as-prepared CuS hollow spheres. The inset in (c) shows an SAED pattern taken on hollow spheres. (d) HRTEM image of as-prepared CuS hollow spheres.

3   Results andDiscussions

Figure 1(a) shows the representative XRD pattern of as-prepared product obtain at 180 ℃ for 24 h. All the diffraction peaks in the pattern can be indexed to the hexagonal CuS with lattice constants $a = 3.972$ Å and $c = 16.344$ Å (space group: P63/mmc (194)), which are consistent with the values in the literature (JCPDS 06 -0464). No peak of other impurities can be detected under the current experimental conditions, indicating the high purity of the products. The size and morphology of as-prepared products were characterized by scanning electron microscopy (SEM) and transmission electron microscopy (TEM). The representative SEM image of as-prepared CuS obtained at 180 ℃ for 24 h is showed in Figure 1(b), inducing that a large quantity of hollow spheres with the diameters in the range from 200 to 400 nm were successfully synthesized under the current experimental conditions. The inset shows a high-magnified SEM image of an individual CuS hollow sphere with broken shells, which clearly reveals the hollow nature. The wall thickness of CuS hollow sphere is about 70 nm. Figure 1(c) shows the typical TEM image of CuS hollow spheres, in which we can find the CuS spheres with hollow interior clearly. The SAED pattern taken from the hollow spheres shows clear diffraction rings (inset in Figure 1C), indicating polycrystalline characteristic of the product. Figure 1(d) clearly shows the interlayer spacing measured to be about 0.322 nm, which agrees well with the separation between (101) lattice planes of CuS.

The morphologies of the products strongly depend on the synthetic parameters, such as reaction time, temperature, and L - cysteine. Figure 2(a) displays a typical SEM image of as-prepared product obtained at 180 ℃ for 4 h. A large quantity of spherical CuS particles with hollow interiors can be clearly observed. With careful observation, the broken shell of as-prepared CuS spheres also confirms that the interior of the particles is hollow. When the reaction time was prolonged to 8 h, the product was mainly composed of uniform hollow spherical structure, as shown in Figure 2(b). However, with increasing the reaction temperature, the product obtained at 200 ℃ is the mixture of hollow spheres and irregular particles, as shown in Figure 2(c). It is generally believed that the high temperature causes the release of sulfide anions from L - cysteine and destroys the structure stability of L - cysteine. In particular, L - cysteine seems to play a crucial role in the formation of hollow spheres. Figure 2(d) shows a typical SEM image of as-prepared product synthesized with the absence of

Figure 2 The SEM images of products obtained at different time or temperature
(a) 180 ℃ for 4 h, (b) 180 ℃ for 8 h, (c) 200 ℃ for 24 h, (d) with the absence of L – Cysteine at 180 ℃ for 24 h

L – cysteine at 180 ℃ for 24 h. Herein, we can distinctly observe the product composed of many lamellar structures. We assume that L – cysteine serves as a biomolecular template for the formation of CuS hollow spheres based on the Ostwald ripening process, which is consistent with our earlier reports.[3] Firstly, L – cysteine is dispersed in deionized water and simultaneously capped by copper ions due to the coordination of functional groups in the L – cysteine molecules, which has a strong tendency to coordinate with inorganic cations. Subsequently the freshly copper coordination compound is unstable due to the high surface energy and tend to aggregate into spherical product, which can then convert into hollow spheres with the elevation of reaction time and temperature. Furthermore, thiourea also serves as a coordinate agent to adjust the morphology of as-prepared CuS hollow spheres and provided alkaline environment to induce the decomposition of L – cysteine.

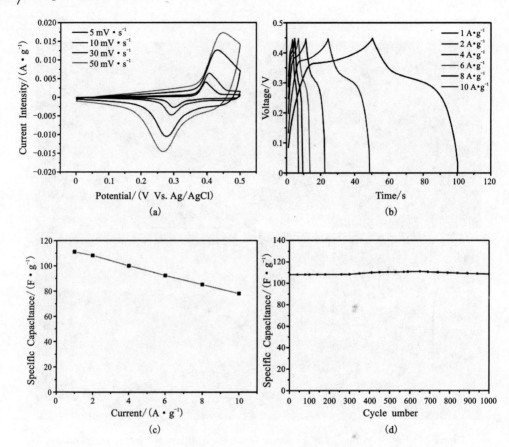

Figure 3 Electrochemical performances of as-prepared CuS hollow spheres

(a) CV curves at different scan rates of 5, 10, 30, and 50 mV/s, (b) Charge-discharge behavior at different current densities, (c) the corresponding specific capacitance as a function of current density, and (d) Specific capacitance versus cycle number of as-prepared CuS hollow spheres at a galvanic charge and discharge current density of 1 A/g.

The electrochemical behavior of CuS hollow spheres in 2 M KOH aqueous electrolyte were investigated using cyclic voltammetry (CV) and galvanic charge-discharge cycling. Figure 3(a) shows the representative CV curves of CuS hollow spheres at scan rates ranged from 5 to 50 mV/s. The well-defined redox peaks within 0–0.5 V appear in all CV curves, which exhibits that the electrochemical capacitance of CuS hollow spheres is derived from pseudocapacitive behavior of reversible Faradaic redox reactions. Furthermore, the increasing scan rate caused the increase of the internal diffusion resistance in the pseudoactive material, which induced the increase of current and the oxidation/reduction peaks shifted to a more positive/negative positions.

Figure 3(b) shows the galvanic charge-discharge curves of CuS hollow spheres measured between 0 and 0.45 V at various current densities. In the whole potential region, the charge curves exhibit symmetry to corresponding discharge counter parts. Figure 3(c) shows the specific capacitances of as-prepared CuS hollow spheres, which were measured from charge-discharge curves from Figure 3(b). The specific capacitance of as-prepared CuS hollow spheres can be calculated to be 111.2 F/g at 1 A/g, 108.2 F/g at 2 A/g, 100 F/g at 4 A/g, 92.3 F/g at 6 A/g, 85.2 F/g at 8 A/g, and 78 F/g at 10 A/g, respectively. This suggests that when the charge-discharge rate is increased from 1 to 10 A/g, there is about 70.1% of the capacitance still retaining, reflecting the good ion diffusion and electron transport ability. To evaluate the electrocatalytic stability of as-prepared CuS hollow spheres, the cycle life was studied by charge-discharge measurements. As shown in Figure 3(d), with the increase of cycle times, the capacitance increases to a maximum. This may be due to the fact that active materials have not been fully used at the initial stage. At a cycle number of 600, the specific capacitance of CuS hollow nanospheres increases to the maximum of 111.2 F/g. After 1000 cycles, the capacitance value still retains at about 97.8%, suggesting good cycling performances of CuS hollow spheres. Compared with plate-like morphology, CuS hollow spheres exhibited the higher specific capacitance and excellent cycling performance.[11]

4 Conclusion

In summary, we have developed a convenient biomolecule-assisted hydrothermal synthetic method for the fabrication of uniform CuS hollow spheres. The morphologies of the final products could be controlled by adjusting the synthetic parameters. The synthetic strategy may provide an effective route to synthesize other transition-metal sulfide hollow spheres. The as-prepared CuS hollow spheres showed specific capacitances and excellent cycling stability in supercapacitors. The CuS hollow spheres may find potential use in catalysts, sensors, electrochemical energy storage, and light weight structural materials.

References

[1] (a) L. F. Shen, L. Yu, X.-Y. Yu, X. G. Zhang, X. W. Lou. Self-templated formation of uniform $NiCo_2O_4$ hollow spheres with complex interior structures for lithium-ion batteries and

supercapacitors [J]. Angewandte Chemie International Edition, 2015, 54(6): 1868 – 1872. (b) D. Zhang, J. Y. Zhu, N. Zhang, T. Liu, L. M. Chen, X. H. Liu, R. Z. Ma, H. T. Zhang, G. Z. Qiu. Controllable fabrication and magnetic properties of double-shell cobalt oxides hollow particles [J]. Scientific reports, 2015, 5: 8737. (c) J.-M. Jeong, B. G. Choi, S. C. Lee, K. G. Lee, S.-J. Chang, Y.-K. Han, Y. B. Lee, H. U. Lee, S. Kwon, G. Lee, C.-S. Lee, Y. S. Huh. Hierarchical hollow spheres of Fe_2O_3@ polyaniline for lithium ion battery anodes [J]. Advanced Materials, 2013, 25(43): 6250 – 6255. (d) R. R. Shi, G. Chen, W. Ma, D. Zhang, G. Z. Qiu, X. H. Liu. Shape-controlled synthesis and characterization of cobalt oxides hollow spheres and octahedra [J]. Dalton Transactions, 2012, 41(29): 5981 – 5987. (e) B. Wang, J. S. Chen, H. B. Wu, Z. Wang, X. W. Lou. Quasiemulsion-templated formation of α-Fe_2O_3 hollow spheres with enhanced lithium storage properties [J]. Journal of the American Chemical Society, 2011, 133(43): 17146 – 17148. (f) W. Liu, B. J. Li, C. L. Gao, Z. Xu. A magnetically separable heterogeneous catalyst Co@CoO hollow spheres: synthesis and catalytic performance for the selective oxidation of alcohol [J]. Chemistry Letters, 2009, 38 (11): 1110 – 1111.

[2] (a) D. Kim, E. Kim, J. Lee, S. Hong, W. Sung, N. Lim, C. G. Park, K. Kim. Direct synthesis of polymer nanocapsules: self-assembly of polymer hollow spheres through irreversible covalent bond formation [J]. Journal of the American Chemical Society, 2010, 132(28): 9908 – 9919. (b) Y. Zhao, L. Jiang. Hollow micro/nanomaterials with multilevel interior structures [J]. Advanced Materials, 2009, 21(36): 3621 – 3638. c) Z.-R. Shen, J.-G. Wang, P.-C. Sun, D.-T. Ding, T.-H. Chen. Fabrication of lanthanide oxide microspheres and hollow spheres by thermolysis of pre-molding lanthanide coordination compounds [J]. Chemical Communications, 2009, 13: 1742 – 1744. (d) X. W. Lou, L. A. Archer, Z. Yang. Hollow Micro-/Nanostructures: Synthesis and Applications [J]. Advanced Materials, 2008, 20(21): 3987 – 4019. (e) J. N. Gao, Q. S. Li, H. B. Zhao, L. S. Li, C. L. Liu, Q. H. Gong, L. M. Qi. One-pot synthesis of uniform Cu_2O and CuS hollow spheres and their optical limiting properties [J]. Chemistry of Materials, 2008, 20(19): 6263 – 6269. (f) H. B. Zeng, W. P. Cai, P. S. Liu, X. X. Xu, H. J. Zhou, C. Klingshirn, H. Kalt. ZnO-based hollow nanoparticles by selective etching: elimination and reconstruction of metal semiconductor interface, improvement of blue emission and photocatalysis [J]. ACS Nano, 2008, 2(8): 1661 – 1670. (g) S. H. Im, U. Jeong, Y. N. Xia. Polymer hollow particles with controllable holes in their surfaces [J]. Nature Materials, 2005, 4(9): 671 – 675.

[3] (a) G. Chen, X. H. Liu, W. Ma, H. M. Luo, J. H. Li, R. Z. Ma, G. Z. Qiu. Hollow spherical rare-earth-doped yttrium oxysulfate: A novel structure for upconversion [J]. Nano Research, 2014, 7(8): 1093 – 1102. (b) G. Chen, W. Ma, X. H. Liu, S. Q. Liang, G. Z. Qiu, R. Z. Ma. Controlled fabrication and optical properties of uniform CeO_2 hollow spheres [J]. RSC Advances, 2013, 3(11): 3544 – 3547. (c) W. Ma, Y. F. Guo, X. H. Liu, D. Zhang, T. Liu, R. Z. Ma, K. C. Zhou, G. Z. Qiu. Nickel dichalcogenide hollow spheres:

Controllable fabrication, structural modification, and magnetic properties [J]. Chemistry-A European Journal, 2013, 19(46): 15467 – 15471.

[4] (a) W. B. Rogers, V. N. Manoharan. Programming colloidal phase transitions with DNA strand displacement [J]. Science, 2015, 347(6222): 639 – 642. (b) P. W. Wu, Y. Yu, C. E. McGhee, L. H. Tan, Y. Lu. Applications of synchrotron-based spectroscopic techniques in studying nucleic acids and nucleic acid-functionalized nanomaterials [J]. Advanced Materials, 2014, 26(46): 7849 – 7872. (c) A. Jatsch, E. K. Schillinger, S. Schmid, P. Bauerle. Biomolecule assisted self-assembly of π-conjugated oligomers [J]. Journal of Materials Chemistry, 2010, 20(18): 3563 – 3578. (d) B. X. Li, Y. Xie, Y. Xue. Controllable synthesis of CuS nanostructures from self-assembled precursors with biomolecule assistance [J]. The Journal of Physical Chemistry C, 2007, 111(33): 12181 – 12187. (e) A. P. Alivisatos, K. P. Johnsson, X. Peng, T. E. Wilson, C. J. Loweth, M. P. Bruchez, Jr. P. G. Schultz. Organization of nanocrystal molecules' using DNA [J]. Nature, 1996, 382(6592): 609 – 611.

[5] (a) Y. Liu, Y. Qiao, W. X. Zhang, Z. Li, X. L. Hu, L. X. Yuan, Y. H. Huang. Coral-like α-MnS composites with N-doped carbon as anode materials for high-performance lithium-ion batteries [J]. Journal of Materials Chemistry, 2012, 22(45): 24026 – 24033. (b) C. W. Kung, H. W. Chen, C. Y. Lin, K. C. Huang, R. Vittal, K. C. Ho. CoS acicular nanorod arrays for the counter electrode of an efficient dye-sensitized solar cell [J]. ACS Nano, 2012, 6 (8): 7016 – 7025. (c) X. M. Zhao, S. W. Zhou, L. P. Jiang, W. H. Hou, Q. M. Shen, J. J. Zhu. Graphene-CdS nanocomposites: facile one-step synthesis and enhanced photoelectrochemical cytosensing [J]. Chemistry-A European Journal, 2012, 18(16): 4974 – 4981. (d) X. Huang, M. Wang, M. G. Willinger, L. D. Shao, D. S. Su, X. M. Meng. Assembly of three-dimensional hetero-epitaxial ZnO/ZnS core/shell nanorod and single crystalline hollow ZnS nanotube arrays [J]. ACS Nano, 2012, 6(8): 7333 – 7339. (e) S. J. Bao, Y. B. Li, C. M. Li, Q. L. Bao, Q. Lu, J. Guo. Shape evolution and magnetic properties of cobalt sulfide [J]. Crystal Growth and Design, 2008, 8(10): 3745 – 3749.

[6] (a) D.-H. Ha, A. H. Caldwell, M. J. Ward, S. Honrao, K. Mathew, R. Hovden, M. K. A. Koker, D. A. Muller, R. G. Hennig, R. D. Robinson. Solid-solid phase transformations induced through cation exchange and strain in 2d heterostructured copper sulfide nanocrystals [J]. Nano Letters, 2014, 14(12): 7090 – 7099. (b) Y. Xie, A. Riedinger, M. Prato, A. Casu, A. Genovese, P. Guardia, S. Sottini, C. Sangregorio, K. Miszta, S. Ghosh, T. Pellegrino, L. Manna. Copper sulfide nanocrystals with tunable composition by reduction of covellite nanocrystals with Cu^+ ions [J]. Journal of the American Chemical Society, 2013, 135 (46): 17630 – 17637. (c) L. Qian, X. Tian, L. Yang, J. Mao, H. Yuan, D. Xiao. High specific capacitance of CuS nanotubes in redox active polysulfide electrolyte [J]. RSC Advances, 2013, 3(6): 1703 – 1708. (d) X. Feng, Y. Li, H. Liu, Y. Li, S. Cui, N. Wang, L. Jiang, X. Liu, M. Yuan. Controlled growth and field emission properties of CuS nanowalls [J]. Nanotechnology, 2007, 18(14): 145706. (e) X. L. Yu, C. B. Cao, H. S. Zhu, Q. S. Li,

C. L. Liu, Q. H. Gong. Nanometer-sized copper sulfide hollow spheres with strong optical-limiting properties [J]. Advanced Functional Materials, 2007, 17(8): 1397 - 1401.

[7] (a) L. Zhou, W. Li, Z. W. Chen, E. Ju, J. S. Ren, X. G. Qu. Growth of hydrophilic cus nanowires via dna-mediated self-assembly process and their use in fabricating smart hybrid films for adjustable chemical release [J]. Chemistry-A European Journal, 2015, 21(7): 2930 - 2935. (b) M. T. Mayer, Z. I. Simpson, S. Zhou, D. W. Wang. Ionic-Diffusion-Driven, low-temperature, solid-state reactions observed on copper sulfide nanowires [J]. Chemistry of Materials, 2011, 23(22): 5045 - 5051. (c) Q. Y. Lu, F. Gao, D. Y. Zhao. One-step synthesis and assembly of copper sulfide nanoparticles to nanowires, nanotubes, and nanovesicles by a simple organic amine-assisted hydrothermal process [J]. Nano Letters, 2002, 2(7): 725 - 728. (d) J. F. Mao, Q. Shu, Y. Q. Wen, H. Y. Yuan, D. Xiao, M. M. F. Choi. Facile fabrication of porous CuS nanotubes using well-aligned [Cu(tu)] $Cl_{1/2}^{-}H_2O$ nanowire precursors as self-sacrificial templates [J]. Crystal Growth and Design, 2009, 9(6): 2546 - 2548. (e) M. Kruszynska, H. Borchert, A. Bachmatiuk, M. H. Rummeli, B. Buchner, J. Parisi, J. Kolny-Olesiak. Size and shape control of colloidal copper (I) sulfide nanorods [J]. ACS Nano, 2012, 6(7): 5889 - 5896. (f) H. B. Wu, W. Chen, Synthesis and reaction temperature-tailored self-assembly of Copper Sulfide Nanoplates [J]. Nanoscale, 2011, 3(12): 5096 - 5102. (g) X. H. Liao, N. Y. Chen, S. Xu, S. B. Yang, J. J. Zhu. A microwave assisted heating method for the preparation of copper sulfide nanorods [J]. Journal of Crystal Growth, 2003, 252(4): 593 - 598.

[8] (a) D. X. Wu, J. F. Duan, Y. S. Lin, Z. G. Meng, C. Y. Zhang, H. T. Zhu. Photo-thermal conversion of copper sulfide hollow structures with different shape and thickness [J]. Journal of Nanoscience and Nanotechnology, 2015, 15(4): 3191 - 3195. (b) H. T. Zhu, J. X. Wang, D. X. Wu. Fast synthesis, formation mechanism, and control of shell thickness of CuS hollow spheres [J]. Inorganic Chemistry, 2009, 48(15): 7099 - 7104.

[9] (a) L. R. Guo, D. D. Yan, D. F. Yang, Y. J. Li, X. D. Wang, O. Zalewski, B. F. Yan, W. Lu. Combinatorial photothermal and immuno cancer therapy using chitosan-coated hollow copper sulfide nanoparticles [J]. ACS Nano, 2014, 8(6): 5670 - 5681. (b) L. R. Guo, I. Panderi, D. D. Yan, K. Szulak, Y. J. Li, Y.-T. Chen, H. Ma, D. B. Niesen, N. Seeram, A. Ahmed, B. F. Yan, D. Pantazatos, W. Lu. A comparative study of hollow copper sulfide nanoparticles and hollow gold nanospheres on degradability and toxicity [J]. ACS Nano, 2013, 7(10): 8780 - 8793. (c) S. Ramadan, L. R. Guo, Y. J. Li, B. F. Yan, W. Lu. Hollow copper sulfide nanoparticle-mediated transdermal drug delivery [J]. Small, 2012, 8(20): 3143 - 3150.

[10] (a) G. Z. Chen, F. Rosei, D. L. Ma. Template engaged synthesis of hollow ceria-based composites [J]. Nanoscale, 2015, 7(13): 5578 - 5591. (b) J. N. Zhang, J. Z. Lin, B. Wu, B. B. Yang, M. Y. Wu, W. L. Yang, Q. Y. Wu, J. J. Yang. Facile fabrication of hybrid hollow microspheres via in situ pickering miniemulsion polymerization [J]. Chemistry

Letters, 2012, 41(9): 970 – 971. (c) H. L. Xu, W. Z. Wang. Template synthesis of multishelled Cu_2O hollow spheres with a single-crystalline shell wall [J]. Angewandte Chemie International Edition, 2007, 46(9): 1489 – 1492.

[11] C. J. Raj, B. C. Kim, W.-J. Cho, W.-G. Lee, Y. Seo, K.-H. Yu. Electrochemical capacitor behavior of copper sulfide (CuS) nanoplatelets [J]. Journal of Alloys and Compounds, 2014, 586: 191 – 196.

☆ Y. J. Zhu, T. Liu, W. Ma, D. Zhang, N. Zhang, L. M. Chen, X. H. Liu, G. Z. Qiu. Biomolecule-assisted Hydrothermal Synthesis and Electrochemical Properties of Copper Sulfide Hollow Spheres [J]. Chemistry Letters, 2015, 44(10): 1321 – 1323.

Chapter VI

Other Hollow Spheres

Hollow Metallic Microspheres: Fabrication and Characterization

Abstract: Hollow metallic nickel spheres with an average diameter of 1.8 μm have been successfully synthesized via a decomposition and reduction route by using hollow nickel hydroxide spheres as precursor. Microsized hollow bimetallic (Ni/Au, Ni/Ag, Ni/Pt and Ni/Pd) and noble metal (Pt and Pd) spheres have also been selectively synthesized by adjusting the amount of reactants of replacement reaction in which corresponding noble metal compound and hollow metallic nickel spheres were used as starting materials. The magnetic properties of hollow metallic nickel spheres have been investigated. It is found that the coercivity of hollow Ni spheres is much enhanced compared with their bulk counterparts. The study on catalytic activities of as-prepared hollow noble metal spheres for hydrogen generation reveals that Ni/Pt hollow bimetallic spheres exhibit favorable catalytic activities which have potential applications in portable hydrogen generation systems.

1 Introduction

During the past decades, the impetus for the considerable research into the design and preparation of hollow spheres emanates from their exceptional properties and important technological applications in the fields of catalysis, biological sensing, and optoelectronics.[1-3] Among the target materials, hollow metallic spheres with nanometer to micrometer dimensions have been especially interesting in recent years due to their fascinating mechanical, electrical, and magnetic properties, as well as their promising applications in drug release and catalytic fields.[4,5] Considerable effort has been devoted to the fabrication of hollow metallic spheres during the past few years. Many methods have also been investigated to fabricate hollow metallic spheres, in which the most versatile way is based on the template-directed synthesis. Various hard and soft sacrificial templates, such as polystyrene spheres,[6] silica spheres,[7] liquid droplets,[8] and microemulsion droplets,[9] etc., are employed to fabricate hollow spheres. Xia et al. have demonstrated the use of amorphous selenium (a-Se) spherical colloids as templates in forming platinum hollow spheres.[10] Hyeon and co-workers have synthesized palladium hollow spheres using silica spheres as templates, and demonstrated the palladium hollow spheres owned good catalytic activities in Suzuki cross-coupling reactions and can be reused many times without loss of catalytic activity.[11] What is worth noting is that the above-mentioned hollow metallic spheres have been based on template-based synthesis. Unfortunately, the products of such synthesis often involved the complicated process and are difficult to purify. The development of the facile methods for fabricating hollow metallic spheres still remains a highly sophisticated challenge.

Metallic nickel materials have received enormous attention due to their unique magnetic and electronic properties, as well as their widely potential applications.[12-14] In particular, surface area is an especially important characteristic of metallic nickel, as the catalytic activity, for example, is greatly enhanced with increasing nickel surface area.[15] Metallic nickel spheres with hollow interiors are maybe of benefit to such applications. However, there have been only a few reports on the fabrication of hollow metallic nickel spheres. Xu and co-works reported the synthesis of hollow nickel spheres in micrometer and nanometer dimensions.[5, 16] Moreover, hierarchical porous hollow nickel microspheres with nickel nanoparticles have been fabricated by Zhu et al.[17] Recently, hierarchical structured hollow nickel sphere arrays and nickel chain assembled by hollow spheres also have been obtained using through-pore template and PVP as hard and soft templates, respectively.[18, 19]

Noble metalmaterials with hollow interiors have been shown to exhibit a range of interesting properties superior to those of their solid counterparts.[11, 20] Several methods have been successfully applied in the synthesis of hollow noble metal spheres. Besides the synthetic methods mentioned above, seed-mediated method with silver seeds can be employed to prepare hollow gold spheres.[21] Recently an alternative route has been reported to prepare hollow noble metal materials with different shapes by a transmetallation reaction in which suitable metal ion nanoparticle reacts with a sacrificial partner and finally lead to the formation of hollow spheres inheriting the morphology of the sacrificial partner, and urchinlike hollow metallic and bimetallic nanospheres have been rapidly synthesized by using in situ produced Ag as sacrificial templates.[20, 22, 23]

Herein, we demonstrate a facile synthesis strategy for fabricating hollow metallic nickel microspheresvia a decomposition and reduction route using nickel hydroxide hollow spheres as precursors. Hollow metallic nickel spheres can be obtained by H_2 reduction of nickel oxide hollow spheres which were prepared through thermal decomposition of nickel hydroxide hollow spheres. The size and morphologies are perfectly inherited from nickel oxide hollow spheres to hollow metallic nickel spheres. The decomposition and reduction route provides a new method for large-scale synthesis of hollow metallic nickel spheres and other metallic hollow spheres. on the base of replacement reaction, hollow bimetallic (Ni/Au, Ni/Ag, Ni/Pt, and Ni/Pd) and noble metal (Pt and Pd) spheres have also been obtained as corresponding noble metal compound interact with as prepared hollow metallic nickel spheres. The investigation on magnetic properties of hollow metallic nickel spheres and catalytic activities of hollow noble metal and bimetallic spheres reveals the enhanced coercivity (Hc) of hollow metallic nickel spheres and superior catalytic activities of Ni/Pt hollow bimetallic spheres for hydrogen generation, respectively. This method can be extended for the preparation of other pure and multilayered metallic hollow spheres.

2 Experimental Section

All chemicals were of analytical grade from Beijing Chemical Reagents Factory and used as starting materials without further purification.

2.1 Preparation of hollow $Ni(OH)_2$ and NiO spheres. Hollow $Ni(OH)_2$ microspheres were prepared according to a modified literature procedure.[24] In a typical procedure, 5.0 mmol of $NiCl_2 \cdot 6H_2O$, 2.0 g of glycine, and 2.0 g of Na_2SO_4 salt were put into a Teflon-lined stainless steel autoclave of 50 mL capacity and dissolved in 30 mL of deionized water. Then 10 mL of NaOH solution (5 M) was added into the autoclave under vigorous stirring. The solution was stirred vigorously for 10 min and sealed and maintained at 180 ℃ for 24 h in an electric oven. Subsequently, the system was allowed to cool to room temperature naturally. The resulting green precipitate was collected by filtration and washed with absolute ethanol and distilled water in sequence for several times. The final product was dried in a vacuum box at 50 ℃ for 4 h. As-prepared hollow $Ni(OH)_2$ spheres were calcined to produce hollow NiO spheres in air at 400 ℃ and 600 ℃ for 2 h, respectively.

2.2 Preparation of hollow metallic Ni spheres. Hollow metallic Ni spheres can be obtained by the reduction of as-prepared hollow NiO spheres with a 5% H_2/N_2 mixture at 500 ℃ for 2 h. The flow rate and heating rate were controlled at 20 mL/min and 10 ℃/min, respectively.

2.3 Preparation of bimetallic hollow spheres (Ni/Au, Ni/Ag, Ni/Pt, Ni/Pd) and hollow noble metal (Pt and Pd) spheres. As-prepared hollow metallic Ni spheres

(1 mmol) were suspended in deionized water by ultrasonic treatment. Afterward, a freshly prepared HCl solution (5 wt%) was added slowly to remove the oxidation layer on hollow metallic Ni spheres. The upper solution was decanted immediately when continuous bubbles were observed, which indicated metallic Ni started to react with HCl to release H_2 and removal process of oxidation layer was complete. The Ni powders were washed with distilled water several times by centrifugation and then freshly reduced Ni powders were added to a certain amount of corresponding noble metal salt or acid solution ($HAuCl_4 \cdot 4H_2O$, $AgNO_3$, $H_2PtCl_6 \cdot 6H_2O$, and $PdCl_2$) according to the designed molar ratio of Ni:M (10:1 and 1:1, M = Au, Ag, Pt, Pd). The replacement reaction was allowed to proceed for 6 h without stirring and shaking. Finally, the obtained products were collected and washed with distilled water for several time and then dried in vacuum box at 50 ℃ for 2 h.

2.4 Catalyst Study. The hydrolysis of ammonia borane (NH_3BH_3) was carried out at room temperature and concentration. 0.0012 g of catalyst (hollow metallic Ni, Pt and Pd spheres and Ni/Pt and Ni/Pd bimetallic hollow spheres) was kept in a round-bottom flask with opening connected to an inverted, water-filled and graduated burette. The reaction was started and proceeded under constant stirring with the addition of aqueous NH_3BH_3 solution (50 mL, 0.5 wt.%). The volume of generated H_2 was recorded per minute.

2.5 Characterization. The obtained samples were characterized on a Rigaku Dmax - 2000 X-ray powder diffractometer (XRD) with Cu K_α radiation (λ = 1.5418 Å). The operation voltage and current were kept at 40 kV and 40 mA, respectively. The size and morphology of as-synthesized products were determined at 20 kV by a XL30 S - FEG scanning electron microscope (SEM). Energy-dispersive X-ray spectroscopy (EDS) was taken on the SEM. Magnetic properties of samples were measured with a vibrating sample magnetometer (VSM).

3 Results and Discussion

X-ray diffraction (XRD) was carried out to determine the chemical composition and crystallinity of as-prepared products. Figure 1 shows the typical XRD pattern of hollow Ni spheres reduced by H_2 atmosphere with hollow NiO spheres as precursors. Three characteristic diffraction peaks can be indexed as (111), (200), and (222) crystal planes of face-centered cubic Ni, in good accordance with reported data [space group: Fm3m (225), JCPDS Card 04 - 0850, a = 3.5238 Å]. No peaks of any impurities were observed, which indicates that metallic face-centered cubic Ni has been successfully obtained via H_2 reduction route with face-centered cubic NiO as precursors at 500 ℃ for 2 h.

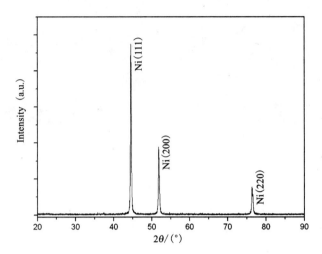

Figure 1 XRD pattern of as-prepared hollow Ni microspheres prepared via H_2 reduction route at 500 ℃ for 2 h, using hollow NiO spheres as precursors

Scanning electron microscopy (SEM) analysis shows that hollow Ni spheres were synthesized in high yield with use of NiO calcined at 600 ℃ as precursors [Figure 2(a)]. Several broken spheres were found in Figure 2(a), suggesting the hollow nature of as-prepared samples. More details of those hollow spheres can be learnt from high magnification SEM image. A broken sphere with hollow interior structure was shown in Figure 2(b). The average diameter of hollow spheres was estimated to be approximately 1.8 μm, which show the hollow Ni spheres have the tendency to shrink from their precursors after H_2 reduction. Furthermore, careful observation shows the building units of hollow Ni spheres changed to nanoparticles with diameters of 150 nm to 250 nm from nanosheets which assembled into hollow Ni(OH)$_2$ and NiO spheres. The result of the SEM analysis above suggests that the hollow nature was perfectly inherited from Ni(OH)$_2$ spheres to NiO spheres and finally to metallic Ni spheres. The result of EDS analysis [Figure 2(c)] reveals the sample mainly consists of Ni. A slight peak of O indicates the existence of oxidation layer on the surface of hollow Ni spheres, resembling that of the snowflakelike metallic Co microcrystal in our previous work, and the peak of C is caused by the C substrate.[25] The magnetic properties of as-prepared

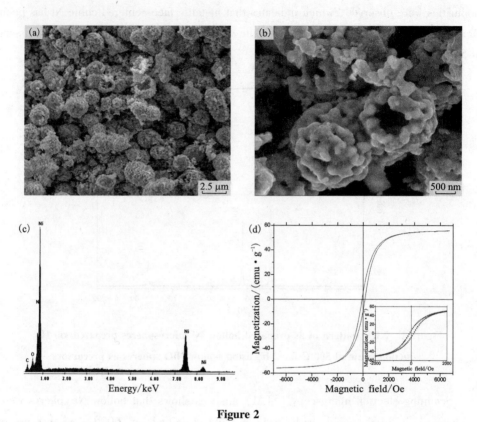

Figure 2

SEM images (a) and (b), EDS spectrum (c), and hysteresis loop (d) of as-prepared hollow Ni microspheres. The inset in Figure (d) is an enlarged hysteresis loop of as-prepared samples at 300 K.

hollow Ni spheres were investigated with a vibrating sample magnetometer at 300 K. As shown in Figure 2(d), the coercivity (Hc), saturation magnetization (Ms) and remanent magnetization (Mr) of as-prepared hollow Ni spheres are 116.6 Oe, 55.6 emu/g and 7.55 emu/g, respectively. Compared with bulk Ni (about 0.7 Oe), the coercivity of microsized hollow Ni spheres is much enhanced. But it is lower than that of hierarchical hollow Ni microspheres which are composed of nanopaticles with diameters of 40 nm to 60 nm,[14] smaller than those of nanopariticles ranging from 150 nm to 250 nm as building units of our samples. As the magnetic properties of nanosized materials are associated with their size and mircostructure, the difference between Hc values may be ascribed to the different size and unique morphologies of corresponding samples.

The hollow noble metal spheres were obtained by replacement reaction between hollow metallic Ni spheres and corresponding metal compound. The driving force of those reaction comes from the large standard reduction potential gap between Ni^{2+}/Ni redox pair (-0.25 V vs standard hydrogen electrode (SHE)) and M^{x+}/M redox pair (1.00 V for $AuCl_4^-/Au$, 0.80 V for Ag^+/Ag, 0.74 V for $PtCl_6^{2-}/Pt$ and 0.83 V for Pd^{2+}/Pd vs (SHE), respectively). Figure 3 shows the XRD patterns of as-prepared Ni/M hollow bimetallic spheres. The characteristic peaks of Au, Ag, Pt and Pd are observed from Figure 3(a) - 3(d) besides those of Ni, respectively, indicating the successful replacement of noble metal.

The morphology, structure, and size of as-prepared samples were characterized with SEM. Figure 4 shows the SEM images of Ni/M hollow bimetallic spheres. It can be found that the samples are built up of numerous aggregated nanoparticles, similar to their precursors hollow Ni spheres. The diameters of these hollow spheres are in the range from 1.5 μm to 2 μm and the broken ones in each image reveal the hollow interior structure clearly, indicating the shape of hollow sphere were kept perfectly from hollow metallic Ni spheres. EDS analysis was also carried out and only the peaks of noble metals are observed with the absence of any impurities within the EDS analysis limit, which are consistent with XRD patterns.

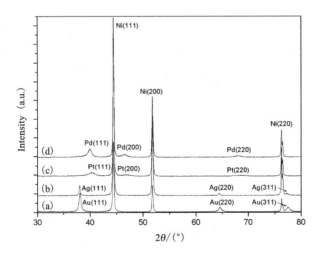

Figure 3 XRD patterns of as-prepared hollow spheres

(a) Ni/Au, (b) Ni/Ag, (c) Ni/Pt, (d) Ni/Pd

It is know that the bimetallic nanoparticles prepared using solid templates could be alloy nanoparticles in which the two constituent metals are mixed at the atomic level[26-28] or core/shell nanoparticles in which the two components are separated by distinct phase boundaries,[29, 30] or even core/shell nanoparticles with an alloy shell and a pure core.[31] However, the employment of hollow templates-hollow metallic Ni microspheres in our work made much difference. All the peaks of noble metals can be observed and the peaks of Ni shows no shift (Figure 3) compared with those of pure Ni (Figure 1), distinct from the XRD pattern of an alloy which exhibits only diffraction peaks of Ni shifting to lower angles caused by the lattice expanding due to the substitution of the larger Pt atoms for the smaller Ni atoms.[15] Meanwhile, many pinholes, in addition to some broken shells, in the surface of hollow Ni microspheres [Figure 2(b)] provide convenient transportation channels for M^{x+} cations to penetrate into the shell and thus M^{x+} cations may react with Ni atoms homogeneously not only on the outer surface but also on the inner surface and within shell of Ni microspheres, different from the outer surface of templates as starting reaction interface in the case of the formation of the core/shell structure. We suggest that noble metals are homogeneously mixed with Ni particles after replacement reaction based on the discussion above. EDS mapping of Ni/Au hollow bimetallic spheres has been conducted as an example to verify this hypothesis. The distribution of Au atoms are relatively even in Ni substrates except for the slightly lower density in the center deriving from the hollow nature of Ni/Au bimetallic spheres, indicating Au atoms exist as a mixture with Ni particles throughout the shell.

Figure 5 displays XRD patterns of as-prepared metallic hollow Pt and Pd spheres. Prominent peaks in the XRD patterns correspond to the reflections of the face-centered cubic structure of Pt with lattice constants $a = 3.923$ Å and Pd with lattice constants $a = 3.890$ Å, consistent with reported values (JCPDS Card 04 – 0802 and 46 – 1043) apart. However, a nominal amount of Ni (marked by asterisk) is found in the pattern of Pd sample, which may be caused by the oxide layer on the surface of hollow Ni spheres, not removed thoroughly during acid treatment, preventing the further replacement of Pd.

The SEM images of as-prepared Pt and Pd samples are shown in Figure 6. Figure 6(a) and 6(b) are the SEM images of Pt products, in which four broken spheres can be clearly seen, exhibiting the hollow structure of the obtained Pt spheres. Furthermore, the building units are also nanopaticles, resembling those of hollow metallic Ni spheres. The EDS spectra of hollow Pt spheres, in which Ni was found with small peak area compared with that of Pt. The peaks of C come from the C substrate. The morphology of obtained Pd samples is depicted in Figure 6(c) and 6(d). Several spheres with diameters ranging from 1.5 μm to 2 μm are seen in Figure 6(c), whereas

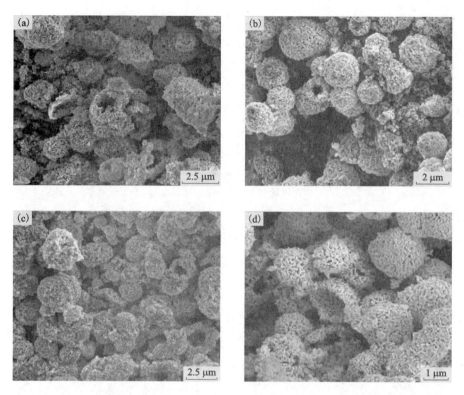

Figure 4 SEM images of as-prepared Ni/M hollow bimetallic spheres
(a) Ni/Au; (b) Ni/Ag; (c) Ni/Pt; (d) Ni/Pd

a broken one in Figure 6(d) reveals its hollow nature. It is worth noting that the morphology of Pd shell structure is quite different from those of other hollow spheres presented in this paper. Hollow Pd spheres are assembled by nanopaticles much smaller [Figure 6(d)] than those of the rest and the thickness of wall is also relatively small. The strong peaks of Pd and weak peaks of Ni agree well with XRD records. The presence of C is derived from C substrate.

The catalytic activities of as-prepared hollow spheres for hydrogen generation from ammonia borane (NH_3BH_3) were investigated and the results are shown in Figure 7. Hollow metallic Ni spheres show the lowest activity and hollow Pd spheres exhibit higher catalytic performance. The activity of Ni/Pd hollow bimetallic spheres is much higher than that of pure hollow Pd spheres. In comparison, the catalytic activities of products containing Pt are favorable, among which Ni/Pt hollow bimetallic spheres show the highest activity. Note the linear relationship with little deviation between the amount of H_2 and time, indicating the uniform continuity of H_2 generated by various catalysts in

334 / Inorganic Functional Materials with Hollow Interiors

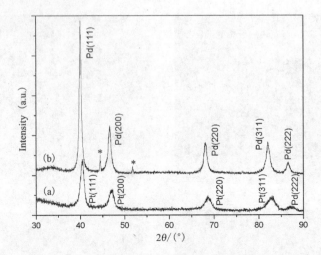

Figure 5   XRD patterns of as-prepared hollow noble metal spheres
(a) Pt; (b) Pd

Figure 6   SEM images of as-prepared hollow noble metal spheres
(a), (b) Pt; (c), (d) Pd

3 h observation. It is noteworthy that both the H_2/NH_3BH_3 ratio reached close 3.0 when the hydrogen release process was completed with Pt and Ni/Pt hollow bimetallic spheres as catalysts, respectively. Similar to previous work,[15, 32, 33] catalysts containing Pt and Pd exhibit the much higher activities than Ni does regarding the aqueous NH_3BH_3 system. The reason why Ni/Pt and Ni/Pd are more active than pure Pt and Pd may be that firmly combined bimetallic nanoparticles on the surface and within Ni/Pt and Ni/Pd hollow spheres can provide great quantities of catalytic sites,[15] and parts of hollow Pt and Pd spheres have inevitably collapsed after complete replacement reaction, making the surface area decrease to some extent and further weakening the catalytic activities. Due to their relatively higher catalytic activities and much lower cost compared to hollow Pt spheres, Ni/Pt hollow bimetallic spheres may have potential applications in portable hydrogen generation systems.

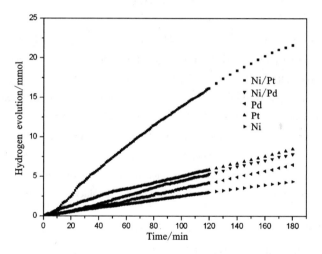

Figure 7 Hydrogen release from the hydrolysis of aqueous NH_3BH_3 solution (50 mL, 0.5 wt.%) in the presence of various as prepared catalysts (12 mg)

4 Conclusion

In summary, hollow metallic nickel spheres with an average diameter of 1.8 μm have been successfully synthesized via a decomposition and reduction route by using hollow nickel hydroxide spheres as precursors. Hollow metallic nickel spheres can be obtained by H_2 reduction of hollow nickel oxide spheres which were prepared through thermal decomposition of nickel hydroxide hollow spheres. Various hollow metallic spheres have also been obtained via a replacement reaction route using hollow metallic nickel spheres as sacrificial templates. Among those spheres, hollow bimetallic (Ni/Au, Ni/Ag, Ni/Pt and Ni/Pd) and noble metal (Pt and Pd) spheres can be selectively prepared by adjusting the amount of noble metal compound and hollow metallic nickel spheres. The magnetic properties of hollow metallic nickel spheres and catalytic activities of as-prepared hollow noble metal spheres for hydrogen generation have also been investigated. The results shows the much enhanced magnetic properties compared to bulk Ni and favorable catalytic performance of Ni/Pt hollow bimetallic spheres which have potential applications in portable fuel cells, respectively. This strategy provides an effective method for the fabrication of other pure and multilayered hollow spheres.

References

[1] S. Mandal, D. Roy, R. V. Chaudhari, M. Sastry. Pt and Pd nanoparticles immobilized on amine-functionalized zeolite: excellent catalysts for hydrogenation and heck reactions [J]. Chemistry of Materials, 2004, 16(19): 3714 - 3724.

[2] T. A. Taton, C. A. Mirkin, R. L. Letsinger. Scanometric DNA array detection with nanoparticle probes [J]. Science, 2000, 289(5485): 1757 - 1760.

[3] M. L. Brongersma, J. W. Hartman, H. A. Atwater. Spectrally selective splitters with metal-dielectric-metal surface plasmon waveguides [J]. Applied Physics B, 2009, 95(4): 807 - 812.

[4] Y. Sun, Y. Xia. Relative parameter contributions for encapsulating silica-gold nanoshells by poly (N-isopropylacrylamide-co-acrylic acid) hydrogels [J]. Macromolecular Research, 2009, 17 (5): 307 - 312.

[5] J. C. Bao, Y. Y. Liang, Z. Xu, L. Si. Facile synthesis of hollow nickel submicrometer spheres [J]. Advanced Materials, 2003, 15(21): 1832 - 1835.

[6] F. Caruso, R. A. Caruso, H. Mohwald. Nanoengineering of inorganic and hybrid hollow spheres by colloidal templating [J]. Science, 1998, 282(5391): 1111 - 1114.

[7] K. P. Velikov, A. van Blaaderen. Synthesis and characterization of monodisperse core-shell colloidal spheres of zinc sulfide and silica [J]. Langmuir, 2001, 17(16): 4779 - 4786.

[8] X. Gao, J. Zhang, L. Zhang. Hollow sphere selenium nanoparticles: their in-vitro anti hydroxyl radical effect [J]. Advanced Materials, 2002, 14(4): 290 - 293.

[9] L. Qi, J. Li, J. Ma. Biomimetic morphogenesis of calcium carbonate in mixed solutions of surfactants and double-hydrophilic block copolymers [J]. Advanced Materials, 2002, 14(4): 300 - 303.

[10] B. Mayers, X. Jiang, D. Sunderland, B. Cattle, Y. Xia. Hollow nanostructures of platinum with controllable dimensions can be synthesized by templating against selenium nanowires and

colloids [J]. Journal of the American Chemical Society, 2003, 125(44): 13364 – 13365.
[11] S. W. Kim, M. Kim, W. Y. Lee, T. Hyeon. Fabrication of hollow palladium spheres and their successful application to the recyclable heterogeneous catalyst for Suzuki coupling reactions [J]. Journal of the American Chemical Society, 2002, 124(26): 7642 – 7643.
[12] M. P. Zach, R. M. Penner. Nanocrystalline nickel nanoparticles [J]. Advanced Materials, 2000, 12(12): 878 – 883.
[13] R. K. Thauer. Nickel to the fore [J]. Science, 2001, 293(5533): 1264 – 1265.
[14] F. Liu, J. Y. Lee, W. J. Zhou. Segmented Pt/Ru, Pt/Ni, and Pt/RuNi nanorods as model bifunctional catalysts for methanol oxidation [J]. Small, 2006, 2(1), 121 – 128.
[15] F. Y. Cheng, H. Ma, Y. M. Li, J. Chen. $Ni_{1-x}Pt_x$ ($x = 0 – 0.12$) Hollow spheres as catalysts for hydrogen generation from ammonia borane [J]. Inorganic Chemistry, 2007, 46(3), 788 – 794.
[16] Q. Liu, H. J. Liu, M. Han, J. M. Zhu, Y. Y. Liang, Z. Xu, Y. Song. Nanometer-sized nickel hollow spheres [J]. Advanced Materials, 2005, 17(16): 1995 – 1999.
[17] Y. Wang, Q. S. Zhu, H. G. Zhang. J. Fabrication and magnetic properties of hierarchical porous hollow nickel microspheres [J]. Journal of Materials Chemistry, 2006, 16(13): 1212 – 1214.
[18] G. T. Duan, W. P. Cai, Y. Y. Luo, Z. G. Li, Y. J. Lei. Hierarchical structured Ni nanoring and hollow sphere arrays by morphology inheritance based on ordered through-pore template and electrodeposition [J]. The Journal of Physical Chemistry B, 2006, 110(32): 15729 – 15733.
[19] N. Wang, X. Cao, D. S. Kong, W. M. Chen, L. Gao, C. P. Chen. Nickel chains assembled by hollow microspheres and their magnetic properties [J]. The Journal of Physical Chemistry C, 2008, 112(17): 6613 – 6619.
[20] H. Liang, H. Zhang, J. Hu, Y. Guo, L. Wan, C. Bai. Pt hollow nanospheres: facile synthesis and enhanced electrocatalysts [J]. Angewandte Chemie International Edition, 2004, 43(12): 1540 – 1543.
[21] X. C. Xu, C. M. Shen, C. W. Xiao, T. Z. Yang, H. R. Zhang, J. Q. Li, H. L. Li, H. J. Gao. Wet chemical synthesis of gold nanoparticles using silver seeds: a shape control from nanorods to hollow spherical nanoparticles [J]. Nanotechnology, 2007, 18(11): 115608.
[22] Y. Sun, B. Mayers, Y. Xia. Metal nanostructures with hollow interiors [J]. Advanced Materials, 2003, 15(7 – 8): 641 – 646.
[23] S. J. Gao, S. J. Dong, E. K. Wang. A general method for the rapid synthesis of hollow metallic or bimetallic nanoelectrocatalysts with urchinlike morphology [J]. Chemistry – A European Journal, 2008, 14(15): 4689 – 4695.
[24] Y. Wang, Q. S. Zhu, H. G. Zhang. Fabrication of β – Ni(OH)$_2$ and NiO hollow spheres by a facile template-free process [J]. Chemical Communications, 2005, 41: 5231 – 5233.
[25] X. H. Liu, R. Yi, Y. T. Wang, G. Q. Zhou, N. Zhang, X. G. Li. Highly ordered

snowflakelike metallic cobalt microcrystals [J]. The Journal of Physical Chemistry C, 2007, 111(1): 163 – 167.

[26] M. P. Mallin, C. J. Murphy. Solution-phase synthesis of sub – 10 nm Au – Ag alloy nanoparticles [J]. Nano Letters, 2002, 2(11): 1235 – 1237.

[27] R. G. Sanedrin, D. G. Georganopoulou, S. Park, C. A. Mirkin. Seed-mediated growth of bimetallic prisms [J]. Advanced Materials, 2005, 17(8): 1027 – 1031.

[28] I. Lee, S. W. Han, K. Kim. Production of Au – Ag alloy nanoparticles by laser ablation of bulk alloys [J]. Chemical Communications, 2001, 18: 1782 – 1783.

[29] Y. W. Cao, R. Jin, C. A. Mirkin. DNA-modified core-shell Ag/Au nanoparticles [J]. Journal of the American Chemical Society, 2001, 123(32): 7961 – 7962.

[30] M. Tsuji, N. Miyamae, K. Matsumoto, S. Hikino, T. Tsuji. Rapid formation of novel Au core-Ag shell nanostructures by a microwave-polyol method [J]. Chemistry Letters, 2005, 34 (11): 1518 – 1519.

[31] Q. B. Zhang, J. P. Xie, J. Y. Lee, J. X. Zhang, C. Boothroyd. Synthesis of Ag@ AgAu metal core/alloy shell bimetallic nanoparticles with tunable shell compositions by a galvanic replacement reaction [J]. Small, 2008, 4(8): 1067 – 1071.

[32] Q. Xu, M. Chandra. Catalytic activities of non-noble metals for hydrogen generation from aqueous ammonia-borane at room temperature [J]. Journal of Power Sources, 2006, 163(1): 364 – 370.

[33] M. Chandra, Q. J. Xu. Room temperature hydrogen generation from aqueous ammonia-borane using noble metal nano-clusters as highly active catalysts [J]. Journal of Power Sources, 2007, 168(1): 135 – 142.

☆ R. Yi, R. R. Shi, G. H. Gao, N. Zhang, X. M. Cui, Y. H. He, X. H. Liu. Hollow Metallic Microspheres: Fabrication and Characterization [J]. The Journal of Physical Chemistry C, 2009, 113(4): 1222 – 1226.

Selective Synthesis and Properties of Monodisperse Zn Ferrite Hollow Nanospheres and Nanosheets

Abstract: Monodisperse Zn ferrite hollow nanospheres and nanosheets could be selectively synthesized via a facile solvothermal method. The shape, structure and size of Zn ferrites were investigated by XRD, TEM, SEM, and HRTEM. The magnetic and microwave absorption properties of hollow Zn ferrite nanospheres and nanosheets were also investigated in detail. Magnetic studies revealed that both the saturation magnetization (Ms) and coercivity (Hc) of hollow nanospheres are drastically higher than those of nanosheets. Electromagnetic and resulting microwave adsorption properties showed hollow nanospheres have four dips and the maximum magnitudes of the dips is −31.44 dB at 10.48 GHz. The probable formation mechanism of hollow nanospheres was discussed on the basis of the experimental results.

1 Introduction

The fabrication of spinel ferrites nanoparticles has been intensively investigated in recent years due to their unique physical and chemical properties, as well as their technological applications in ferrofluids,[1] high-density magnetic recording media,[2] biomedicine,[3] and radar-absorbent materials (RAM).[4] However, many intrinsic properties and applications of these magnetic particles are mainly determined by their shape, size, and chemical composition. For example, hollow spheres have remarkable advantage used in biomedicine especially in target drug delivery when compared with solid counterparts because of their low densities resulted from this structure, which enhances the stability of suspensions. In principle, one could control any one of these parameters to fine-tune the properties of this nanostructure.

As an important member of ferrite family, Zn ferrite has attracted significant research interest based on its fascinating magnetic and electromagnetic properties. Recently, many methods have been applied to prepare uniform Zn ferrite nanoparticles with controllable composition and size.[5-6] Although hollow Zn ferrite nanostructures have also been prepared through a template-method,[7] the control of the morphology and size still remains a highly sophisticated challenge. Herein we demonstrated a one-step, economical, and efficient solution phase chemical method for the large-scale fabrication of monodisperse hollow Zn ferrite nanospheres and nanosheets. The shape of these particles could be readily controlled by changing reactive conditions and the nature of the protective agent. This work will result in a general, economic and environmentally friendly synthetic strategy for obtaining various monodisperse ferrites nanoparticles and provided an opportunity to further apply these promising materials.

2 Experimental section

All chemicals in this work, such as hydrated iron chloride ($FeCl_3 \cdot 6H_2O$), zinc chloride ($ZnCl_2$), ammonium acetate (NH_4Ac), and ethylene glycol, were analytical grade regents from the Beijing Chemical Reagents Factory and used as starting materials without further purification.

2.1 Synthesis. Monodisperse Zn ferrite nanocrystals were obtained via a facile solvothermal synthetic route. The typical preparation procedure of hollow Zn ferrite nanospheres is as follows. $FeCl_3 \cdot 6H_2O$ (2 mmol) and $ZnCl_2$ (1 mmol) were added into ethylene glycol (30 mL) to form clear solution. Then a protective agent such as NH_4Ac (15 mmol) was added into the solution to form a mixture under vigorous stirring at room temperature. Subsequently, the mixture was put into a Teflon-lined stainless steel autoclave of 50 mL capacity and sealed and maintained respectively at 170 – 200 ℃ for 24 – 48 h. Finally, the system was allowed to cool to roomtemperature naturally. The resulting black precipitate was collected by filtration and washed with absolute ethanol and distilled water in sequence for several times. The final product was dried in a vacuum box at 50 ℃ for 4 h. Zn ferrite nanosheets were synthesized using KAc (15 mmol) as protective agent at 180 ℃ for 24 h. The other procedures were the same to those for hollow nanospheres.

2.2 Characterization. The obtained samples were characterized on a Rigaku Dmax – 2000 X-ray powder diffractometer (XRD) with Cu K_α radiation (λ = 1.5418 Å). The operation voltage and current were kept at 40 kV and 40 mA, respectively. The size and morphology of as-synthesized products were determined at 20 kV by a XL30 S-FEG scanning electron microscope (SEM). Energy-dispersive X-ray spectroscopy (EDS) was taken on the SEM. Magnetic properties of the samples were carried out by using a LDJ – 9600 vibrating sample magnetometer (VSM). The complex permittivity and permeability and resulting microwave adsorption property have been studied by HP-8720ET vector network analyzer.

3 Results and discussion

Figure 1(a) shows the typical SEM image of Zn ferrite hollow nanospheres obtained using NH_4Ac as protective agent at 180 ℃ for 24 h and indicates the large quantity and good uniformity of Zn ferrite hollow nanospheres achieved using this approach. These hollow nanospheres had a mean size of about 250 nm, and there exist many broken spheres, which shows that they are hollow inside. The inset in Figure 1(a) is a higher magnification SEM image obtained from a selected area of

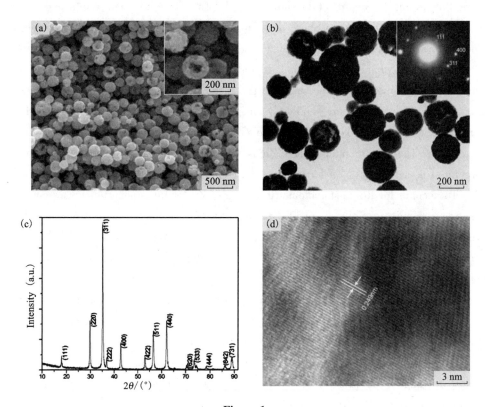

Figure 1

Low-magnification (a) and high-magnification (the inset) SEM images of hollow Zn ferrite nanospheres synthesized using NH_4Ac as protective agent at 180 ℃ for 24 h. (b) A TEM image of the same batch of sample. The inset shows the SAED pattern taken from an individual hollow sphere. (c) XRD pattern of the same batch of sample. (d) High-resolution TEM image of a partial area of individual hollow Zn ferrite nanosphere.

Figure 1(a). Herein, the hollow inside of Zn ferrite can be clearly observed. Figure 1(b) presents typical TEM images of hollow Zn ferrite nanospheres, which have diameters ranging from 150 nm to 400 nm and a shell thickness of about 50 nm. The inset of Figure 1(b) is the selected area electron diffraction (SAED) of an individual hollow Zn ferrite nanosphere, which shows that the hollow Zn ferrite nanospheres are single-crystalline structures. Figure 1(c) shows the typical XRD pattern of the as-prepared Zn ferrite hollow nanosphere. All diffraction peaks in this pattern can be indexed as the pure cubic phase [space group: $Fd\bar{3}m$ (227)] of Zn ferrite with cell constants $\alpha = 8.446$ Å (PDF 79-1150). No impurity peak was observed, which indicated that the high purity of Zn ferrite crystalline was successfully synthesized by using this approach. Figure 1(d) shows the HRTEM image obtained from a partial area of an individual hollow nanosphere and further reveals their single-crystalline nature. The distance between two adjacent planes is measured to be 0.249 nm, corresponding to (222) lattice planes in the spinel Zn ferrite. EDS was also measured to determine the chemical composition of the sample. The result from EDS spectra shows that the as-prepared hollow nanosphere hollow nanosphere contain Fe, O and Zn, and no contamination elements is detected (the C peak in the pattern is caused by the C substrate). The atomic ratio of Fe: Zn is about 40:1, indicating that the chemical formula of as-synthesized Zn ferrite is unstoichiometric nature.

To account for the unstoichiometric nature of Zn ferrite nanoparticles, we consider the strong chemical affinity of specific cations like Zn^{2+} to the tetrahedral A site and the metastable cation distribution in nanoscale range.[8-9] Based on these features, we propose the following mechanism: The nucleation and particle growth during solvothermal synthesis is a process that involves absorption/desorption of cations and the occupancy of their specific sites. In the present system, several cations, including Fe^{3+}, Fe^{2+} (the product of the redox reaction of Fe^{3+} and ethylene glycol[10]), and Zn^{2+} are involved. During the formation of Zn ferrite nuclei, Zn^{2+} will preferentially occupy the tetradedral site and force the Fe^{3+} to the octahedral site.[8] Even in nanoscale range, the preferential occupancy of Zn^{2+} dominates over the metastable cation distribution. The restriction of Zn^{2+} to occupy only the tetrahedral sites results in the formation of the unstoichiometric Zn ferrite, since the absorption of either Zn^{2+} or other cations by the nuclei will depend on the availability of their preferential sites.

The shape and diameter of the products can be tuned by altering the reaction conditions or the nature of protective agents. For example, if the temperature was increased from 180 ℃ to 200 ℃, the product was mainly dominated by hollow spheres with the dimension of about 370 nm [Figure 2(a)]. However, when the temperature was reduced to 170℃, solid-structured ones with the mean size of 250 nm were obtained [Figure 2(b)]. If the reaction time was prolonged from 24 h to 48 h, with careful observation, the as-synthesized was still dominated by uniform hollow spheres, but the average dimensions of Zn ferrite particleswere increased to about 280 nm with a narrower size distribution [Figure 2(c)]. Figure 2(d) shows TEM image of the sample that was synthesized with the same amount of potassium acetate (15.0 mmol KAc) as a protective agents under the similar reaction conditions. The large quantity of uniform Zn ferrite nanosheets with average dimension of 90 nm was achieved. We extended the reaction process for the hollow Zn ferrite nanospheres to synthesize ternary Zn ferrites using a metal substitution method by adding a different metal precursor to the mixture of $FeCl_3 \cdot 6H_2O$, $ZnCl_2$ and $NH_4 \cdot Ac$ as well. For example, when a 1 : 1 molar ratio of MCl_2 (M = Mn, Cu) to $ZnCl_2$ was introduced into the reaction system (according to the 1 : 2 molar ratio of the total molar amount of Zn^{2+} and M^{2+} to Fe^{3+}, the according amount of $FeCl_3$ was also added at the same time), hollow-structured Mn – Zn ferrite [Figure 2(e)] and Cu – Zn ferrite [Figure 2(f)] nanospheres were also obtained, respectively.

The mechanism leading to the formation of the different shapes of Zn ferrite nanoparticles under the different reaction conditions is not yet clear. However, our own experimental evidences have led us to believe that the nature of the cation of protective agents plays an important role in the formation of the different shapes of Zn ferrites nanoparticles. For example, the NH_4^+ cation was found to be critical for the formation of hollow nanospheres. As is well known, most of the methods for preparation of hollow spheres are template-based ones, in which preparation of sacrificial templates, either hard or soft ones,[11-13] coating of the template with a shell, and removal of the template to create the hollow center are successively conducted. We assume that NH_4Ac could be decomposed into NH_3 because of the heat-instability of ammonium salts at higher reaction temperature[14] (such as 180℃). In the sealed solvothermal reaction system, NH_3 could produce a great deal of bubble which serves as "soft" template to prepare the hollow spheres. Based on above hypothesis, we can further explain the effects of the reaction conditions on the structures (hollow verse solid) and sizes of Zn ferrites nanoparticles prepared using NH_4Ac as the protective reagent. When the reaction temperature was increased from 180℃ to 200℃, the decomposition reaction of NH_4Ac surely occurred and lead to the hollow spheres. At the lower temperature (such as 170℃), however, the reaction might be not occur. Without the "soft" template,

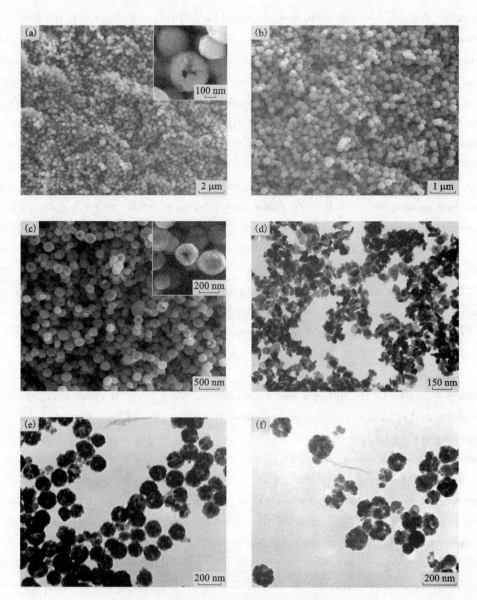

Figure 2 SEM or TEM images of Zn ferrites and substituted Zn ferrites prepared under different conditions

(a) and (b) The same as in Figure 1, except that the temperature was increased from 180℃ to 200℃ or reduced to 170℃, respectively. (c) The same as in Figure 1, except that growth time was prolonged from 24 h to 48 h. (d) The same as in Figure 1, except that the protective agent of NH_4Ac was substituted with KAc. (e) and (f) Hollow-structured MnZn and CuZn ferrite nanospheres synthesized with NH_4Ac as protective agent at 180℃ for 24 h, respectively.

i.e., bubble, the solid-structured nanospheres was certainly the reasonable consequence. When the reaction time was prolonged from 24 h to 48 h, the improvement of both dimension and uniformity of hollow Zn ferrite spheres synthesized at the 180 ℃ indicates that the initial form larger hollow nanospheres grow at the cost of the smaller ones (Ostwald ripening).

The magnetic properties of as-prepared Zn ferrites nanoparticles were investigated with a vibrating sample magnetometer at 300 K. The magnetic hysteresis loops for the different samples are shown in Figure 3. Figure 3(a) and 3(b) correspond to hollow Zn ferrites nanosphere and sheet-like counterparts, respectively. Herein, it is obvious that both the saturation magnetization and coercivity of sheet-like nanoparticles are drastically lower than the hollow ones. The saturation magnetization values Ms of as-prepared hollow Zn ferrites nanosphere is 80.2 emu/g, and the coercivity values Hc for the samples are 103 Oe. However, the saturation magnetization values Ms of the as-prepared nanosheets is 45 emu/g, and the coercivity values Hc for the samples are 49 Oe.

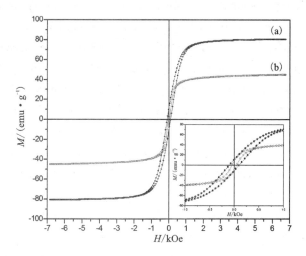

Figure 3 Hysteresis loops of different shapes of Zn ferrite nanoparticles at the room temperature

(a) hollow nanospheres, (b) nanosheets. The inset is an enlarged hysteresis loop of as-prepared Zn ferrite nanoparticles.

It is known that typical bulk $ZnFe_2O_4$ of the normal spinel structure is paramagnetic above Neel temperature. However, in nanocrystalline Zn ferrite, the cation distribution has changed from normal to mixed spinel type. Thus a significant fraction of the Zn^{2+} occupy the octahedral sites and force the Fe^{3+} to the tetrahedral sites, resulting in a larger magnetization. For example, maximum magnetization, M_s, was increased from bulk value of ~ 5 emu/g[15] to nanoscale one of 8 emu/g for 3.1 nm Zn ferrite represented by chemical composition $ZnFe_2O_4$ at room-temperature.[16] However, this does not explain why, in the present study, M_s for any one of Zn ferrites nanocrystals is far larger than these values above. We propose that the off-stoichiometric chemical composition of Zn ferrite particles may further lead to achange of the structure of as-prepared particles from normal to mixed spinel type. The rich Fe^{3+} in our samples (the molar ratio of Fe/Zn is about 40) may result in more Fe^{3+} ions in the A site, making the materials more ferrimagnetic and of higher saturation magnetization. Furthermore, magnetic properties of as-prepared samples are also directly related to their size and shape. Similar results have been reported by Yao et al.[17] In their study, the Ms of an off-stoichiometric zinc ferrite sample with 10 nm synthesized via the thermal decomposition is about 45 emu/g at room temperature.

We have found that, as a potential electromagnetic shielding material, the shapes of Zn ferrites have a complicated effect on electromagnetic properties, including complex permittivity and permeability. Complex permittivity ($\varepsilon_r = \varepsilon' - j\varepsilon''$) and permeability ($\mu_r = \mu' - j\mu''$) represent the dielectric and dynamic magnetic properties of magnetic materials. The real parts (ε', μ') of complex permittivity and permeability symbolize the storage capability of electric and magnetic energy. The imaginary ones (ε'', μ'') represent the loss of electric and magnetic energy. The real and imaginary parts of complex permittivity (ε', ε'') and permeability (μ', μ'') of Zn ferrite particles were measured using an 8720ET vector network analyzer working at the 2 – 18 GHz band (the samples were molded into the hollow pipe of rectangular cavities having a dimension of 10.2 mm × 2.9 mm × 2.5 mm). Figure 4 shows the complex permittivity and permeability of different shapes of Zn ferrites nanoparticles. As shown in the Figure, the shapes of Zn ferrites have an obvious effect on all of the four electromagnetic parameters (ε', ε'', μ' and μ''). For example, both of the real and imaginary parts of permittivity of sheet-like Zn ferrite crystals are larger than those of hollow one in most frequency range, as shown in Figure 4(a) and 4(b), suggesting that the sheet-like particles have higher electric conductivity than hollow spheres. Figure 4(c) and 4(d) show the real and imaginary parts of complex permeability for the two shapes of Zn ferrites particles, respectively. It is observed that the real parts of permittivity (μ') of sheet-like Zn ferrite spheres are larger than those of hollow ones in higher frequency region (f > 5.5 GHz). The imaginary parts (μ''), however, are lower than the values of hollow one in most frequency range, indicating that the hollow Zn ferrite spheres have higher loss of magnetic energy than sheet-like ones.

Microwave absorption properties of Zn ferrite particles were calculated from a computer simulation using the values of electromagnetic parameters (ε', ε'', μ' and μ'') and other parameters, such as the speed of light (c) and the thickness of the samples (t). The calculation process referred to the work of Yusoff.[18] The reflection loss minimum or the dip in plot is equivalent to the occurrence of minimal reflection of the microwave absorber. The thickness of the samples (t) is one of important parameters which influence the intensity and the position of the frequency at the reflection loss minimum. For example, increase of the thickness of absorbers may lead to the dip

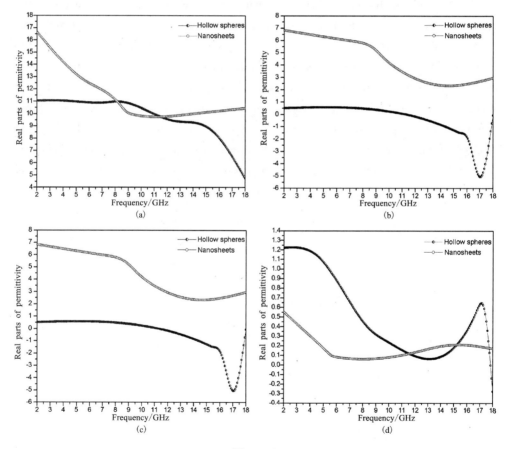

Figure 4

Frequency dependence on (a) real and (b) imaginary parts of permittivity, and (c) real and (d) imaginary parts of permeability for Zn ferrite nanoparticles.

frequency shift toward lower ranges.[18] In addition, the thickness of the absorbers should well match with the material electromagnetic parameters such as complex permittivity and permeability. Therefore, we prepared samples with the same thicknesses in order to eliminate the influence originated from the thickness of samples. Figure 5 shows the frequency dependence of the reflection loss for the Zn ferrites with different shapes at the desired thicknesses (t = 2.5 mm). Figure 5(a) and 5(b) respectively correspond to hollow Zn ferrites nanospheres and nanosheets. Four dips can be observed in the different frequency ranges for hollow nanospheres [Figure 5(a)]. Figure 5(b) shows that there is only one wide and shallow dip around 10 GHz. The magnitudes and situations of the dips vary with the shapes of as-prepared samples. The maximum magnitudes of the dips are −17.92 dB at 10 GHz for sheet-like sample and −31.44 dB at 10.48 GHz for hollow spheres, respectively. Further study is needed to fully probe the mechanism leading to the difference of microwave absorption properties of Zn ferrites nanoparticles with the different shapes. However, the surface structure of sheet-like Zn ferrites nanoparticles is greatly different from the hollow-structured nanospheres. Compared to the curved topology of hollow nanospheres, the flat surfaces of sheet-like nanoparticles enables the surface metal cations to posses a more symmetric coordination, and thus few missing coordinating oxygen ions exist.[19] This difference of surface structure, together with other factors such as sizes, may be the crucial causes

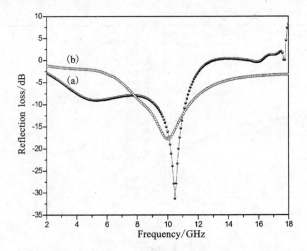

Figure 5 Absorption characteristics of different shapes of Zn ferrite nanoparticles at radar band from 2 GHz to 18 GHz

(a) hollow nanospheres; (b) nanosheets

resulting in the different microwave absorption properties of Zn ferrites nanoparticles with the different shapes.

4 Conclusion

In conclusion, monodisperse Zn ferrite hollow nanospheres and nanosheets could be selectively prepared via simple solvothermal method. Magnetic studies revealed that both M_s and H_c of hollow Zn ferrite nanospheres are drastically higher than those of the sheet-like ones. As a potential microwave absorber, the shapes of Zn ferrites have a complicated effect on electromagnetic parameters (ε', ε'', μ', μ'') and resulting microwave absorption properties. For hollow spheres, four dips can be observed in the different frequency ranges and the maximum magnitudes of the dips is −31.44 dB at 10.48 GHz. There is, however, only one wide and shallow dip (the maximum magnitude of −17.92 dB) around 10 GHz for the sheet-like sample.

References

[1] K. Raj, R. Moskowitz. Commercial applications of ferrofluids [J]. Journal of Magnetism and Magnetic Materials, 1990, 85(1): 233 − 245.

[2] A. Moser, K. Takano, D. T. Margulies, M. Albrecht, Y. Sonobe, Y. Ikeda, S. Sun, E. E. Fullerton. Magnetic recording: advancing into the future [J]. Journal of Physics D: Applied Physics, 2002, 35(19): R157 − R167.

[3] J. Bai, J. P. Wang, High-magnetic-moment core-shell-type FeCo − Au/Ag nanoparticles [J]. Applied Physics Letters, 2005, 87(15): 152502 − 152502 − 3.

[4] R. C. Che, L. M. Peng, X. F. Duan, Q. Che, X. L. Liang. Microwave absorption enhancement and complex permittivity and permeability of Fe encapsulated within carbon nanotubes [J]. Advanced Materials, 2004, 16(5): 401 − 405.

[5] M. D. Shultz, M. J. Allsbrook, E. E. Carpenter. Control of the cation occupancies of MnZn ferrite synthesized via reverse micelles [J]. Journal of Applied Physics, 2007, 101(9): 09M518 − 1 − 09M518 − 3.

[6] C. Upadhyay, H. C. Verma, V. Sathe, A. V. Pimpale. Effect of size and synthesis route on the magnetic properties of chemically prepared nanosize $ZnFe_2O_4$ [J]. Journal of Magnetism and Magnetic Materials, 2007, 312(2): 271 − 279.

[7] Y. Q. Zhang, Z. B. Huang, F. Q. Tang, J. Ren, Ferrite hollow spheres with tunable magnetic properties [J]. Thin Solid Films, 2006, 515(4): 2555 − 2561.

[8] R. Valenzuela. Magnetic ceramics [M]. Cambridge: Cambridge University Press, 2005.

[9] C. Rath, N. C. Mishra, S. Anand, R. P. Das, K. K. Sahu, C. Upadhyaya, H. C. Verma.

Appearance of superparamagnetism on heating nanosize $Mn_{0.65}Zn_{0.35}Fe_2O_4$ [J]. Applied Physics Letters, 2000, 76(4): 475 – 477.

[10] H. Deng, X. L. Li, Q. Peng, X. Wang, J. P. Chen, Y. D. Li. Monodisperse magnetic single-crystal ferrite microspheres [J]. Angewandte Chemie International Edition, 2005, 44 (18): 2782 – 2785.

[11] X. Xu, S. A. Asher. Synthesis and utilization of monodisperse hollow polymeric particles in photonic crystals [J]. Journal of the American Chemical Society, 2004, 126(25): 7940 – 7945.

[12] Z. Zhong, Y. Yin, B. Gates, Y. Xia. Preparation of mesoscale hollow spheres of TiO_2 and SnO_2 by templating against crystalline arrays of polystyrene beads [J]. Advanced Materials, 2000, 12(3): 206 – 209.

[13] Y. Ding, Y. Hu, X. Jiang, L. Zhang, C. Yang. Polymer-monomer pairs as a reaction system for the synthesis of magnetic Fe_3O_4 – polymer hybrid hollow nanospheres [J]. Angewandte Chemie International Edition, 2004, 43(46): 6369 – 6372.

[14] C. Oommen, S. R. Jain. Ammonium nitrate: a promising rocket propellant oxidizer [J]. Journal of Hazardous Materials, 1999, 67(3): 253 – 281.

[15] T. M. Clark, B. J. Evans. Enhanced magnetization and cation distributions in nanocrystalline $ZnFe_2O_4$: a conversion electron mossbauer spectroscopic investigation [J]. IEEE Transactions on Magnetics, 1997, 33(5): 3745 – 3747.

[16] R. D. K. Misra, S. Gubbala, A. Kale, W. F. Egelhoff Jr. A comparison of the magnetic characteristics of nanocrystalline nickel, zinc, and manganese ferrites synthesized by reverse micelle technique [J]. Materials Science and Engineering: B, 2004, 111(2): 164 – 174.

[17] C. Yao, Q. Zeng, G. F. Goya, T. Torres, J. Liu, H. Wu, M. Ge, Y. Zeng, Y. Wang, J. Z. Jiang. $ZnFe_2O_4$ nanocrystals: synthesis and magnetic properties [J]. The Journal of Physical Chemistry C, 2007, 111(33): 12274 – 12278.

[18] A. N. Yusoff, M. H. Abdullah, S. H. Ahmad, S. F. Jusoh, A. A. Mansor, S. A. A. Hamid, Electromagnetic and absorption properties of some microwave absorbers [J]. Journal of Applied Physics, 2002, 92(2): 876 – 882.

[19] C. R. Vestal, Z. J. Zhang. Effects of surface coordination chemistry on the magnetic properties of $MnFe_2O_4$ spinel ferrite nanoparticles [J]. Journal of the American Chemical Society, 2003, 125(32): 9828 – 9833.

☆ A. G. Yan, X. H. Liu, R. Y. Yi, R. R. Shi, N. Zhang, G. Z. Qiu. Selective synthesis and properties of monodisperse Zn ferrite hollow nanospheres and nanosheets [J]. The Journal of Physical Chemistry C, 2008, 112(23): 8558 – 8563.

Controllable Fabrication of SiO_2/Polypyrrole Core-Shell Particles and Polypyrrole Hollow Spheres

Abstract: An in situ polymerization route has been developed to synthesize SiO_2/polypyrrole core-shell particles and polypyrrole hollow spheres with controllable shell thickness. In this process, pyrrole monomers adsorbed on the surface of SiO_2 spheres due to electrostatic interactions, SiO_2/polypyrrole core-shell particles could be subsequently obtained in large quantities via in situ polymerization under hydrothermal conditions. Furthermore, polypyrrole hollow spheres could be obtained after the removal of the core via HF etched method. The shell thickness of core-shell particles and hollow spheres could be easily controlled by adjusting the process parameters such as monomer concentration and hydrothermal temperature. The morphology, size and structure of the final products were investigated in detail by transmission electron microscopy (TEM), scanning electron microscopy (SEM), and Fourier-transform infrared spectra (FTIR). The probable formation mechanism of core-shell particles and hollow spheres was proposed on the basis of the experimental results.

1 Introduction

During the past decades, the impetus for the considerable research into the fabrication of the core-shell particle with unique and tailored properties emanates from their intense scientific and technological interest.[1] These composite particles provide a robust platform for incorporating process ability and diverse functionalities because of fine-tuning the chemical composition, structure, and sizes of the cores or shells.[2] Hollow spheres could be obtained via removal of the core of the core-shell particles.[3] Hollow spheres have received enormous attention in recent years due to their unique properties and their wide variety of potential applications in various fields. Hollow spheres can be used as delivery vehicle for the removal of contaminated waste and the controlled release of substances such as drugs, cosmetics, dyes, and inks or as a carrier of catalyst.[4-8]

Polypyrrole (PPy) is one of the most promising conducting polymers due to unique properties, excellent environmental stability and potential application in electronic devices.[9-11] PPy hollow spheres may provide some immediate advantages over their solid counterparts for applications in various fields due to their relatively low densities. During the past decades, various synthetic methods are continually improved for the preparation of PPy hollow spheres, however, the most versatile procedure has been based on template assisted synthesis, e. g., polystyrene latex spheres,[12] SiO_2 spheres,[13] metal[14] and metal ramification as template,[15] PPy hollow spheres could be subsequently obtained via removal of the templates.[16-18] What is worth noting is that the above-mentioned template assisted synthesis is generally based on surface-modified process of the template. However, the products of such synthesis are often involved in a complicated process due to the surface-modified layer.

In this paper, we demonstrated a facile synthetic method to fabricate SiO_2/PPy core-shell particles and PPy hollow spheres with controllable shell thickness. This method does not require the prior surface-modified process of the template. We used mono-dispersed SiO_2 spheres as the template without any modification. Pyrrole monomers adsorbed on the surface of SiO_2 spheres due to electrostatic interactions, SiO_2/polypyrrole core-shell particles could be subsequently prepared via in situ polymerization procedure without using any polymeric stabilizers under hydrothermal conditions. PPy hollow spheres could be obtained after the removal of the SiO_2 core. Compared with the other methods, the synthetic strategies were facile and the resultant product is purer PPy hollow sphere.

2 Experimental Section

All chemicals in this work, such as tetraethylsiloxane (TEOS), aqueous ammonia (32 wt. %), pyrrole monomers, oxidant $(NH_4)_2S_2O_8$ (APS), and HF (37 wt. %) were all of analytical grade and used without further purification.

2.1 Synthesis of mono-dispersed SiO_2 spheres. Mono-dispersed SiO_2 spheres were synthesized by base-catalyzed hydrolysis of TEOS, as described by Stöber et al.[19] Generally, 5 mL of aqueous ammonia was added into a solution containing 40 mL of ethanol and 5 mL of deionized water. Three millilitres of TEOS was dropped into the as-prepared mixture at room temperature under vigorous stirring. The reaction mixture was stirred continuously for 6 h to fabricate the mono-dispersed SiO_2 spheres.

2.2 Synthesis of SiO_2/PPy core-shell particles. A typical preparation of

SiO$_2$/PPy core-shell particles was carried out as follows: 0.1 g SiO$_2$ spheres were dispersed in 20 mL deionized water by ultrasound. Then 0.1 – 0.3 mL of pyrrole monomer was dropped into the mixture under stirring. The mixture was transferred to a Teflon autoclave after treated by ultrasound for about 1 h and then APS solution prepared by adding 0.1g APS into 10 mL deionized water was dropped slowly into the mixture. The mixture was sealed and maintained at 100 – 140℃ for 8 h. Subsequently, the autoclave was allowed to cool down to room temperature naturally after heat treatment. Finally, the resulting dark-gray precipitates were collected by filtration, washed with distilled water and absolute ethanol in sequence for several times and then dried at 50℃ for 6 h in air.

2.3 Synthesis of PPy hollow spheres. SiO$_2$/PPy core-shell structure composites could be converted to PPy hollow spheres by soaking composites in an aqueous solution of 10 wt.% HF for 24 h and then the product was washed by distilled water and ethanol.[20] Finally, the PPy hollow spheres were dried at 50℃ for 6 h in air.

2.4 Characterization. The size and morphology of as-synthesized products werecharacterized by a JEM – 200CX transmission electron microscope (TEM) at 160 kV. Samples were prepared for TEM analysis by placing several drops of suspension on a carbon-coated copper grid. Scanning electron microscopy (SEM) images were obtained on a XL30 S – FEG microscope operated at 20 kV. Fourier-transform infrared spectra (FTIR) of the samples were recorded on a Bruker VECTOR22 spectrometer.

3 Results and Discussion

Figure 1(a) and 1(b) show the typical SEM images of SiO$_2$/polypyrrole core-shell particles of product 1 obtained at 140℃ for 8 h using 0.2 mL pyrrole monomer and indicate large quantity and good uniformity core-shell particles were achieved using this approach. The diameter of SiO$_2$/polypyrrole core-shell particles is about 350 – 700 nm. Figure 1(b) is a higher magnification SEM image obtained from a selected area of Figure 1(a). The morphology of composite particles can be clearly observed, and the PPy shell composed granular structures is evident, which may be due to pyrrole monomer adsorbed on the surface of SiO$_2$ spheres and in situ polymerized to form granular PPy shell structures. TEM images provided further insight into the structure of SiO$_2$/polypyrrole core-shell particles. Figure 1(c) shows a typical TEM image of SiO$_2$/polypyrrole core-shell particles. Herein, the SiO$_2$ cores surrounded by polypyrrole shell

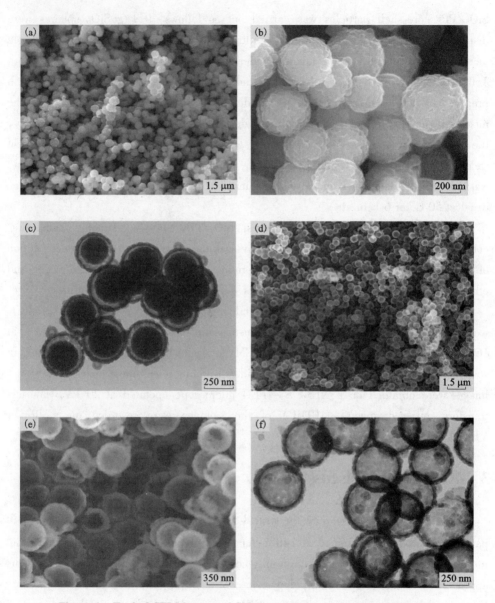

Figure 1 Typical SEM images of SiO_2/polypyrrole core-shell particles of product 1 obtained at 140℃ for 8 h using 0.2 mL pyrrole monomer

(a) low and (b) higher resolution SEM images in which the surface morphology of core-shell particle can be observed; (c) TEM image of the core-shell particles with PPy shell. (d) low and (e) higher resolution SEM image of hollow PPy spheres of product 1′ obtained by using 10 wt.% HF aqueous solution etching core-shell particles, some broken-shell spheres can be observed form (e). (f) TEM image of PPy hollow spheres of product 1′ further proved the hollow inside.

can be clearly observed. The particles with the shell thickness of 50 nm and core diameter of 250 – 300 nm are uniform in size and about 500 nm in diameter.

PPy hollow spheres of product 1' could be obtained by removed the SiO_2 cores using hydrofluoric acid aqueous solution (HF, 10 wt.% content). Figure 1(d) indicate that large quantity and good uniformity PPy hollow spheres were achieved using this approach. These hollow spheres had a mean diameter of about 500 nm with a little deviation. SEM observations also indicated that almost 100% of as-prepared products are uniformity PPy hollow spheres. The hollow structure could be further confirmed via some broken spheres. Figure 1(e) is a higher magnification SEM image obtained from a selected area of Figure 1(d), which clearly show the PPy spheres with hollow inside. The shell thickness is about 50 nm with a mean diameter of about 500 nm. Figure 1(f) shows typical TEM image of PPy hollow spheres of product 1' with shell thickness of about 50 nm. The sharp contrast between the dark edge and the pale center in the TEM image further proved the hollow inside.

In our experiment, the influences of volume of pyrrole monomer on the morphology and size of products were explored. The product 2 and 3 were obtained at 140℃ for 8 h using 0.1 mL and 0.3 mL pyrrole monomer, respectively. The PPy hollow spheres of product 2' and 3' could be obtained by removed the corresponding cores of core-shell particles of product 2 and 3 using 10 wt.% HF aqueous solution. Figure 2(a) shows a typical TEM image of product 2. Herein, it is presented that the large quantity of core-shell particles with shell thickness of about 40 nm. Figure 2(b) shows a typical TEM image of PPy hollow spheres with shell thickness of about 40 nm of product 2', which clearly show the PPy spheres with hollow inside, and the thickness of shell is thinner than that of product 1'. Figure 2(c) is the low resolution SEM image of products 2', which indicate that large quantity and good uniformity PPy hollow spheres were achieved. When the volume of pyrrole monomer is lifted to 0.3 mL, product 3 has a rough surface overlayer compared to product 1 and 2, as shown in Figure 2(d). The shell thickness is about 70nm. Figure 2(e) shows PPy hollow spheres of product 3' with hollow inside. Figure 2(f) is the high resolution SEM image of product 3', which clearly shows a broken-shell sphere. The broken-shell sphere future confirmed the hollow interior of the PPy hollow spheres.

Influence of temperature on the morphology and size of composites was investigated by preparing core-shell structure composites at different reaction temperatures. Figure 3(a) shows the TEM image of product 4 obtained under the same conditions as product 3 except the reaction temperature at 100℃. The shell thickness is about

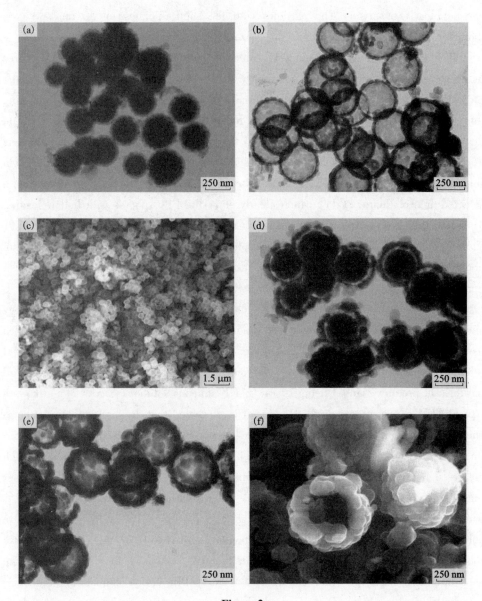

Figure 2

(a) TEM image of SiO_2/polypyrrole core-shell particles of product 2 prepared at 140 ℃ for 8 h with 0.1 mL pyrrole monomer. (b) TEM and (c) SEM images of hollow spheres of products 2′. (d) TEM image of core-shell particles of product 3 prepared at 140 ℃ for 8 h with 0.3 mL pyrrole monomer. (e) TEM and (f) high resolution SEM images of products 3′ obtained by using 10 wt.% HF aqueous solution etching core-shell particles of products 3.

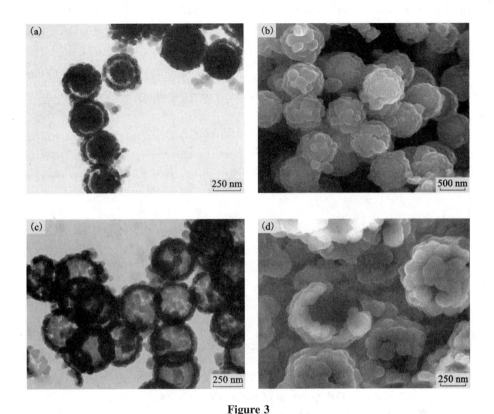

Figure 3

(a) TEM image of product 4 obtained at 100℃ for 8 h using 0.3 mL pyrrole monomer. (b) SEM image of product 4. (c) TEM and (d) high resolution SEM images of product 4' obtained by by using 10 wt. % HF aqueous solution etching core-shell particles of products 4.

70 nm, and the surface of composites is rough. Figure 3(b) shows the SEM image of product 4, which is rougher than product 2. We can also find that the PPy shell was composed of many small PPy particles in Figure 3(b). And the product 4' [in Figure 3(c)] obtained by using HF solution to etch the core of product 4 is so similar with product 3'. However, the temperature below glass transition temperature (160 – 170℃) of PPy, which consequently form a rough overlayer.[21] Figure 3(d) is the high resolution SEM image of product 4', herein the broken spheres can be observed.

The molecular structure of the PPy hollow spheres was characterized by Fourier-transform Infrared (FTIR) spectroscopy, as shown in Figure 4. No obvious differences were found between the FTIR spectra of the PPy hollow sphere samples and those of PPy obtained by common methods. The characteristic bands of the PPy hollow spheres were the pyrrole ring fundamental vibration at 1596 cm^{-1}, =C—H in-plane vibration at 1383 cm^{-1} and 1039 cm^{-1}, —NH$_2$ vibration at 1105 cm^{-1}, C—N stretching vibration in the ring at 1443 cm^{-1}, and a C—H in plane vibration at 1105 cm^{-1}.[22] 1509 cm^{-1} band and 1209 cm^{-1} band are observed, which are most likely related to the different ions doped and their degree of doping.[23-24]

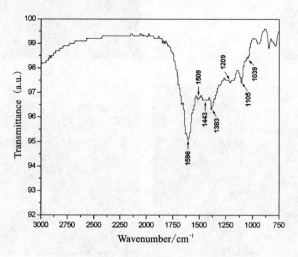

Figure 4 FTIR spectrum of hollow PPy spheres of product 1′ obtained at 140℃ for 8 h using 0.2 mL pyrrole monomer

The formation mechanism of SiO$_2$/PPy core-shell structure particles and PPy hollow spheres is represented in Scheme 1. The mono-dispersed SiO$_2$ spheres mixed with pyrrole monomer were first dispersed in deionized water under stirring/ultrasonic treatment for appropriate time at room temperature. Pyrrole monomers subsequently adsorbed on the surface of SiO$_2$ spheres due to electrostatic interactions. After initiator (APS) was added, in situ polymerization occurred and SiO$_2$/PPy core-shell structure spheres were formed under hydrothermal conditions. Finally, uniform PPy hollow spheres could be obtained after the removal of the core of core/shell particles using HF etching solutions.

Sufficient stirring and ultrasonic treatment of the mixture solution containing monomer and SiO$_2$ spheres are necessary to prevent flocculation of excessive irregular PPy particles in the preparation process of SiO$_2$/PPy core-shell structure particles. Furthermore, relatively low concentrations of monomer and initiator are also important

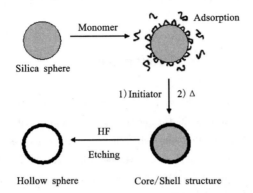

Scheme 1. The formation mechanism of SiO_2/polypyrrole core-shell structure particles and PPy hollow spheres.

for production SiO_2/PPy core-shell structure particles with uniform shell. It could be attributed to the slowing down of polymerization rate that favors the adsorption of monomer on the surface of SiO_2 spheres followed by subsequent preferential polymerization, which is in good accordance with literature results.[25] Because SiO_2 spheres are yielded from aqueous solution, the surface of SiO_2 spheres was surrounded by OH^- and they undoubtedly possess unique hydrophilic abilities.

4 Conclusion

In summary, SiO_2/PPy core-shell structure particles were fabricated via a simple solution-based in situ polymerization approach. PPy hollow spheres could be obtained via HF etched method. The size of SiO_2/PPy core-shell particles and PPy hollow spheres are dependent on the size of SiO_2 template. The shell thickness of core-shell particles and hollow spheres is continuously improved with the elevation of addition volume of pyrrole monomer, and which could be controlled ranging from tens to hundreds of nanometers by adjusting the addition volume of pyrrole monomer. This approach allows the fabrication of hollow spheres with predetermined diameters and size distribution and doping another kind of ions in the shell of the hollow spheres. We think that this facile method is versatile and can be extended many other templates to fabricate inorganic/organic core-shell structures and other hollow conducting polymer materials. We are currently furthering our studies using mono-dispersed SiO_2 spheres as templates to prepare colloids coated with a variety of inorganic/organic materials, with the aim of introducing different properties and functions, thereby making them suitable for various technological applications.

References

[1] J. F. Gohy, N. Willet, S. Varshney, J. X. Zhang, R. Jérôme. Core-shell-corona micelles with a responsive shell [J]. Angewandte Chemie International Edition, 2001, 40(17): 3214 – 3216.

[2] J. Jang, J. Ha, B. Lim. Synthesis and characterization of monodisperse silica-polyaniline core-shell nanoparticles [J]. Chemical Communications, 2006, 15: 1622 – 1624.

[3] F. Caruso. Hollow capsule processing through colloidal templating and self-assembly [J]. Chemistry-A European Journal, 2000, 6(3): 413 – 419.

[4] P. Jiang, F. J. Cizeron, V. L. Colvin. Preparation of macroporous metal films from colloidal crystals [J]. Journal of the American Chemical Society, 1999, 121(34): 7957 – 7958.

[5] H. Fau, S. Herminghaus, P. Lenz, R. Lipowsky. Liquid morphologies on structured surfaces: from microchannels to microchips [J]. Science, 1999, 283(5398): 46 – 49.

[6] F. Caruso, M. Spaova, V. Salguesirno-Maceria. Multilayer assemblies of silica-encapsulated gold nanoparticles on decomposable colloid templates [J]. Advanced Materials, 2001, 13(14): 1090 – 1094.

[7] Y. Zhang. S. Wang, Y. Qian, Z. Zhang. Complexing-reagent assisted synthesis of hollow CuO microspheres [J]. Solid State Sciences, 2006, 8(5): 462 – 466.

[8] D. G. Shuchukin, G. B. Sukhorukov. Nanoparticle synthesis in engineered organic nanoscale reactors [J]. Angewandte Chemie International Edition, 2003, 42(37): 4472 – 4475.

[9] D. G. Shchukin, G. B. Suknorukov. Smart inorganic/organic nanocomposite hollow microcapsules [J]. Angewandte Chemie International Edition, 2003, 42(37): 4472 – 4475.

[10] J. Stejskal, P. Kratochvíl, A. D. Jenkins. The formation of polyaniline and the nature of its structures [J]. Polymer, 1996, 37(2): 367 – 369.

[11] A. Kitani, M. Kaya, K. Sasak. Performance study of aqueous polyaniline batteries [J]. Journal of The Electrochemical Society, 1986, 133(6): 1069 – 1073.

[12] Y. Yang, Y. Chu, F. Y. Yang, Y. P. Zhang. Uniform hollow conductive polymer microspheres synthesized with the sulfonated polystyrene template [J]. Materials Chemistry and Physics, 2005, 92(1): 164 – 171.

[13] L. Y. Hao, C. L. Zhu, C. N. Chen, P. Kang, Y. Hu, W. C. Fan, Z. Y. Chen. Fabrication of silica core-conductive polymer polypyrrole shell composite particles and polypyrrole capsule on monodispersed silica templates [J]. Synthetic Metals, 2003, 139(2): 391 – 396.

[14] S. M. Marinakos, J. P. Novak, L. C. Brousseau. Gold particles as templates for the synthesis of hollow polymer capsules. Control of capsule dimensions and guest encapsulation [J]. Journal of the American Chemical Society, 1999, 121(37): 8518 – 8522.

[15] D. M. Cheng, H. Xia, H. S. Chan. Facile fabrication of AgCl@ polypyrrole-chitosan core-shell nanoparticles and polymeric hollow nanospheres [J]. Langmuir, 2004, 20(23): 9909 – 9912.

[16] Z. Zhong, Y. D. Yin, B. Gate, Y. N. Xia. Preparation of mesoscale hollow spheres of TiO_2 and SnO_2 by templating against crystalline arrays of polystyrene beads [J]. Advanced Materials, 2000, 12(3): 206 – 209.

[17] X. D. Wang, W. L. Yang, Y. Tang, J. Wang, S. K. Fu, Z. Gao. Fabrication of compact silver nanoshells on polystyrene spheres through electrostatic attraction [J]. Chemical Communications, 2002, 4: 350 – 351.

[18] N. Kawahashi, H. Shiho. Copper and copper compounds as coatings onpolystyrene particles and as hollow spheres [J]. Journal of Materials Chemistry, 2000, 10(10): 2294 – 2297.

[19] W. Stöber, A. Fink, E. Bohn. Controlled growth of monodisperse silica spheres in the micron size range [J]. Journal of Colloid and Interface Science, 1968, 26(1): 62 – 69.

[20] S. W. Kim, M. Kim. Fabrication of hollow palladium spheres and their successful application to the recyclable heterogeneous catalyst for suzuki coupling reactions [J]. Journal of the American Chemical Society, 2002, 124(26): 7642 – 7643.

[21] M. Biswas, A. Roy. Thermal, stability, morphological, and conductivity characteristics of polypyrrole prepared in aqueous medium [J]. Journal of Applied Polymer Science, 1994, 51(9): 1575 – 1580.

[22] A. Dhanabalan, S. V. Mello, O. N. Oliveira. Preparation of langmuir-blodgett films of soluble polypyrrole [J]. Macromolecules, 1998, 31(6): 1827 – 1832.

[23] S. T. Selvan. Novel nanostructures of gold-polypyrrole composites [J]. Chemical Communications, 1998, 3: 351 – 352.

[24] S. T. Selvan, V. T. Truong. Spectroscopic study of thermo-oxidative degradation of polypyrrole powder by FT-IR [J]. Synthetic Metals, 1997, 89(2): 103 – 109.

[25] R. Gangopadhyay, A. De. Conducting polymer nanocomposites: a brief overview [J]. Chemistry of Materials, 2000, 12(3): 608 – 622.

☆ X. H. Liu, H. Y. Wu, F. L. Ren, G. Z. Qiu, M. T. Tang. Controllable fabrication of SiO_2/polypyrrole core-shell particles and polypyrrole hollow spheres [J]. Materials Chemistry and Physics, 2008, 109(1): 5 – 9.

图书在版编目(CIP)数据

中空结构无机功能材料/刘小鹤,马仁志,李星国著.
—长沙:中南大学出版社,2015.7
ISBN 978 – 7 – 5487 – 1740 – 9

Ⅰ.中... Ⅱ.①刘... ②马... ③李... Ⅲ.中空纤维 – 纳米材料 – 无机材料 – 功能材料　Ⅳ.TB321

中国版本图书馆 CIP 数据核字(2015)第 159791

中空结构无机功能材料

刘小鹤　马仁志　李星国　著

□责任编辑	刘颖维	
□责任印制	易建国	
□出版发行	中南大学出版社	
	社址:长沙市麓山南路	邮编:410083
	发行科电话:0731-88876770	传真:0731-88710482
□印　　装	长沙超峰印务有限公司	
□开　　本	720×1000　1/16　□印张 24　□字数 477 千字	
□版　　次	2015 年 7 月第 1 版　　□印次　2015 年 7 月第 1 次印刷	
□书　　号	ISBN 978 – 7 – 5487 – 1740 – 9	
□定　　价	115.00 元	

图书出现印装问题,请与经销商调换